Biocommunication of Archaea

Guenther Witzany
Editor

Biocommunication of Archaea

Springer

Editor
Guenther Witzany
Telos-Philosophische Praxis
Bürmoos, Salzburg
Austria

ISBN 978-3-319-65535-2 ISBN 978-3-319-65536-9 (eBook)
DOI 10.1007/978-3-319-65536-9

Library of Congress Control Number: 2017949167

Printed on acid-free paper

This Springer imprint is published by Springer Nature
The registered company is Springer International Publishing AG
The registered company address is: Gewerbestrasse 11, 6330 Cham, Switzerland

Foreword

Biocommunication of Archaea is the last book in this exceptional new series on communication as prerequisite for all biotic coordination processes that apply biocommunication theory to all organismic kingdoms or domains of life, respectively.

In the first book "Biocommunication and Natural Genome Editing" (2010), the whole theoretical concept was outlined, followed by "Biocommunication in Soil Microorganisms" (2011), "Biocommunication of Plants" (2012 with co-editor Frantisek Baluska), "Biocommunication of Fungi," and "Viruses: Essential Agents of Life." In 2014, "Biocommunication of Animals" followed and in 2016 also the investigation on a species of unicellular eukaryotes "Biocommunication of Ciliates" (2016, together with co-editor Mariusz Nowacki).

Theory of biocommunication is the first and only theory that integrates all domains of life empiristically in a non-reductionistic and non-mechanistic way. This theory of biocommunication strictly avoids metaphysical narratives (e.g., holism, mechanism, physicalism) and opens a groundbreaking understanding of life.

The biocommunicative approach investigates both communication processes within and among cells, tissues, organs and organisms as sign-mediated interactions, and nucleotide sequences as code, i.e., language-like text, which follows in parallel three (3) kinds of rules: combinatorial (syntactic), context-sensitive (pragmatic), and content-specific (semantic).

Key levels of biocommunication research therefore integrate (1) sensing, memory, and interpretation of abiotic indices, (2) intraorganismic communication (intra- and intercellular), (3) interorganismic communication, and (4) transorganismic communication.

Biocommunication theory translates empirical data into a coherent perspective on the functions within and between biological organisms and arguably leads to a sustainable integrative biology.

Bürmoos, Austria   Guenther Witzany

Preface

Communication means interactions that are mediated by signals in contrast to purely physicochemical interactions where no signals are present. Additionally these sign-mediated interactions need rules on how the signals may be combined to transport more complex informational content. Last but not least, communicative interactions mediated by signs depend essentially on living agents that are able to follow such rules of sign use. Additionally it must be mentioned that sign-mediated rule-governed interactions represent a kind of social interaction, i.e., communication processes are social events, involving groups of interacting agents that share the rules on how to use signals, and this means that group identity is essential in living nature.

Whenever living cells, whether akaryotic or eukaryotic, coordinate their behavior, communication processes are necessary to reach coordinative goals. Each activity on all levels of biocommunication (intraorganismic, interorganismic, transorganismic, and generating response behavior to abiotic influences) is usually achieved by means of specialized signaling. If these signaling processes are disturbed, damaged or incomplete, the activity remains incomplete or rudimentary or is deleted.

The prerequisites of communication are its social character, its dependence on the use of signals according to three levels of rules, the primacy of context (pragmatics) which determines the meaning/function (semantics) of the used signs and—last but not least—the inherent capability of sign-using agents to change these rules of sign use according to environmental or adaptational needs. None of which is found in inanimate nature. No signs, no semiotic rules, and no socially interacting living agents are present when, e.g., water freezes to ice.

The change of sign-using rules, which gives signals and sign sequences new meanings that never existed before and are not the result of a recombination of former ones, is an inherent feature of living agents competent to use a natural language or a natural code. The generation of new sign sequences is essentially not the result of the selection of beneficial mutations out of an abundance of defective variants. In contrast to this passive derivation narrative of positive selective forces, the generation of new sequences is an active process in which natural code-using agents produce new sequences in vivo.

This short description of the essentials of biocommunication contradicts former opinions of communication in natural sciences, information theory, systems theory, mechanistic and reductionistic approaches which rely on a mathematical theory of language, i.e., concepts of quantifiable sets of signs. The history of the philosophy of science clearly demonstrates a variety of such approaches in which signals are molecules which are subject to quantitative investigations and comparisons usually based on investigations of the molecular syntax structure of the natural codes. This means that molecular syntax as a result of chance mutations (error replications) and selection represents the material reality of the physicochemical world which can best be represented by mathematical equations. In this perspective, material reality is the only reality because it can be objectivized, measured, and empirically investigated. The molecular syntax of natural codes is therefore the information-bearing content out of which its functions and its meaning can be distilled.

Yet this paradigm was falsified in the 1980s by pragmatic philosophy and sociology which empirically tested that meaning was not represented by the syntax structure of natural codes, but by the context in which sign sequences are used by in vivo interactions of living agents. In contrast to the former narrative, pragmatics (context) determines the meaning of sign sequences, with the consequence that identical sequences of signs may transport different meanings, even contradictory ones. This makes sense in light of energy costs: It is not necessary to represent an ontological entity or event by unequivocal representations. One sign sequence can designate multiple meanings according to contextual needs. A sign- and rule-sharing population only needs a limited number of signals and a limited number of rules to produce multiple variant communications; even de novo generation, although rare, is possible in principle.

This means it is not the syntax which is the relevant information for extracting the meaning of signals used to coordinate and organize behavior. To avoid the term "meaning," Francis Crick wanted to use "Information" only in a strict mechanistic way ("*'Information' in the DNA, RNA, protein sense is merely a convenient shorthand for the underlying causal effect. As to "meaning" ... I would keep away from the term.*").

In contrast to this perspective, it is the context in which social interactions occur, i.e., that in which signs and sign sequences are used. This means the sociological aspect is essential for deciphering the meaning of natural codes.

Biocommunication processes have been documented meanwhile on the whole area of living nature, i.e., plants, animals, fungi, akaryotes, viruses, and even RNA consortia. One missing publication on communication of archaea is presented here. The published works on biocommunication are not the end but the start of a coherent process of investigations and data mining regarding communicative actions within cells, between cells and between non-related organisms during the whole biosphere, and could lead to a better understanding of the principles governing living nature and a better picture of life on this planet to restructure the behavioral motifs of humankind in its relation to non-human living nature.

The benefit of this new type of research which integrates empirically derived knowledge about archaea physics and chemistry with pragmatic action theory is its more coherent explanatory power. It complements the current knowledge about the physiology of archaea and motifs on each level of archaeal life with the available habitats and contexts in which archaea species live. In contrast to pure reductionistic biology, it can integrate the basic motifs of archaea signaling within varying contexts with the knowledge about all physiological interactions. In contrast to mechanistic biology—it only recalls on the outdated narrative of "information" transfer—the biocommunication approach focuses on the real-life situations in which signaling directs the various forms of interaction. In this context, communication is not restricted to information transfer but predominantly acts as a kind of social interaction.

Finally, the editor hopes that Biocommunication of Archaea will integrate a diversity of research goals on the function, taxonomy, and genetics of archaea, representing their main principles of life, evolution, and developmental stages. Understanding the full range of archaea life will have repercussions for the understanding of life and its evolution in general.

Bürmoos, Austria
May 2017

Guenther Witzany

Acknowledgements

I am grateful to all authors that contributed to the success of the biocommunication book series:

Biocommunication in Soil Microorganisms (2011)

Stephen T. Abedon, Robert Armon, Susumu Asakawa, Mina Aziz, Omar Bagasra, Eshel Ben-Jacob, Emilie Chapelle, Alan J. Castle, Miguel A. Cevallos, Brian Cheetham, Yves Dessaux, Denis Faure, Mengsheng Gao, Margaret E. Katz, Dror Y. Kenett, Mi-Seong Kim, Makoto Kimura, S. Krishnaswamy, Susan M. Lehman, Asaf Madi, Massimo Merighi, Paul Murugan, Natsuko Nakayama, Ralf Oelmüller, Katherine M. Pappas, D. Gene Pace, Paul W. Paré, Ilona Pfeiffer, Meghna Pohani, Sathish Rajamani, Margaret A. Riley, Dwayne Roach, Jayne Robinson, Dalit Roth, Xin Shen, Irena Sherameti, Peter Sholberg, Neeraj Shrivastava, Shashibala Sing, K.V. Srividhya, Antonet M. Svircev, Max Teplitski, Michael Ting, Kailash C. Upadhayaya, Ajit Varma, Guanghua Wang, Kurt E. Williamson, Gabriele Whittle, Xitao Xie, Huiming Zhang, Jinlin Zhang.

Biocommunication of Plants (2012)

Juan M. Alba, Silke Allmann, Dirk Balmer, Frantisek Baluska, David Mc.K.Bird, Janet Braam, Ana Victoria Carcia, Amélie Charrier, E.Wassim Chehab, Peter M. DiGennaro, Sylvie Dinant, Jurgen Engelberth, Balaji Enugutti, Nancy A. Fujishige, Joris J. Glas, Heribert Hirt, Ann M. Hirsch, Angela Hodge, Marco Kai, Ulrich Melcher, Merijin R. Kant, Daniel K. Manter, Brigitte Mauch-Mani, Miyo Terao Moirta, Moritaka Nakamura, Moritz K. Nowack, Yadira Olvera-Carillo, Birgit Piechulla, Gadi V.P. Reddy, Bernardus C.J. Schimmel, Yuliya Salanenka, Adam

Schikora, Eleni A. Spyropoulou, Marije Stoops, Akifumi Sugiyama, Kay Schneitz, Georg J. Seifert, Paula Suarez López, Masao Tasaka, Anthony Trewawas, Claudio Valverde, Carlos Villarroel, Jorge M. Vivanco, Katrin Wenke, Teresa Weise, Rene Warnke, Dierk Wanke, Krystof Wieczorek.

Biocommunication of Fungi (2012)

Katharyn J. Affeldt, Audrey L. Atkin, Andrey A. Averyanov, Drion G. Boucias, Jeremy Bruenn, Katherine A. Borkovich, Tatiana A. Belozerskaya, Maria C. Bertolini, Eli J. Borrego, Renato M. de Paula, Asharie J. Campbell, Robert Cichewicz, Fernanda B. Cupertino, Aurelie Deveau, Anna A. Fabbri, Corrado Fanelli, Fernanda Z. Freitas, Pascale Frey-Klett, N. Louise Glass, Natalie N. Gessler, Rodrigo D. Goncalves, Jessica C. Hargarten, Sahar Hasim, Elizabeth A. Hutchison, Alexander Idnurm, Nancy P. Keller, Michael V. Kolomiets, Valerie Legue, Verena U. Litze, Francis Martin, John P. Morrissey, Jonathan M. Plett, Tatiana Potapova, Zdena Palkova, Libuse Vachova, Philippe Silar, Silvia Polaino, Kenneth W. Nickerson, Ruvini Pathirana, Massimo Reverberi, Rusty J. Rodriguez, Marilyn Roossinck, Jacqueline A. Severin, David R. Soll, Danielle M. Troppens, Peter Teal.

Viruses: Essential Agents of Life (2012)

Chantal Abergel, Alessia Armezzani, Frédérick Arnaud, Nina S. Atanasova, Omar Bagasra, Dennis H. Bamford, Pierre-Adrien Bolze, Harald Brüssow, Domitille Chalopin, Jean-Michel Claverie, Elena Capel, Angela Ciuffi, Michael Chandler, Philippe Colson, Andrew B. Conley, Sébastien Desfarges, Jonathan Fileé, Sandro Franco, Patrick Forterre, Delphine Galiana, Matthew Haynes, Matti Jalasvuori, I. King Jordan, Mart Krupovic, Hanna M. Oksanen, Donald G. Pace, Massimo Palmarini, Mariona Parera, Philippe Pérot, Maija K. Pietilä, Elina Roine, Francois Mallet, Miguel Angel Martinez, Gloria Martrus, Lita Murphy, Maria Nevot, Didier Raoult, Forest Rohwer, Marilyn J. Roossinck, Ana Sencilo, Muhammad Sheraz, Thomas E. Spencer, Marta Tomaszkiewicz, Rachael E. Tarlinton, Jean-Nicolas Volff, Luis P. Villarreal, Merry Youle.

Biocommunication of Animals (2014)

Piero Amodio, Gustavo Arriaga, Stefan M. Brudzynski, Ana Maria Costa-Leonardo, Shermin de Silva, Tamas Farago, Camila R. Ferrara, Graziano Fiorito, Jacqueline C. Giles, Ives Haifig, Kenji Hara, Lynne D. Houck, Yen-Ping

Hsueh, Mary Lee A. Jensvold, Gerald Kuchling, Walter Lechner, Daniel H.W. Leighton, Pierre Madl, John M. Marzluff, Marc L. Miller, Marc Naguib, Irene Pepperberg, Giovanna Ponte, Jan Randall, Friedericke Range, Katharina Riebel, Laela Sayigh, Savannah M. Schulze, Ulricke Siebeck, Paul W. Sternberg, Angela Stöger, Simon Townsend, Elena Tricarico, Richard C. Vogt, Lisa Wilding.

Biocommunication of Ciliates (2016)

Takahiko Akematsu, Claudio Alimenti, Sarah Allen, John B. Bracht, Kathryn A. Bracht, Yelena Bisharyan, Wendy Ashlock, Takahiko Akematsu, Eric S. Cole, Jean Clobert, Alexis S. Chaine, Deanna Cho, Wie-Jen Chang, Theodore Clark, György Csaba, Emily M. Ferraro,, Arno Germond, Masahiro Fujishima, Stafan Jacob, Yuuki Kodama, Laszlo Köhidai, Delphine Legrand, Christopher S. Lobban, Pierangelo Luporini, Mariusz Nowacki, Toshiyuki Nakajima, Helmut Plattner, Bill Pedrini, Ronald Pearlman, Birgit H. Satir, Peter Satir, Nicolas Schtickzelle, Maria Schefter, Bettina Sonntag, Adriana Vallesi, Thomas Weisse, Kassandra E. Zaila, Rebecca A. Zufall.

Biocommunication of Archaea (2017)

Stephen T. Abedon, Sonja-Verena Albers, Fayez Aziz, Manuel Becerra, Charlotte Beloe, Carol Bernstein, Harris Bernstein, Christian Brininger, Brendan P. Burns, Derek Caetano-Anollés, Gustavo Caetano-Anollés, Kelsey Caetano-Anollés, James C. Charlesworth, Lori Cobani, Daniel Colman, Juan-José Escuder-Rodríguez, Caryn Evilia, Xu Feng, J. Peter Gogarten, María-Isabel González-Siso, Wenyuan Han, Peter J. Kennelly, Kyung Mo Kim, Ibrahim Koç, Eugene V. Koonin, Yuchen Liu, Zuhong Lu, Kira S. Makarova, Bushra F. Minhas, Jay E. Mittenthal, Tomohiro Mochizuki, Fizza Mughal, Arshan Nasir, Alvaro Orell, David Penny, Tessa E.F. Quax, Lennart Randau, Esther Rodríguez-Belmonte, Simone Schopf, Khuram Shahzad, Qunxin She, Shannon M. Soucy, Savannah Spradlin, D. Swati, Guy Tal, Masaharu Takemura, Mario Vera, Shengqin Wang, Cara Watters, Cuiping Zhao.

Contents

Introduction: Keylevels of Biocommunication of Archaea 1
Guenther Witzany

The Cooccurrence of Archaea and Bacteria Among Diverse Globally Distributed Ecosystems 17
Daniel R. Colman

Direct Interspecies Electron Transfer Between Archaea and Bacteria 27
Cuiping Zhao and Yuchen Liu

Why Archaea Are Limited in Their Exploitation of Other, Living Organisms 41
Stephen T. Abedon

Archaeal Surface Structures and Their Role in Communication with the Extracellular Environment 67
Sonja-Verena Albers and Tessa E.F. Quax

Archaeal Biocommunication in Hot Springs Revealed by Metagenomics 85
María-Eugenia DeCastro, Juan-José Escuder-Rodríguez, Manuel Becerra, Esther Rodríguez-Belmonte and María-Isabel González-Siso

Sexual Communication in Archaea, the Precursor to Eukaryotic Meiosis 103
Harris Bernstein and Carol Bernstein

Quorum Sensing in Archaea: Recent Advances and Emerging Directions 119
James C. Charlesworth, Charlotte Beloe, Cara Watters and Brendan P. Burns

Biofilm Lifestyle of Thermophile and Acidophile Archaea 133
Alvaro Orell, Simone Schopf, Lennart Randau and Mario Vera

The Compressed Vocabulary of the Proteins of Archaea 147
Gustavo Caetano-Anollés, Bushra F. Minhas, Fayez Aziz, Fizza Mughal, Khuram Shahzad, Guy Tal, Jay E. Mittenthal, Derek Caetano-Anollés, Ibrahim Koç, Arshan Nasir, Kelsey Caetano-Anollés and Kyung Mo Kim

KaiC-like ATPases as Signal Transduction Hubs in Archaea 175
Kira S. Makarova and Eugene V. Koonin

Archaea Were Trailblazers in Signaling Evolution: Protein Adaptation and Structural Fluidity as a Form of Intracellular Communication 195
Savannah Spradlin, Lori Cobani, Christian Brininger and Caryn Evilia

Protein Phosphorylation-Dephosphorylation and Signal Processing in the Archaea 213
Peter J. Kennelly

Secondary Metabolites in Archaea and Extreme Environments 235
Shengqin Wang and Zuhong Lu

Evolutionary Interaction Between Archaeal-Eukaryal Cell Lineages and Viruses 241
Masaharu Takemura and Tomohiro Mochizuki

Inteins as Indicators of Bio-Communication 265
Shannon M. Soucy and J. Peter Gogarten

Riboswitches: Regulatory ncRNAs in Archaea 277
D. Swati

DNA Damage Repair in Archaea 305
Qunxin She, Xu Feng and Wenyuan Han

Archaeal Lipids as an Adaptation to Higher Temperatures? 319
David Penny

Contributors

Stephen T. Abedon Department of Microbiology, The Ohio State University, Mansfield, OH, USA

Sonja-Verena Albers Molecular Biology of Archaea, Institute for Biology II, Faculty of Biology, University of Freiburg, Freiburg, Germany

Fayez Aziz Department of Crop Sciences, University of Illinois, Urbana, IL, USA

Manuel Becerra Grupo EXPRELA, Centro de Investigacións Científicas Avanzadas (CICA), Departamento de Bioloxía Celular e Molecular, Facultade de Ciencias, Universidade da Coruña, A Coruña, Spain

Carol Bernstein Department of Cellular and Molecular Medicine, College of Medicine, University of Arizona, Tucson, AZ, USA

Harris Bernstein Department of Cellular and Molecular Medicine, College of Medicine, University of Arizona, Tucson, AZ, USA

Charlotte Beloe School of Biotechnology and Biomolecular Sciences, University of New South Wales, Sydney, Australia; Australian Centre for Astrobiology, University of New South Wales, Sydney, NSW, Australia

Christian Brininger Department of Chemistry, Idaho State University, Pocatello, ID, USA

Brendan P. Burns School of Biotechnology and Biomolecular Sciences, University of New South Wales, Sydney, Australia; Australian Centre for Astrobiology, University of New South Wales, Sydney, NSW, Australia

Derek Caetano-Anollés Department of Evolutionary Genetics, Max-Planck-Institut Für Evolutionsbiologie, Plön, Germany

Gustavo Caetano-Anollés Department of Crop Sciences, University of Illinois, Urbana, IL, USA; Illinois Informatics Institute, University of Illinois, Urbana, IL, USA

Kelsey Caetano-Anollés Department of Agricultural Biotechnology and Research Institute for Agriculture and Life Sciences, Seoul National University, Seoul, Republic of Korea

James C. Charlesworth School of Biotechnology and Biomolecular Sciences, University of New South Wales, Sydney, Australia; Australian Centre for Astrobiology, University of New South Wales, Sydney, NSW, Australia

Lori Cobani Department of Chemistry, Idaho State University, Pocatello, ID, USA

Daniel R. Colman Department of Microbiology & Immunology, Montana State University, Bozeman, MT, USA

María-Eugenia DeCastro Grupo EXPRELA, Centro de Investigacións Científicas Avanzadas (CICA), Departamento de Bioloxía Celular e Molecular, Facultade de Ciencias, Universidade da Coruña, A Coruña, Spain

Juan-José Escuder-Rodríguez Grupo EXPRELA, Centro de Investigacións Científicas Avanzadas (CICA), Departamento de Bioloxía Celular e Molecular, Facultade de Ciencias, Universidade da Coruña, A Coruña, Spain

Caryn Evilia Department of Chemistry, Idaho State University, Pocatello, ID, USA

Xu Feng Department of Biology, University of Copenhagen, BioCenter, Copenhagen N, Denmark

J. Peter Gogarten University of Connecticut, Storrs, CT, USA

María-Isabel González-Siso Grupo EXPRELA, Centro de Investigacións Científicas Avanzadas (CICA), Departamento de Bioloxía Celular e Molecular, Facultade de Ciencias, Universidade da Coruña, A Coruña, Spain

Wenyuan Han Department of Biology, University of Copenhagen, BioCenter, Copenhagen N, Denmark

Peter J. Kennelly Department of Biochemistry, Virginia Polytechnic Institute and State University, Blacksburg, VA, USA

Kyung Mo Kim Division of Polar Life Sciences, Korea Polar Research Institute, Incheon, Republic of Korea

Ibrahim Koç Department of Crop Sciences, University of Illinois, Urbana, IL, USA; Molecular Biology and Genetics, Gebze Technical University, Kocaeli, Turkey

Eugene V. Koonin National Center for Biotechnology Information, National Library of Medicine, Bethesda, MD, USA

Yuchen Liu Department of Biological Sciences, Louisiana State University, Baton Rouge, LA, USA

Zuhong Lu State Key Lab of Bioelectronics, School of Biological Science and Medical Engineering, Southeast University, Nanjing, China

Kira S. Makarova National Center for Biotechnology Information, National Library of Medicine, Bethesda, MD, USA

Bushra F. Minhas Department of Crop Sciences, University of Illinois, Urbana, IL, USA; Illinois Informatics Institute, University of Illinois, Urbana, IL, USA

Jay E. Mittenthal Department of Cell and Developmental Biology, University of Illinois, Urbana, IL, USA

Tomohiro Mochizuki Earth-Life Science Institute, Tokyo Institute of Technology, Tokyo, Japan

Fizza Mughal Department of Crop Sciences, University of Illinois, Urbana, IL, USA; Illinois Informatics Institute, University of Illinois, Urbana, IL, USA

Arshan Nasir Department of Crop Sciences, University of Illinois, Urbana, IL, USA; Department of Biosciences, COMSATS Institute of Information Technology, Islamabad, Pakistan

Alvaro Orell Max Planck Institute for Terrestrial, Marburg, Germany

David Penny Institute of Fundamental Sciences, Massey University, Palmerston North, New Zealand

Tessa E.F. Quax Molecular Biology of Archaea, Institute for Biology II, Faculty of Biology, University of Freiburg, Freiburg, Germany

Lennart Randau Max Planck Institute for Terrestrial, Marburg, Germany

Esther Rodríguez-Belmonte Grupo EXPRELA, Centro de Investigacións Científicas Avanzadas (CICA), Departamento de Bioloxía Celular e Molecular, Facultade de Ciencias, Universidade da Coruña, A Coruña, Spain

Simone Schopf Institute of Biosciences, Environmental Microbiology, Technische, Universität Bergakademie Freiberg, Freiberg, Germany

Khuram Shahzad Illinois Informatics Institute, University of Illinois, Urbana, IL, USA

Qunxin She Department of Biology, University of Copenhagen, BioCenter, Copenhagen N, Denmark

Shannon M. Soucy Dartmouth College, Hanover, USA

Savannah Spradlin Department of Chemistry, Idaho State University, Pocatello, ID, USA

D. Swati Department of Physics and Bioinformatics, MMV, Banaras Hindu University, Varanasi, India

Masaharu Takemura Laboratory of Biology, Department of Liberal Arts, Faculty of Science, Tokyo University of Science, Tokyo, Japan

Guy Tal Department of Crop Sciences, University of Illinois, Urbana, IL, USA

Mario Vera Institute for Biological and Medical Engineering, Schools of Engineering, Medicine and Biological Sciences, Department of Hydraulic and Environmental Engineering, Pontificia Universidad Católica de Chile, Santiago, Chile

Shengqin Wang College of Life and Environmental Science, Wenzhou University, Wenzhou, China

Cara Watters School of Biotechnology and Biomolecular Sciences, University of New South Wales, Sydney, Australia; Australian Centre for Astrobiology, University of New South Wales, Sydney, NSW, Australia

Guenther Witzany Telos-Philosophische Praxis, Buermoos, Austria

Cuiping Zhao Department of Biological Sciences, Louisiana State University, Baton Rouge, LA, USA

Introduction: Keylevels of Biocommunication of Archaea

Guenther Witzany

1 Introduction

In his last interview Carl Woese said "Evolution is a process. It is the process we now call biology which is very static. Evolution, however, is dynamic. And we have to understand what rules that dynamic follows" (Woese 2015).

The biocommunication method was developed to increase the explanatory power about the rules that lead biological organization and to better understand evolutionary dynamics. Under the biocommunicative perspective any step in biological organisation must be coordinated between parts that are involved. Referring back to the rules of communicative logics provides an opportunity to answer questions of evolutionary logic and dynamics as questions of interaction logic and dynamics (Witzany 1993, 2000).

Biocommunication means any kind of biological rule-governed, sign-mediated interactions. This means interactions mediated by a natural language or code used by concrete sign using agents which follow contextual (pragmatic), combinatorial (syntactic) and content-sensitive (semantic) rules (Witzany 2005).

If we look at recent advances in biocommunication research we find that the investigations of sign-mediated interactions between cell tissues, organs and organisms in all kingdoms have increased especially in the last two decades.

It is common knowledge now that if biological entities want to coordinate their behavior (or in more specialized cases the cells of organs which share a genetically conserved identity) it needs not only the generation of biomolecules that serve as signaling tools to transport messages of any kind (and their use in different contextual circumstances) but interactional motifs to reach certain individual or common goals (Witzany 2010).

G. Witzany (✉)
Telos-Philosophische Praxis, Vogelsangstr. 18c, 5111 Buermoos, Austria
e-mail: witzany@sbg.at

G. Witzany (ed.), *Biocommunication of Archaea*,
DOI 10.1007/978-3-319-65536-9_1

Communication is essential to reach such goals and the investigations on biocommunication tell us whether communication functions as the ultimate pre-condition to reach such goals or it does not function for one or several reasons and the messages are deformed or damaged. In the case of cell coordination or organization of organs or group behavior in organisms this might lead to disease or similar problems.

Archaea are single celled organisms without eukaryotic nucleus and therefore represent "prokaryotes" although the term akaryot seems to be more appropriate (Forterre 2013). Most of them live at high temperatures, concentrated saline solutions or extremes of pH, and are autotrophs (methanogens, halophiles, thermoacidophiles). But unlike bacteria archaea are not model parasites although they may be harmful to other organisms such as Eukarya, Bacteria and other Archaea (Abedon 2013; Raymann et al. 2015). On the other side they built symbiotic interactions with members of these three domains also.

Archaea share some molecular biological features with eukarya that are not found in any bacteria such as ATP production, protein secretion, cell division and vesicles formation, and protein modification pathways (Woese and Fox 1977; Woese et al. 1990; Forterre 1997). Archaea therefore seem to share a strong evolutionary relationship to the eukarya (Garrett and Klenk 2007). In contrast to the other domains eukarya and bacteria the empirical data on archaea especially their communicative interactions, i.e. interactions based on signal transfer, remains poorly investigated. To update this was the reason to edit this book on biocommunication of archaea.

2 The Concept of Biocommunication

When we consider the biocommunication of archaea, we must first become familiar with the current terms of communication (and with the signaling system, which we would term language or code, the essential tool for communication).

When we speak about language and communication we usually think of humans that talk to each other and communicate to organize common goals and to coordinate common behavior. However, since Karl von Frisch received the Nobel Prize 1973 for detection and investigation of bee languages and dialects it is evident that even non-human animals may communicate to construct complex behavioral patterns. The Nobel Laureate Manfred Eigen insisted nearly at the same time that when we speak about the genetic code we are speaking about a real language, not just a metaphor (Eigen and Winkler 1983; Witzany 1995).

Concerning these fundamental insights I developed the biocommunicative approach which investigates both communication (1) and language (2) as universal requirements for life. The first such investigation was published in 1993 (Witzany 1993) followed by an updated version in 2000 (Witzany 2000) as a draft on the *Theory of Communicative Nature*. The results in virology and the role of viruses in evolution and developmental processes in particular exemplified a variety of the proposed pre-assumptions therein (Witzany 2009; Villarreal and Witzany 2010, 2013a, b, 2015). The *Theory of Biocommunication* was outlined first as a program in 2007 updated in 2010 (Witzany 2010) concerning bees, corals, plants, fungi, bacteria, viruses and subviral RNAs. Several more detailed exemplifications followed in a series of books I edited between 2011 and 2014 with leading researchers in their field (Witzany 2011, 2012a, b, c, 2014; Witzany and Baluska 2012a; Witzany and Nowacki 2016).

2.1 Mechanistic Narratives Used to Explain Communication Are Outdated

The method for analysing any part of a machine in detail to get a picture of the whole functional blueprint, which can then be used to reproduce or manipulate it, or to produce an even more perfect example is still useful if we are dealing with machines.

In contrast, communication between cells, cellular components, tissues, organs and organisms is far from being a procedure that can be reduced to mechanistic input/output or cause/reaction descriptions. It is evident that communication processes between living organisms include a variety of non-mechanistic circumstances and competences that must be satisfied in parallel if biotic communicative interactions are to have successful consequences; for example, innovating common coordination to adapt to new environmental conditions.

Machines cannot create new programs out of a functional blueprint, which is in contrast to the abilities of living organisms that are able to communicate with each other. The universal Turing machine and the self-reproducing machines of von Neumann still remain at the conceptual stage. However, no single self-reproducing machine has ever been produced within the last 80 years. There are good reasons for this, because it is, in principle, impossible that an artificial machine could reproduce itself (Witzany and Baluska 2012b). In contrast to the artificial machines which cannot reproduce themselves, living cells and organisms can reproduce themselves and additionally generate an abundance of behavioral motifs for which no algorithm can be constructed, such as de novo generation of coherent nucleotide sequences.

2.2 Communication Is Interaction Mediated by Sign(al)s

Coherently with current knowledge about natural communication processes communication is defined as the sign-mediated interaction between at least two living agents which share a repertoire of signs that are combined (according to syntactic rules) in varying contexts (according to pragmatic rules) to transport content (according to semantic rules). This means monological concepts such as the sender-receiver narrative, in which the (private) sender codes information and the (private) receiver decodes it according to inherited programs as outlined by the mathematical theories of communication (Shannon and Weaver 1949) cannot explain the emergence of commonly shared meanings.

Contrary to former concepts the importance of this result is that these three levels (syntax, pragmatics, semantics) of semiotic rules (*semeion* = sign) are complementary parts of any natural language or code. If one level is missing, according to Charles Morris, we cannot seriously speak of language- or signal-mediated communication. Therefore, the most recent definition of communication is: sign-mediated and rule-governed interactions, i.e. interactions that depend on a shared repertoire of signs and rules (Witzany 2010). However, these features are lacking in abiotic interactions; no semiotic rules are necessary if water freezes to ice.

Additionally, we know that mathematical and mechanistic theories of language are less helpful in investigating natural language and real-life communication processes, because such theories cannot explain typical features of communication between living entities, which are not formal (i.e. for which no algorithm is available), such as (1) different and even contradictory meanings of identical syntactic sequences/sentences for commonly interacting agents and (2) the de novo generation of coherent sentences or sequences. This means that no natural language or code speaks or codes itself, but requires living agents that are competent in such languages or codes (Witzany 2010). The emergence of natural codes is inherently a social event.

In the biology of the twentieth century, the physiology of all manner of cells, tissues, organs and organisms was the mainstream direction of biological research and experiments. In the 1970s, an increasing use of "communication" as a metaphor also arose in biology. During the last decade of the twentieth century, interest in communication (no longer used as a metaphor) within, and between, organisms overtook that of the purely physiological understanding of organisms. This was owed to concrete communication processes designating varying contexts in real-life circumstances. Cell-to-cell communication now dominates contemporary cell biology, including knowledge of a great variety of signaling pathways, serving for both organization and coordination of production, release, uptake, interpretation and processing of context-dependent "information" (content) within and between cells. Context dependency determines the crucial fact that it is not the syntax (grammar) of a sequence of signs (information) which determines the meaning (semantics), but the context (pragmatics) in which the concrete use of the sequence occurs.

In parallel, the use of "language" as a metaphor has increased since the middle of the twentieth century with the improved knowledge of the genetic code. Most of the processes that evolve, constitute, conserve and rearrange the genetic storage medium (DNA) are terms that were originally used in linguistics. For example: nucleic acid language, genetic code, "code without commas" (F. Crick), coding, copying, translation, transcription, "genetic text" (F. Jacob), sequence homology, etc. Meanwhile, the linguistic approach has also lost its metaphorical character, and the similarity between natural languages and codes and the genetic storage medium of

DNA have not only been accepted but adopted in epigenetics, comparative genomics, bioinformatics, biolinguistics, biocommunication theory and biosemiotics.

2.3 Primacy of Context: Pragmatics

Natural communication assembles the full range of signal-mediated interactions that are necessary in order to organize all evolutionary and developmental coordination within, and between, cells, tissues, organs and organisms. To identify biocommunication of organisms we therefore have to look at the interaction motifs in the real lifeworld context of the organisms (Witzany 2010).

1. No single organism is able to communicate as an emerging property; it must be part of a community, society or swarm of organisms that share an identity and have the ability to sense whether others are part of this identity or not (self/non-self differentiation competence), even if this competence is solely shared genetically. For communication it is necessary for organisms to have assets that serve as signs, signals or symbols, such as chemical molecules, either produced directly by the organism or as secondary metabolites, or even molecules in the surroundings, which can be manipulated according to the organism's needs. In the case of animals, especially complex ones, visible and audible sign repertoires have evolved.
2. Organisms must also share the ability to use these signs in a coherent manner, which means using them in a strictly temporal and spatial context. In most cases it is not just one sign but several that are combined in a specific manner to transport messages or information. This represents a common feature of sign use in communication processes, and is termed the correct combination or syntax.
3. Organisms are part of the natural habitat in which they live, together with similar organisms of the same or related species, but usually also with an abundance of unrelated organisms. This historically developed context fully represents the natural history of swarms or communities and the way in which they have evolved certain abilities and are able to mount appropriate response behaviors to enable their survival. These competencies, which include sensing, monitoring, learning and memory, are preconditions for faster adaptation.
4. Signaling molecules, which serve as signs, transport messages with meanings (semantics). The informational (semantic) content which is transported triggers certain response behaviors in the same, related, or even unrelated, organisms. Interestingly, the signal sequence or content does not necessarily depict a strict meaning, i.e. a function, but can vary according to different situational contexts. This means that identical signs can transport a variety of different messages according to different contextual needs. The different uses of identical signs, or sequences, enable the generation of dialects within the same species that can transport messages which are microecosphere-specific. This includes very sensitive (self/non-self) recognition between slightly differently adapted populations of the same species (Witzany 2000).

Although sign-mediated interactions (i.e. communication processes) are very reliable in most cases, they do not function mechanistically in a strict sense. Syntax (combination), pragmatics (context) and semantics (content) must function in parallel to ensure and optimize the coordination and thus survival of group members.

These three levels of semiotic rules (syntax, pragmatics and semantics) do not function mechanistically but can be varied, deleted, or, in certain circumstances, and in contrast to the capabilities of machines, generated de novo. Additionally, semiotic rules do not function by themselves but need semiotic subjects, i.e. living organisms that utilize such rules. If no living organism is present, semiotic rules, signs and communication are absent. Although highly conserved semiotic rules are modifiable, environmental circumstances, such as stress, can trigger adaptive responses. In such cases, signals may transport new messages which previously did not exist, broadening the communicative competences of organisms, i.e. broadening their evolutionary capabilities.

3 In Vitro Analyses Lack Context-Dependent Behaviors of Real-Life Habitats

In vitro investigations focus on ecological setups, which do not represent the entire interactional context in which an organism is involved in vivo. In a mechanistic and reductionistic perspective on living organisms as machine-like entities this makes sense. But the evolution and development of each organism depends on the in vivo habitat with its inter-, intra-, and transorganismic triggers of genetic reading patterns, which are absent from in vitro setups.

Especially if we think on the fact that in vivo habitats of all organisms known today are embedded into a virosphere which outnumbers cells of organisms ten times. This means that the real life world of any organism is under continued pressure of genetic parasites and any organism has to activate immune functions constantly (Ryan 2009). Therefore, it is likely that isolated organisms in laboratory setups lack a variety of features which would be triggered in in vivo habitats by natural phenomena such as symbiotic and parasitic microorganisms. This may lead to limited conclusions regarding their intra- and interorganismic biocommunicative capabilities.

4 The Biocommunication Method Applied to Archaea

The advantage of a methodical adaptation of communication and linguistic terminology is that it provides appropriate tools for differentiation at specific levels, which are otherwise difficult to describe in non-reductive terms with pure physiology. The result of this is, that language-like structures and communication processes occur, at the simplest levels of nature.

Language and communication are not the evolutionary inventions of humans, nor are they anthropomorphic adaptations for describing non-human interactions, but they are the key tools to coordinate and organize in and between biotic agents.

It is an empirical fact that all coordination and organization within, and between cells, tissues, organs, and organisms need signs; i.e., molecules that serve as signals in messages, or serve as vital indicators of environmental conditions. Since no natural code can encode itself, in the way that no natural language can speak itself, these signs must be generated, sensed and interpreted in the correct way by biological agents; i.e., there must be subjects of sign production and sign interpretation (i.e., comparison with and evaluation of memorized experiences). The consequence of this is that generation, sensing, as well as interpretation and memory storage, may fail. This can result in inappropriate behavior, or even have letal consequences.

The biocommunication approach can integrate empirical data into an integrative perspective of an organism and its interactional patterns at various levels.

4.1 Designation of the Semiochemical Vocabulary

Many archaeal species produce metabolic pathways that are not found in other domains of life. Investigations, therefore, may start with the identification of the used semiochemicals, i.e., chemicals such as secondary metabolites that serve as signs within interactions of various motifs, such as reproduction, mating, community (identity) organisation, attack or defense, etc. In this respect, it is important to interconnect the semiochemicals with the concrete interactional motif (Pertseva and Shpakov 2009; Bräsen et al. 2014; Charlesworth and Burns 2015). Identical signals may be used in several interaction motifs with quite different meanings, and this indicates that interaction context determines meaning of signals.

4.2 Sensing, Interpretation, Memory of Abiotic Indices and Coordination of Response

The next level to be identified in this perspective is the empirical data on how archaea sense their abiotic environment, i.e., pH level, temperature, radiation, moisture, gravity, etc. Such sensing is crucial for the survival and interactional (e.g., symbiotic) motifs of these organisms because it may decide their behavior. Sensing may lead to correct interpretation, which means the comparison with lasting experiences in this field are memorized. Epigenetic markings are essential tools in this respect (Ammar et al. 2012; Oliverio and Katz 2014). The benefit of correct interpretation is that it may lead to faster and more appropriate reactions.

4.3 Intraorganismic Communication in Archaea

The various behavioral motifs, as appropriate response behavior to environmental sensing, interpretation, and memory, leads us to various signaling pathways within archaeal organisms (Jung 2007; Schlesner et al. 2009; Pal et al. 2015; Grau-Bové et al. 2015; Esser et al. 2016). Intraorganismic investigations will focus on metabolic signaling pathways and signaling within and between archaeal genomes (Shockley et al. 2003; Coker and DasSarma 2007; Brochier-Armanet et al. 2011; Sato and Atomi 2011; Atomi et al. 2012; Morris et al. 2013; Kono et al. 2017), and —additionally—the signaling pathways between archaeal plasmids and infectious agents such as viruses, to generate available resources for the generation of signaling molecules of the archaea to other organisms (Bräsen et al. 2014; Quemin et al. 2016).

One of the keylevels of archaeal intraorganismic communication certainly will focus on RNA processing pathways (Mura et al. 2013). More recently, the roles of circular RNAs, especially in a variety of coordinated splicing activities, have been investigated (Tang et al. 2002; Petkovic and Müller 2015). In addition, the repair pathways are of special interest in archaea because of their extreme high temperature exposure (Grogan 2004; Ishino ad Narumi 2015).

The special kind of signaling which is needed for that is of exceptional interest and leads us to the interactions between archaea and persistent viruses and viral derived "defectives", such as the variety of mobile genetic elements in archaea (Fig. 1).

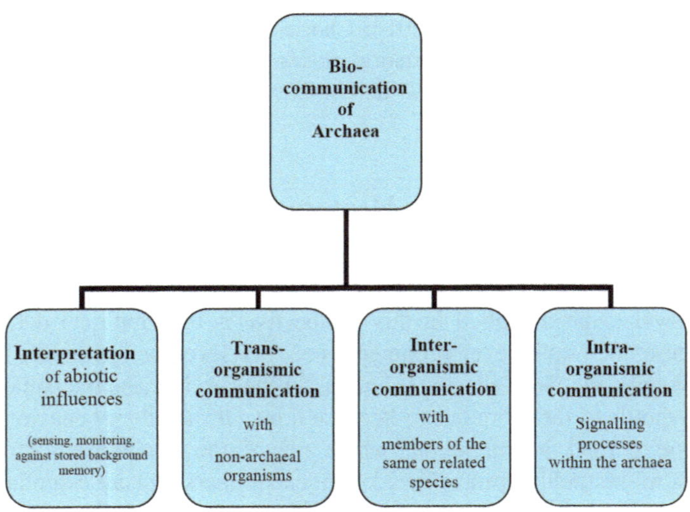

Fig. 1 Key levels of biocommunication of archaea

4.3.1 The Role of Persistent Viruses

As mentioned in recent years, the lytic consequences of viral infection are a special case if viruses are not able to develop a sessile lifestyle without harming the host (Villarreal 2005, 2009). In most cases, viruses living within organisms—or integrated as "defectives"—help to ward off competing (in most cases, close related) parasites, from the host and become part of its evolutionary history (Ortmann et al. 2006; Krupovic et al. 2011; Krupovic and Koonin 2016). Interestingly, such defectives may represent mobile genetic elements, such as transposons or introns, which are not very frequently found in archaea but are still present (Filée et al. 2007; Bagasra and Pace 2011; Makarova et al. 2014; Wang et al. 2015; Simon et al. 2016). The recycling and recombination of such defective viral parts ("defectives") represents a constant resource for host genetic identity (Villarreal 2016). It is obvious that such extreme environments may support recombination events between the RNA consortia, RNA viruses and DNA viruses (Snyder et al. 2015; Rensen et al. 2016; Stedman 2013, 2015); i.e. access of the ancient RNA world to the ancient and evolutionary polyphyletic DNA viruses (Koonin et al. 2006; Villarreal 2011; Villarreal and Witzany 2015).

Viruses have long been accepted only as disease-causing, epidemic phenomena, with lytic and, therefore, dangerous consequences for infected organisms. However, new research has corrected this picture (Forterre 2010; Prangishvili 2013). Viruses are part of the living world; in most cases, integrated in the cytoplasm or the nucleoplasma of cells, without harming the host (Adriaenssens et al. 2016). Viruses are on their way to representing the best examples of symbiotic relationships, because there is no living being since the start of life that has not been colonized by them, most often in the form of multiple colonizations (Ryan 2007; Villarreal 2005, 2009; Witzany 2012c).

Viruses infecting archaea, as described so far, all carry either single stranded or double stranded DNA genomes. Interesingly, archaeal viruses are rather unique, have an exceptional virion architecture, and can be found in the other domains of life at not more than 1% (Prangishvili 2015; Mochizuki et al. 2012; Iranzo et al. 2016)

Persistent viruses are decisive for species diversity and host genome editing (Villarreal 2012). The persistent status is most often reached by a competing genetic parasite that counterregulates their competences, now co-opted by the host organism for regulating all replication-relevant pathways and intron excision (Witzany 2014; Villarreal 2015, 2016). The recently found communicative capabilities of bacteriophages (Erez et al. 2017) can be predicted to be valid also for archaea.

Especially, the ability to generate really new sequences (not out of previous ones) allows such groups to constantly infect other nucleic sequence-based agents, whether virus-like or cellular genomes, such as archaea (Prangishvili 2013, 2015). The generation of such new sequences by phages (Villarreal 2005) or by co-operating RNA stem loop groups, leads to identity groups, such as viruses that represent toxic codes and even counteracting antitoxins. Persistently (non-lytic), infected host organisms are the preferred habitat, where former competing agent groups are unified in the basic behavioural motif of 'addiction modules' that can be identified as Toxin-Antitoxin, Restriction-Modification, Insertion-Deletion modules. All of the former competing groups are now unified to form stable/unstable modules that are counter-regulated and also provide immunity against related genetic parasites (Villarreal, 2009, 2011, 2015). Additionally, the CRIPRs/Cas adaptive immunity in archaea can be investigated under this perspective (Makarova et al. 2014; Villarreal and Witzany 2015—supplementary file 3, Krupovic et al. 2016; Béguin et al. 2016).

The benefit for host organisms, colonized successfully by persistent viruses and/or by their defectives that remain as co-adapted regulatory tools, is the immune function against similar genetic colonizers (Marrafini and Sontheimer 2010; Villarreal 2011; Sharama et al. 2016; Seligmann and Raoult 2016).

In this respect, there are some indices and good arguments which suggest that archaea infecting large dsDNA viruses may be an evolutionary forerunner of the eukaryotic nucleus (Forbes et al. 1983; Villarreal and DeFilippis 2000; Bell 2006; Villarreal 2005, 2009; Chaikeeratisak et al. 2017).

4.4 *Interorganismic Communication in Archaea*

Sign-mediated interactions between archaea can be investigated by comparing signaling and interconnected behavioral motifs. In biofilm formation, archaea produce and exchange a variety of signals that are sensed and interpreted (Koerdt et al. 2010; Fröls 2013; Bennighoff et al. 2017). Additionally, the coordinated behavior between two archaea depends on correct uptake of the signals by the receptors and transportation to the intracellular domain of the archaea for further information processing (Orell et al. 2014). Quorum sensing and quorum quenching in archaea, is an important tool to coordinate common behavioral motifs in this respect (Mackin 2011; Zhang et al. 2012; Montgomery et al. 2013), not to forget that each individual has to activate the signaling pathways within its organism to produce appropriate semiochemicals.

4.5 *Transorganismic Communication in Archaea*

If archaea interact with non-archaeal organisms, we can identify several signaling pathways which coordinate this interaction, either in a variety of symbiotic (e.g., sponges, corals) or parasitic lifestyles (Pape et al 2006; Wrede et al. 2012; Bayer et al. 2014). Some methanogenic archaea can be found symbiotically in arthropod guts, rumen of cattles, and human intestine. In nearly all known cases, these are highly specialized associations (Moissl-Eichinger and Huber 2011).

5 The Benefit of the Biocommunication Method

The benefit of the biocommunication approach is that it can be handled as a relatively easy method to identify all biotic features of organisms empiristically, but in a non-mechanistic and non-reductionistic way, and in contrast to all other recent approaches and concepts (Witzany 2015, 2016). In parallel, it strictly avoids (quasi-) metaphysical narratives such as holism, materialism, mechanicism, and atomism, to identify all secreted and produced semiochemicals that are used in all signaling processes within archaea, between archaea, and between archaea and non-archaeal organisms. Additionally, the biocommunication method tries to identify the behavioral motifs that follow such signaling processes. Last, but not least, it is interesting to identify how archaea sense, monitor, interpret, and finally react to changing abiotic influences. Since the biocommunication method is relevant also in other domains (viruses, bacteria, protozoa, fungi, animals and plants), this may lead to an integrative biology (Witzany 2016).

6 Conclusions

An overview of all significant levels of archaea communication shows, that identification of signal-mediated processes in signaling pathways are context-dependent, both within and among archaea, as well as between archaea and other organisms. Depending on the context, semiochemicals (molecular components) are produced and integrated into unique signaling pathways where they are used to transport certain meanings. Such meanings are subject to change, i.e., they rely on the various behavioral contexts, which differ under altered conditions. The interactional context determines the semantic relationship, i.e., its meaning, and the function of the chemical components, and forms a signal-mediated communication pattern in archaea.

After recognizing how versatile archaeal communication competences really are, we can see that one main principle is followed throughout all these signaling processes: archaea coordinate all their behavioral patterns with a core set of chemical molecules. The interactional context and the different modes of coordinating appropriate response behavior in e.g. development, growth, mating, attack, defense, etc. determine the combinations of signals that generate the appropriate meaning function, i.e. the informational content of messages. These generating processes normally function in a very conservative way, but under certain circumstances may fail, or selective pressure may lead to changes that can be a driving force in archaea evolution. Additionally, it must be recognized that the persistent lifestyle of viruses is a driving force in archaea evolution, in that they are the main resource for immunity and genetic identity of these fascinating organisms.

References

Abedon S (2013) Are archaeons incapable of being parasites or have we simply failed to notice? Bioessays 35:501

Adriaenssens EM, Krupovic M, Knezevic P, Ackermann HW, Barylski J, Brister JR, Clokie MR, Duffy S, Dutilh BE, Edwards RA, Enault F, Jang HB, Klumpp J, Kropinski AM, Lavigne R, Poranen MM, Prangishvili D, Rumnieks J, Sullivan MB, Wittmann J, Oksanen HM, Gillis A, Kuhn JH (2016) Taxonomy of prokaryotic viruses: 2016 update from the ICTV bacterial and archaeal viruses subcommittee. Arch Virol 162:1153–1157

Ammar R, Torti D, Tsui K, Gebbia M, Durbic T, Bader GD, Giaever G, Nislow C (2012) Chromatin is an ancient innovation conserved between Archaea and Eukarya. Elife 1:e00078

Atomi H, Imanaka T, Fukui T (2012) Overview of the genetic tools in the Archaea. Front Microbiol 3:337

Bagasra O, Pace DG (2011) Back to the soil: retroviruses and transposons. In: Witzany G (ed) Biocommunication in soil microorganisms. Springer, Heidelberg, pp 161–187

Bayer K, Kamke J, Hentschel U (2014) Quantification of bacterial and archaeal symbionts in high and low microbial abundance sponges using real-time PCR. FEMS Microbiol Ecol 89:679–690

Béguin P, Charpin N, Koonin EV, Forterre P, Krupovic M (2016) Casposon integration shows strong target site preference and recapitulates protospacer integration by CRISPR-Cas systems. Nucleic Acids Res 44:10367–10376

Bell P (2006) Sex and the eukaryotic cell cycle is consistent with a viral ancestry of the eukaryotic nucleus. J Theor Biol 243:54–63

Bennighoff JC, Wingender J, Flemming HC, Siebers B (2017) Biofilms X-treme: composition of extracellular polymeric substances in Archaea. In: Flemming HC, Neu TR, Wingender J (eds) The perfect slime. Microbial extracellular polymeric substances. IWA Publishing, London, pp 301–320

Bräsen C, Esser D, Rauch B, Siebers B (2014) Carbohydrate metabolism in archaea: current insights into unusual enzymes and pathways and their regulation. Microbiol Mol Biol Rev 78:89–175

Brochier-Armanet C, Forterre P, Gribaldo S (2011) Phylogeny and evolution of the archaea: one hundred genomes later. Curr Opin Microbiol 14:274–281

Chaikeeratisak V, Nguyen K, Khanna K, Brilot AF, Erb ML, Coker JK, Vavilina A, Newton GL, Buschauer R, Pogliano K, Villa E, Agard DA, Pogliano J (2017) Assembly of a nucleus-like structure during viral replication in bacteria. Science 355:194–197

Charlesworth JC, Burns BP (2015) Untapped resources: biotechnological potential of peptides and secondary metabolites in archaea. Archaea 4(2015):282035

Coker JA, DasSarma S (2007) Genetic and transcriptomic analysis of transcription factor genes in the model halophilic archaeon: coordinate action of TbpD and TfbA. BMC Genet 8:61

Eigen M, Winkler R (1983) The laws of the game: how the principles of nature govern chance. Princeton University Press, Princeton

Erez Z, Steinberger-Levy I, Shamir M, Doron S, Stokar-Avihail A, Peleg Y, Melamed S, Leavitt A, Savidor A, Albeck S, Amitai G, Sorek R (2017) Communication between viruses guides lysis-lysogeny decisions. Nature 541:488–493

Esser D, Hoffmann L, Pham TK, Bräsen C, Qiu W, Wright PC, Albers SV, Siebers B (2016) Protein phosphorylation and its role in archaeal signal transduction. FEMS Microbiol Rev 40:625–647

Filée J, Siguier P, Chandler M (2007) Insertion sequence diversity in archaea. Microbiol Mol Biol Rev 71:121–157

Forbes DJ, Kirschner MW, Newport JW (1983) Spontaneous formation of nucleus-like structures around bacteriophage DANN microinjected into Xenopus eggs. Cell 34:13–23

Forterre P (1997) Archaea: what can we learn from their sequences? Curr Opin Genet Dev 7:764–770

Forterre P (2010) Defining life: the virus viewpoint. Orig Life Evol Biosph 40:151–160
Forterre P (2013) The common ancestor of archaea and eukarya was not an archaeon. Archaea 2013:372396
Fröls S (2013) Archaeal biofilms: widespread and complex. Biochem Soc Trans 41:393–398
Garrett R, Klenk HP (2007) Archaea: evolution, physiology, and molecular biology. Blackwell Publishing, Maiden
Grau-Bové X, Sebé-Pedrós A, Ruiz-Trillo I (2015) The eukaryotic ancestor had a complex ubiquitin signaling system of archaeal origin. Mol Biol Evol 32:726–739
Grogan DW (2004) Stability and repair of DNA in hyperthermophilic Archaea. Curr Issues Mol Biol 6:137–144
Iranzo J, Koonin EV, Prangishvili D, Krupovic M (2016) Bipartite network analysis of the archaeal virosphere: evolutionary connections between viruses and capsidless mobile elements. J Virol 90:11043–11055
Ishino Y, Narumi I (2015) DNA repair in hyperthermophilic and hyperradioresistant microorganisms. Curr Opin Microbiol 25:103–112
Jung KH (2007) The distinct signaling mechanisms of microbial sensory rhodopsins in Archaea, Eubacteria and Eukarya. Photochem Photobiol 83:63–69
Koerdt A, Goedeke J, Berger J, Thormann KM, Albers SV (2010) Crenarchaeal biofilm formation under extreme conditions. PLoS ONE 5(11):e14104
Koonin EV (2010) On the origin of cells and viruses: primordial virus world scenario. In: Witzany G (ed) Natural genetic engineering and natural genome editing. Wiley, Hoboken, pp 47–64
Koonin E, Senkevich TG, Dolja VV (2006) The ancient virus world and evolution of cells. Biol Dir 1:29
Kono T, Mehrotra S, Endo C, Kizu N, Matusda M, Kimura H, Mizohata E, Inoue T, Hasunuma T, Yokota A, Matsumura H, Ashida H (2017) A RuBisCO-mediated carbon metabolic pathway in methanogenic archaea. Nat Commun 8:14007
Krupovic M, Koonin EV (2016) Self-synthesizing transposons: unexpected key players in the evolution of viruses and defense systems. Curr Opin Microbiol 31:25–33
Krupovic M, Prangishvili D, Hendrix RW, Bamford DH (2011) Genomics of bacterial and archaeal viruses: dynamics within the prokaryotic virosphere. Microbiol Mol Biol Rev 75:610–635
Krupovic M, Shmakov S, Makarova KS, Forterre P, Koonin EV (2016) Recent mobility of casposons, self-synthesizing transposons at the origin of the CRISPR-Cas immunity. Genome Biol Evol 8:375–386
Mackin C (2011) Quorum sensing in Archaea. Honors Scholar Theses, p 206
Makarova KS, Wolf Y, Forterre P, Prangishvili D, Krupovic M, Koonin EV (2014) Dark matter in archaeal genomes: a rich source of novel mobile elements, defense systems and secretory complexes. Extremophiles 18:877–893
Marraffini LA, Sontheimer EJ (2010) CRISPR interference: RNA directed adaptive immunity in bacteria and archaea. Nat Rev Genet 11:181–190
Mochizuki T, Krupovic M, Pehau-Arnaudet G, Sako Y, Forterre P, Prangishvili D (2012) Archaeal virus with exceptional virion architecture and the largest single-stranded DNA genome. Proc Natl Acad Sci U S A 109:13386–13391
Moissl-Eichinger C, Huber H (2011) Archaeal symbionts and parasites. Curr Opin Microbiol 14:364–370
Montgomery K, Charlesworth JC, LeBard R, Visscher PT, Burns BP (2013) Quorum sensing in extreme environments. Life (Basel) 3:131–148
Morris BEL, Henneberger R, Huber H, Moissl-Eichinger C (2013) Microbial syntrophy: interaction for the common good. FEMS Microbiol Rev 37:384–406
Mura C, Randolph PS, Patterson J, Cozen AE (2013) Archaeal and eukaryotic homologs of Hfq: a structural and evolutionary perspective on Sm function. RNA Biol 10:636–651

Oliverio AM, Katz LA (2014) The dynamic nature of genomes across the tree of life. Genome Biol Evol 6:482–488
Orell A, Fröls S, Albers SV (2014) Archaeal biofilms: the great unexplored. Ann Rev Microbiol 67:337–354
Ortmann AC, Wiedenheft B, Douglas T, Young M (2006) Hot crenarchaeal viruses reveal deep evolutionary connections. Nat Rev Microbiol 4:520–528
Pal A, Banerjee R, Mondal UK, Mukhopadhyay S, Bothra AK (2015) Deconstruction of archaeal genome depict strategic consensus in core pathways coding sequence assembly. PLoS ONE 10 (2):e0118245
Pape T, Hoffmann F, Quéric NV, Juterzenka K, Reitner J, Michaelis W (2006) Dense populations of Archaea associated with the demosponge *Tentorium semisuberites* Schmidt, 1870 from Arctic deep-waters. Polar Biol 29:662–667
Pertseva MN, Shpakov AO (2009) The prokaryotic origin and evolution of eukaryotic chemosignaling systems. Neurosci Behav Physiol 39:793–804
Petkovic S, Müller S (2015) RNA circularization strategies in vivo and in vitro. Nucleic Acids Res 43:2454–2465
Prangishvili D (2013) The wonderful world of archaeal viruses. Annu Rev Microbiol 67:565–585
Prangishvili D (2015) Archaeal viruses: living fossils of the ancient virosphere? Ann N Y Acad Sci 1341:35–40
Quemin ER, Chlanda P, Sachse M, Forterre P, Prangishvili D, Krupovic M (2016) Eukaryotic-like virus budding in archaea. MBio 13;7(5). pii: e01439-16
Raymann K, Brochier-Armanet C, Gribaldo S (2015) The two-domain tree of life is linked to a new root for the Archaea. Proc Natl Acad Sci U S A 112:6670–6675
Rensen EI, Mochizuki T, Quemin E, Schouten S, Krupovic M, Prangishvili D (2016) A virus of hyperthermophilic archaea with a unique architecture among DNA viruses. Proc Natl Acad Sci U S A 113:2478–2483
Ryan F (2009) Virolution. Harper Collins, London
Ryan RF (2007) Viruses as symbionts. Symbiosis 44:11–21
Sato T, Atomi H (2011) Novel metabolic pathways in archaea. Curr Opin Microbiol 14:307–314
Schlesner M, Miller A, Streif S, Staudinger WF, Müller J, Scheffer B, Siedler F, Oesterhelt D (2009) Identification of Archaea-specific chemotaxis proteins which interact with the flagellar apparatus. BMC Microbiol 9:56
Seligmann H, Raoult D (2016) Unifying view of stem-loop hairpin RNA as origin of current and ancient parasitic and non-parasitic RNAs, including in giant viruses. Curr Opin Microbiol 31:1–8
Shannon CE, Weaver W (1949) The mathematical theory of communication. University of Illinois Press, Campaign
Shockley KR, Ward DE, Chhabra SR, Conners SB, Montero CI, Kelly RM (2003) Heat shock response by the hyperthermophilic archaeon *Pyrococcus furiosus*. Appl Environ Microbiol 69:2365–2371
Simon DM, Clarke NAC, McNeil BA, Johnson I, Pantuso D, Dai L, Chai D, Zimmerly S (2016) Group II introns in eubacteria and archaea: ORF-less introns and new varieties. Wiley Interdiscip Rev RNA 7:341–355
Snyder JC, Bolduc B, Young MJ (2015) 40 Years of archaeal virology: expanding viral diversity. Virology 479–480:369–378
Stedman K (2013) Mechanisms for RNA capture by ssDNA viruses: grand theft RNA. J Mol Evol 76:359–364
Stedman KM (2015) Deep recombination: RNA and ssDNA virus genes in DNA virus and host genomes. Annu Rev Virol 2:203–217
Tang TH, Rozhdestvensky TS, d'Orval BC, Bortolin ML, Huber H, Charpentier B, Branlant C, Bachellerie JP, Brosius J, Hüttenhofer A (2002) RNomics in Archaea reveals a further link between splicing of archaeal introns and rRNA processing. Nucleic Acids Res 30:921–930
Villarreal LP (2005) Viruses and the evolution of life. ASM Press, Washington

Villarreal LP (2009) The origin of group identity. Viruses, addiction and cooperation. Springer, New York
Villarreal LP (2011) Viruses and host evolution: virus-mediated self identity. In: Lopez-Larrea C (ed) Self and non-self. Landes Bioscience and Springer Science Business Media, Austin, pp 185–217
Villarreal LP (2012) The addiction module as a social force. In viruses: essential agents of life. In: Witzany G (ed) Viruses: essential agents of life. Springer, Dordrecht, pp 107–145
Villarreal LP (2015) Force for ancient and recent life: viral and stem-loop RNA consortia promote life. Ann N Y Acad Sci 1341:25–34
Villarreal LP (2016) Persistent virus and addiction modules: an engine of symbiosis. Curr Opin Microbiol 31:70–79
Villarreal LP, DeFilippis VR (2000) A hypothesis for DNA viruses as the origin of eukaryotic replication proteins. J Virol 74:7079–7084
Villarreal LP, Witzany G (2010) Viruses are essential agents within the roots and stem of the tree of life. J Theor Biol 262:698–710
Villarreal LP, Witzany G (2013a) The DNA habitat and its RNA inhabitants: at the dawn of RNA sociology. Genomics Insights 6:1–12
Villarreal LP, Witzany G (2013b) Rethinking quasispecies theory: from fittest type to cooperative consortia. World J Biol Chem 4:79–90
Villarreal LP, Witzany G (2015) When competing viruses unify: evolution, conservation, and plasticity of genetic identities. J Mol Evol 80:305–318
Wang H, Peng N, Shah SA, Huang L, She Q (2015) Archaeal extrachromosomal genetic elements. Microbiol Mol Biol Rev 79:117–152
Witzany G (1993) Natur der Sprache - Sprache der Natur. Sprachpragmatische Philosophie der Biologie. Würzburg, Königshausen & Neumann
Witzany G (1995) From the "logic of the molecular syntax" to molecular pragmatism. Explanatory deficits in Manfred Eigen's concept of language and communication. Evol Cog 1:148–168
Witzany G (2000) Life: the communicative structure. Libri Books on Demand, Norderstedt
Witzany G (2005) Natural history of life: history of communication logics and dynamics. SEED J 5:27–55
Witzany G (2009) Noncoding RNAs: persistent viral agents as modular tools for cellular needs. Ann N Y Acad Sci 1178:244–267
Witzany G (2010) Biocommunication and natural genome editing. Springer, Dortrecht
Witzany G (ed) (2011) Biocommunication in soil microorganisms. Springer, Dordrecht
Witzany G (ed) (2012a) Biocommunication of fungi. Springer, Dordrecht
Witzany G (ed) (2012b) Viruses: essential agents of life. Springer, Dordrecht
Witzany G (2012c) From molecular entities to competent agents: viral infection-derived consortia act as natural genetic engineers. In: Witzany G (ed) Viruses: essential agents of life. Dordrecht, Netherlands, pp 407–419
Witzany G (ed) (2014) Biocommunication of animals. Springer, Dordrecht
Witzany G (2015) Life is physics and chemistry and communication. Ann N Y Acad Sci 1341:1–9
Witzany G (2016) The biocommunication method: on the road to an integrative biology. Commun Integr Biol 9(2):e1164374
Witzany G, Baluska F (eds) (2012a) Biocommunication of plants. Springer, Heidelberg
Witzany G, Baluska F (2012b) Life's code script does not code itself. The machine metaphor for living organisms is outdated. EMBO Rep 13:1054–1056
Witzany G, Nowacki M (eds) (2016) Biocommunication of ciliates. Springer, Dordrecht
Woese C (2015) Carl Woese evolutions's golden revoltionary. In: Mazur S (ed) The paradigm shifters. Overthrowing the hegemony of the culture of Darwin. Caswell Books, New York
Woese CR, Fox GE (1977) Phylogenetic structure of the prokaryotic domain: the primary kingdoms. Proc Natl Acad Sci U S A 74:5088–5090

Woese CR, Kandler O, Wheelis ML (1990) Towards a natural system of organisms: proposal for the domains Archaea, Bacteria, and Eukarya. Proc Natl Acad Sci U S A 87:4576–4579

Wrede C, Dreier A, Kokoschka S, Hoppert M (2012) Archaea in symbiosis. Archaea 2012:596846

Zhang G, Zhang F, Ding G, Li J, Guo X, Zhu J, Zhou L, Cai S, Liu X, Luo Y, Zhang G, Shi W, Dong X (2012) Acyl homoserine lactone-based quorum sensing in a methanogenic archaeon. ISME J 6:1336–1344

The Cooccurrence of Archaea and Bacteria Among Diverse Globally Distributed Ecosystems

Daniel R. Colman

1 Introduction: Historical Perspective on Archaeal Ecology

Though the discovery of Archaea is commonly attributed to Carl Woese and colleagues, it is important to note that Woese's greatest contribution (and arguably the most important contribution to the study of Archaea so far) was not his culturing of a novel third domain of life, but rather his ability to synthesize available ideas, information and techniques to ultimately discover that the Archaea are a separate biological entity from the Eukaryotes and the Bacteria. Archaea have been isolated and described as early as the late 1800s, when halophilic Archaea (then presumed to be bacteria) were isolated as a means to understand the preservation of salt-packed fish (Kocur and Hodgkiss 1973). Additionally, archaeal methanogens (then also considered a curious type of bacteria) had been commonly isolated from the 1930s onwards. Both of these organismal groups were thought of as strange prokaryotes—methanogens because of their unique biochemistry and the halophiles because of their salinity tolerance. It, however, was not until the efforts in understanding these odd microbial organisms' place in evolutionary trees by Carl Woese et al., that it became clear that methanogenic, halophilic and thermoacidophilic Bacteria actually belonged to a domain of life that was entirely distinct from both Bacteria and eukaryotes (Woese and Fox 1977; Woese et al. 1990).

Following the seminal late 1970s work by Woese and colleagues, members of his collaborative group and his academic progeny advanced the study of Archaea through two different methodological trajectories: culture-based methods and molecular, cultivation-independent techniques. Of equal importance to

D.R. Colman (✉)
Department of Microbiology & Immunology, Montana State University, Bozeman, MT, USA
e-mail: daniel.colman@montana.edu

G. Witzany (ed.), *Biocommunication of Archaea*,
DOI 10.1007/978-3-319-65536-9_2

understanding the phylogenetic relations and diversity of the Archaea is understanding their physiological characteristics in order to inform phylogenetic studies. In this regard, some of the most important early researchers to characterize Archaea, Karl Stetter, and Wolfram Zillig, were both also involved in the initial discovery of the archaeal domain. While Stetter's work has focused on hyperthermophiles, which he was first to discover, his additions to culture collections have helped fill in information gaps concerning the physiological characteristics of Archaea that would be otherwise weighted heavily towards the readily cultured halophiles and mesophilic methanogens. Stetter followed earlier culturing precedents set by Ralph Wolfe (Stetter et al. 1981), who was also influential in the third domain discovery. Using culture conditions that mimic the anoxic, volcanic gas-rich environments of hyperthermophiles, they were able to recover a far greater archaeal diversity from these environments than was previously thought to exist. As of current, the vast majority of isolates within the Thermoprotei, and a significant portion of the hyperthermophilic Euryarchaeota were first cultured and discovered by Stetter, Zillig and their colleagues.

While Stetter, Zillig and others were making great advancements towards elucidating the known diversity of Archaea and providing cultures for physiological study, the molecular biology offshoot of Woese's idealogical descendents were also beginning to develop research programs that would greatly influence the study of Archaea as well as Bacteria. In particular, Norman Pace, a post-doctoral researcher at the University of Illinois at the time of Woese's discovery, began to develop a method incorporating Woese's comparative phylogenetics techniques with environmental microbial communities. As a logical extension to the phylogenetic analysis of strains based on ssu rRNA identities, Pace's group developed a method for fingerprinting natural microbial communities based on the 16S ssu rRNA gene (Olsen et al. 1986). The process relied on extracting environmental DNA, shotgun cloning the DNA to create a vector library, amplifying the library and probing for the presence of 16S rRNA gene sequences. Making use of the newly described dideoxy termination sequencing method developed by Fred Sanger et al., the sequences could be compared against a reference database and the in situ microbial communities could then be inferred based on the database query results. While Pace's approach to ecological analyses did not directly concern the Archaea initially, members of his lab group, as well as those influenced by his methods, quickly began to advance the discovery and study of Archaea without the hindrance of culturing.

In one of the early archaeal studies to use Pace's approach, Edward Delong, a post-doc of Norman Pace's, used archaeal-specific probes to detect archaeal 16S rRNA genes in surface coastal waters off of North America. Surprisingly, he found a particularly abundant ribotype that was highly divergent from other Archaea, but most closely affiliated with the Crenarchaeota (DeLong 1992). Given that the Crenarchaeota were only known to be extreme hyperthermophiles at the time, and other Archaea (aside from the methanogens) were only known to be extremophiles, the finding was surprising and provocative. The discovery of this ribotype was certainly curious and at the time suggested that the previous understanding of

Archaea as strictly extremophilic was not entirely accurate. Importantly, this study marked the beginning of research into what we now know are members of the Thaumarchaeota phylum, which has gained a considerable amount of attention in recent years (discussed later). Importantly, this study also marked the beginning of a more accurate understanding of archaeal diversity outside of the realm of extreme environments.

Throughout the 1990s and 2000s, myriad studies of uncultured 'ribotypes' via environmental 16S rRNA gene surveys led to a surprisingly large amount of known archaeal phylogenetic diversity in a wide variety of ecosystems. In particular, known archaeal diversity was no longer restricted to just extreme environments, but also most environments where Bacteria had been thought to be dominant. More recently, the advent and application of environmental genomics (Tyson et al. 2004) has led to a number of new insights into the functional potential of many of these uncultured archaeal lineages and has been highlighted by expanding roles of Archaea in global carbon, nitrogen and sulfur cycling.

Following from the traditional perspective that Archaea dominate extreme environments to the exclusion of Bacteria, Archaea were largely thought to not significantly overlap ecologically with Bacteria (Valentine 2007). It was not until the seminal environmental studies by Ed Delong et al. which indicated that Archaea were particularly numerous in more mesic environments. The widespread use of cultivation-independent techniques has significantly changed our view of archaeal ecology and, in particular, their ecological co-occurrence with bacterial populations across globally distributed systems. Below, I describe advances in our understanding in the overlap between archaeal and bacterial taxa in numerous environments, highlighting some seminal finds.

2 Methane Oxidizing Archaeal and Bacterial Consortia

One of the most integral discoveries that have been made regarding the interactions between Archaea and Bacteria has been that of anaerobic oxidizing methane (AOM) consortia. AOM are widespread among oceanic environments and comprise diverse groups of euryarchaeal organisms that often cohabitate with bacterial consortia to couple oxidation of methane (CH_4) with the reduction of several oxidants including SO_4^{2-}, NO_3^- and multiple metal oxidants (Knittel and Boetius 2009; Scheller et al. 2010a, b). While the anaerobic oxidation of methane has long been known, the precise microbial mechanisms of this process were only recently discovered. These inter-domain consortia potentially leverage electron transfer conduits in order to surmount energetic barriers to the anaerobic oxidation of methane, which can occur via coupling to a variety of oxidants, including sulfate and nitrate, among others (Beal et al. 2009; Ettwig et al. 2010; Scheller et al. 2010a, b; Lai et al. 2016). A number of anaerobic methane oxidizing euryarchaeotal groups (ANME) are known, which are thought to mediate the anaerobic oxidation of methane in association with bacteria. ANME are affiliated with methanogens in the taxonomic groups of Methanosarcinales, Methanobacteriales, and the Methanococcoides which all belong to the Euryarchaeota phylum. Recently, the isolation of the first anaerobic alkane oxidizing Archaea, *Candidatus* Syntrophobacterium spp. provided important insights into the physiological nature of these organisms (Laso-Perez et al. 2016). Although, somewhat surprisingly, *Candidatus* Syntrophobacterium spp. were responsible for longer chain alkane degradation (butane) via butane activation with the key enzyme in methanogenesis/reverse methanogenesis (McrA).

Of particular global importance, these archaeal and bacterial consortia are responsible for a significant amount of mitigation of global methane flux from the open ocean environments into the atmosphere. For example, 7–25% of the total global methane flux is consumed by AOM, which also nearly for the near total oceanic methane fluxes (Knittel and Boetius 2009). Methane is a potent greenhouse gas, with an order of magnitude greater global warming potential compared to carbon dioxide. Thus, these anaerobic methane oxidizing consortia play a critical role in the biogeochemical cycling of carbon, in addition to the regulation of earth's climatic variations. However, there is still a considerable amount of physiological information that is unknown due to the recalcitrance of these organisms and consortia to cultivation.

3 Ammonia Oxidizing Archaea in Open Oceans and Soils

Perhaps no greater contribution to our current understanding of archaeal ecology was made than by early efforts to understand the role of 'cold Crenarchaeota' that were first discovered by Delong et al. in the early 1990s in cold Antarctic ocean waters (DeLong 1992). These so called 'cold Crenarchaeota' are now known to belong to an entirely distinct phylum of Archaea known as the Thaumarchaeota (Brochier-Armanet et al. 2008), which comprise a diverse group of ammonium oxidizers that are distributed globally in fresh waters, soils, oceans and many other environments where they compete with bacterial assemblages. Early estimates indicated the Thaumarchaeota represented significant fractions of oligotrophic open ocean waters, which implicated them as potentially significant biogeochemical players (DeLong 1992). Interestingly, the first evidence for the physiology of the Thaumarchaeota arose not from free-living planktonic populations, as were discovered by Delong et al., but rather from a metagenomic analysis from a marine sponge which housed a different genus compared to open ocean phylotypes, *Candidatus* Cenarchaeum symbiosum (Preston et al. 1996). Sequencing of the sponge symbiont's genome indicated the capacity for ammonia oxidation, which was then later confirmed in free-living Thaumarchaeota. Considering the ubiquity of free-living thaumarchaeotes among water columns and sediments from globally distributed environments (Francis et al. 2005), their role in global nitrogen cycling is certainly appreciable. Although Ammonia Oxidizing Bacteria (AOB) were thought to be the predominant nitrifiers until the discovery of Ammonia Oxidizing

Archaea (AOA), quantitative cultivation-independent studies (assessing abundance of the key ammonia oxidation gene—ammonia monooxygenase—*amo*) have indicated that AOA are significantly more abundant than AOB in many environments, suggesting they likely play a larger biogeochemical role (Wuchter et al. 2006; Mincer et al. 2007).

Studies of the distribution of AOA and AOB via *amo* gene abundance distributions indicated discrete distributions of the two types of microorganisms along gradients in NH_4^+ availability (Wuchter et al. 2006; Agogue et al. 2008). Later, physiological studies of the first free-living thaumarchaeote isolate, *Candidatus* Nitrosopumilus maritimus (Konneke et al. 2005) provided context for these ecological distributions. In particular, *Ca.* N. maritimus exhibits NH_4^+ affinities that are over two orders of magnitude greater than that of AOB (Martens-Habbena et al. 2009). These data suggests that AOA outcompete AOB at low concentrations of NH_4^+ and thus occupy different ecological niches with regards to NH_4^+ availability. Intriguingly, although AOB were long considered the primary nitrifiers in most environments, the abundance and activity of AOA in oligotrophic environments indicates that AOA may be more globally important, as environments, and particularly the open ocean, tend to be considerably more oligotrophic. In addition to the dominance of AOA in ocean settings, they are also dominant in soil communities (Leininger et al. 2006), where they further contribute to global nitrogen cycling.

As in open oceans, AOA appear to occupy different ecological niches than AOB in other environments, including soils, as the former are generally more prevalent in low NH_4^+-availability environments and also exhibit similar adaptations to low NH_4^+ availabilities. Terrestrial cultivars of the Thaumarchaeota that were recovered after the initial thaumarchaeote isolate (*Ca.* N. maritimus) also exhibit high NH_4^+ affinities (Lehtovirta-Morley et al. 2011), suggesting that they are particularly adapted to natural environments with low nutrient fluxes. Perhaps most conspicuously, the AOA tend to dominate AOB in acidic soils (Nicol et al. 2008) and have provided the missing link to ammonium oxidation in acidic soils, where AOB were not known to be present.

4 Hydrothermal and Acidic Environments

Since the discovery of Archaea as a separate domain, the extreme environments of hydrothermal systems were thought to be mostly dominated by archaeal taxa. Indeed, the prevailing thought was that Archaea predominate in the hottest and most acidic of hydrothermal environments (Reysenbach and Shock 2002). While there is evidence to support their dominance in hydrothermal systems with the highest temperatures and lowest pH (Colman et al. 2017), they also comprise highly diverse and abundant populations in more ‘mesic’ hydrothermal systems (Bowen De Leon et al. 2013; Boyd et al. 2013; Inskeep et al. 2013; Colman 2015). Indeed,

Archaea may contribute significantly to community function within these environments. While less acidic systems have traditionally been squarely in the realm of bacterial dominance, recent cultivation-independent approaches have highlighted a growing role for Archaea in circumneutral hydrothermal environments alongside major bacterial community members (Bowen De Leon et al. 2013; Inskeep et al. 2013; Colman 2015). For instance, numerous uncultured divisions of Archaea have representative genomes that have been reconstructed from circumneutral hydrothermal environments, including members of the candidate divisions 'Bathyarchaeota', 'Hadesarchaea', and the 'Asgard Archaea' (Saw et al. 2015; Baker et al. 2016; Zaremba-Niedzwiedzka et al. 2017). While bacterial divisions in these environments are most likely to be the primary producers (Takacs-Vesbach et al. 2013), it is likely that many of these candidate divisions contribute to carbon cycling within these largely bacterial-dominated environments. For instance, the 'Bathyarchaeota' are thought to be potentially involved in C1 carbon compound cycling (i.e. through methanogenesis or methanotrophy; (Evans et al. 2015)) or potentially acetogenesis (He et al. 2016), although a role in heavier chain alkane degradation may be more likely (Laso-Perez et al. 2016). Regardless, it is clear that these uncultured, recently genomically analyzed, microbial taxa have clear interactions with their bacterial community members in what were once thought to be traditionally bacterially dominated environments.

Perhaps more intriguingly, are the geochemical environments where the shift from bacterial to archaeal dominance occurs within hydrothermal environments. Archaea can be found across the span of major geochemical gradients (i.e. temperature and pH) in hydrothermal systems (Boyd et al. 2013; Inskeep et al. 2013; Xie et al. 2014; Colman 2015). However, Archaea are particularly dominant in hot springs with the lowest pH and the highest temperatures, suggesting that they are uniquely adapted to these extremely stressful environments (Colman et al. 2017). Indeed, others have suggested that Archaea, as a whole, are especially adapted to energetically expensive environments and that energetic limitation is a defining feature of the archaeal domain (Valentine 2007). Though Archaea dominate the highest temperature acidic environments, Bacteria dominate some of the most acidic habitats known, albeit at lower temperatures (Tyson et al. 2004). The observation that the dominance between Archaea and Bacteria shifts along temperature gradients in low pH environments suggests that physical interactions may permit the exchange of genetic information via horizontal gene transfer (HGT) within and among the two domains where their ecological niches overlap at moderate temperatures. Some evidence from comparative genomics exists that genetic transfer has occurred down temperature gradients within acidic habitats within the archaeal domain from the Sulfolobales order taxa that are adapted to higher temperatures to Thermoplasmatales which are adapted to lower temperatures (Ruepp et al. 2000; Futterer et al. 2004; Colman et al. 2017), which may have contributed to the adaptation to acidophily by one or the other of the orders (Colman et al. 2017). However, future research may elucidate whether inter-domain HGT has occurred between Bacteria and Archaea in acidic environments, and whether this HGT has resulted in adaptation to these particularly extreme environments.

5 Ocean Sediment Environments

In addition to the discovery of Archaea as abundant members of open ocean waters, Archaea have long been known to be significant populations of open ocean sediments (Teske and Sorensen 2008; Lloyd et al. 2013). However, it was not until the recent application of environmental metagenomics that their physiologic function has come into focus. In particular, members of the 'Bathyarchaeota' are abundant in open ocean sediments where they cohabitate with organic carbon degrading Bacteria. In one of the first published reports regarding the potential function of 'Bathyarchaeota', it was suggested that they were predominantly responsible for degradation of proteins and other organic carbon metabolites (Lloyd et al. 2013). Later reports have indicated a role in other carbon metabolic pathways including methanogenesis and acetogenesis (discussed above). Regardless, as in other systems, it is becoming increasingly clear that Archaea in these environments, which were once thought to be predominantly bacterial, are important contributors to organic carbon degradation along with their co-occurring bacterial community consortia.

A recent study highlighted the potential function and interaction of thermophilic sediment Archaea with sulfate reducing bacterial consortia at the level of reducing equivalent transfers. As mentioned above, the *Candidatus* genus Syntrophoarchaeum was recently demonstrated to active butane using McrA, the key protein that is involved in activating methane during methanogenesis (Laso-Perez et al. 2016). Syntrophoarchaeum spp. are related to the ANME-1 group of Archaea that carry out methanotrophy via a near reversal of the methanogenesis pathway in concert with bacterial members of the Deltaproteobacteria class that perform sulfate reduction. However, these organisms oxidized butane in concert with co-habitated Deltaproteobacterial organisms in the 'HotSeep-1' clade. A particularly complicated problem for methane (or butane) oxidizing Archaea is that they must get rid of 'reducing equivalents' or excess reductant, which is typically performed in methanogens via the production of methane. Since AOM and butane-oxidizing Archaea run a near reversal of the methanogenesis pathway, an important physiological problem is passing electrons on elsewhere. In AOM consortia, and in Syntrophoarchaeum spp.-bacterial consortia, this problem is solved via passing the reducing equivalents (essentially electrons) through 'nanowires' or pilus-like structures that allow the hydrocarbon degrading Archaea to remove excess reductant, which is then provided to the bacterial partners for SO_4^{2-}, or other oxidant, reduction (Wegener et al. 2015; Laso-Perez et al. 2016). Though these recent analyses have provided much insight into the functioning of sediment and AOM Archaea, they also highlight that much more information is needed regarding the physiological nature and consequent ecological importance of many of these organisms, especially in hydrocarbon degradation, which is of central importance to our understanding of global carbon cycling.

One of the most intriguing questions that remains about the nature of archaeal and bacterial interactions are the nature of these interactions at the level of reductant and/or oxidant transfer. As the above studies highlight, consortia have co-evolved to mitigate the problem of removing reducing equivalents (from the AOM) and providing it to co-occurring partners (i.e. the sulfate reducing bacteria). In many cases, reducing equivalents can be transferred via H_2, which is produced by many organisms as a way to dispose of excess electrons. One of the core observations of archaeal ecology is that sulfate reducing bacteria generally outcompete Archaea for H_2, thus excluding methanogens from environments where these bacteria are present. That consortia have co-evolved around a mechanism of inter-domain electron transfer suggests that many more interactions that await to be discovered may also revolve around syntrophic interactions that rely on shuttling reductants or oxidants among partners in order to increase environmental habitability and/or lower thermodynamic barriers to certain metabolic processes, such as AOM. The recent documentation of 'Cable Bacteria' provides one such example that may be ripe for further understanding inter-domain interactions at the level of direct energy transfer.

6 Conclusions

Though Archaea were once thought to be relegated to only the most extreme environments and were generally presumed to be ecologically distinct from Bacteria, the application of increasingly insightful cultivation-independent analyses have shown otherwise. In particular, Archaea are now known to co-occur and interact with Bacteria in complex, global biogeochemical cycles in a variety of environments in addition to the most extreme environments. It may not be likely that additional microbial diversity will be discovered after the exhaustive efforts of the previous several decades. However, the functional role of most of the recently documented, uncultured Archaea has yet to be fully realized and holds much promise for future insights into their global role. Moreover, the mechanistic functional interactions between many of these Archaea and their co-occurring archaeal and bacterial community members at the intra- and inter-domain level remain to be elucidated and are an important target for future research concerning the interactions between Archaea and Bacteria, and in particular, understanding how these interactions lead to more emergent properties that influence evolutionary trajectories for lineages and global biogeochemical cycles.

References

Agogue H, Brink M, Dinasquet J, Herndl GJ (2008) Major gradients in putatively nitrifying and non-nitrifying Archaea in the deep North Atlantic. Nature 456:788–791

Baker BJ, Saw JH, Lind AE, Lazar CS, Hinrichs K, Teske AP et al (2016) Genomic inference of the metabolism of cosmopolitan subsurface Archaea, Hadesarchaea. Nat Microbiol 1

Beal EJ, House CH, Orphan VJ (2009) Manganese- and iron-dependent marine methane oxidation. Science 325:184–187

Bowen De Leon K, Gerlach R, Peyton BM, Fields MW (2013) Archaeal and bacterial communities in three alkaline hot springs in Heart Lake Geyser Basin, Yellowstone National Park. Front Microbiol 4:330

Boyd ES, Hamilton TL, Wang J, He L, Zhang CL (2013) The role of tetraether lipid composition in the adaptation of thermophilic archaea to acidity. Front Microbiol 4:62

Brochier-Armanet C, Boussau B, Gribaldo S, Forterre P (2008) Mesophilic crenarchaeota: proposal for a third archaeal phylum, the Thaumarchaeota. Nat Rev Microbiol 6:245–252

Colman DR (2015) Diversity of understudied archaeal and bacterial populations of Yellowstone National Park: from genes to genomes. Ph.D. thesis, University of New Mexico, Albuquerque, NM

Colman DR, Poudel S, Hamilton TL, Havig JR, Selensky MJ, Shock EL et al (2017) Geobiological feedbacks and the evolution of thermoacidophiles. *ISMEJ* (in Press)

DeLong EF (1992) Archaea in coastal marine environments. Proc Natl Acad Sci U S A 89:5685–5689

Ettwig KF, Butler MK, Le Paslier D, Pelletier E, Mangenot S, Kuypers MMM et al (2010) Nitrite-driven anaerobic methane oxidation by oxygenic bacteria. Nature 464:543–548

Evans PN, Parks DH, Chadwick GL, Robbins SJ, Orphan VJ, Golding SD et al (2015) Methane metabolism in the archaeal phylum Bathyarchaeota revealed by genome-centric metagenomics. Science 350:434–438

Francis CA, Roberts KJ, Beman JM, Santoro AE, Oakley BB (2005) Ubiquity and diversity of ammonia-oxidizing archaea in water columns and sediments of the ocean. Proc Natl Acad Sci U S A 102:14683–14688

Futterer O, Angelov A, Liesegang H, Gottschalk G, Schleper C, Schepers B et al (2004) Genome sequence of *Picrophilus torridus* and its implications for life around pH 0. Proc Natl Acad Sci U S A 101:9091–9096

He Y, Li M, Perumal V, Feng X, Fang J, Xie J et al (2016) Genomic and enzymatic evidence for acetogenesis among multiple lineages of the archaeal phylum Bathyarchaeota widespread in marine sediments. Nat Microbiol 1

Inskeep WP, Jay ZJ, Tringe SG, Herrgård MJ, Rusch DB, YNP Metagenome Project Steering Committee et al (2013) The YNP Metagenome project: environmental parameters responsible for microbial distribution in the yellowstone geothermal ecosystem. Front Microbiol 4:67

Knittel K, Boetius A (2009) Anaerobic oxidation of methane: progress with an unknown process. Annu Rev Microbiol 63:311–334

Kocur M, Hodgkiss W (1973) Taxonomic status of genus Halococcus Schoop. Int J Syst Bacteriol 23:151–156

Konneke M, Bernhard AE, de la Torre JR, Walker CB, Waterbury JB, Stahl DA (2005) Isolation of an autotrophic ammonia-oxidizing marine archaeon. Nature 437:543–546

Lai CY, Wen LL, Shi LD, Zhao KK, Wang YQ, Yang XE et al (2016) Selenate and nitrate bioreductions using methane as the electron donor in a membrane biofilm reactor. Environ Sci Technol 50:10179–10186

Laso-Perez R, Wegener G, Knittel K, Widdel F, Harding KJ, Krukenberg V et al (2016) Thermophilic archaea activate butane via alkyl-coenzyme M formation. Nature 539:396–401

Lehtovirta-Morley LE, Stoecker K, Vilcinskas A, et al (2011) Cultivation of an obligate acidophilic ammonia oxidizer from a nitrifying acid soil. Proc Natl Acad Sci U S A 108:15892–15897

Leininger S, Urich T, Schloter M, Schwark L, Qi J, Nicol GW et al (2006) Archaea predominate among ammonia-oxidizing prokaryotes in soils. Nature 442:806–809

Lloyd KG, Schreiber L, Petersen DG, Kjeldsen KU, Lever MA, Steen AD et al (2013) Predominant archaea in marine sediments degrade detrital proteins. Nature 496:215–218

Martens-Habbena W, Berube PM, Urakawa H, de la Torre JR, Stahl DA (2009) Ammonia oxidation kinetics determine niche separation of nitrifying Archaea and Bacteria. Nature 461:976–979

Mincer TJ, Church MJ, Taylor LT, Preston C, Karl DM, DeLong EF (2007) Quantitative distribution of presumptive archaeal and bacterial nitrifiers in Monterey Bay and the North Pacific Subtropical Gyre. Environ Microbiol 9:1162–1175

Nicol GW, Leininger S, Schleper C, Prosser JI (2008) The influence of soil pH on the diversity, abundance and transcriptional activity of ammonia oxidizing archaea and bacteria. Environ Microbiol 10:2966–2978

Olsen GJ, Lane DJ, Giovannoni SJ, Pace NR, Stahl DA (1986) Microbial ecology and evolution: a ribosomal RNA approach. Annu Rev Microbiol 40:337–365

Preston CM, Wu KY, Molinski TF, DeLong EF (1996) A psychrophilic crenarchaeon inhabits a marine sponge: *Cenarchaeum symbiosum* gen. nov., sp. nov. Proc Natl Acad Sci U S A 93:6241–6246

Reysenbach AL, Shock E (2002) Merging genomes with geochemistry in hydrothermal ecosystems. Science 296:1077–1082

Ruepp A, Graml W, Santos-Martinez ML, Koretke KK, Volker C, Mewes HW et al (2000) The genome sequence of the thermoacidophilic scavenger *Thermoplasma acidophilum*. Nature 407:508–513

Saw JH, Spang A, Zaremba-Niedzwiedzka K, Juzokaite L, Dodsworth JA, Murugapiran SK et al (2015) Exploring microbial dark matter to resolve the deep archaeal ancestry of eukaryotes. Philos Trans R Soc Lond B Biol Sci 370:20140328

Scheller S, Goenrich M, Boecher R, Thauer RK, Jaun B (2010a) The key nickel enzyme of methanogenesis catalyses the anaerobic oxidation of methane. Nature 465:606–U697

Scheller S, Goenrich M, Mayr S, Thauer RK, Jaun B (2010b) intermediates in the catalytic cycle of methyl coenzyme M reductase: isotope exchange is consistent with formation of a σ-Alkane-Nickel complex. Angew Chem Int Ed 49:8112–8115

Stetter KO, Thomm M, Winter J, Wildgruber G, Huber H, Zillig W et al (1981) Methanothermus-fervidus, sp-nov, a novel extremely thermophilic methanogen isolated from an icelandic hot-spring. *Zentralblatt Fur Bakteriologie Mikrobiologie Und Hygiene I Abteilung Originale C-Allgemeine Angewandte Und Okologische Mikrobiologie* 2:166–178

Takacs-Vesbach C, Inskeep WP, Jay ZJ, Herrgard MJ, Rusch DB, Tringe SG et al (2013) Metagenome sequence analysis of filamentous microbial communities obtained from geochemically distinct geothermal channels reveals specialization of three aquificales lineages. Front Microbiol 4

Teske A, Sorensen KB (2008) Uncultured archaea in deep marine subsurface sediments: have we caught them all? ISME J 2:3–18

Tyson GW, Chapman J, Hugenholtz P, Allen EE, Ram RJ, Richardson PM et al (2004) Community structure and metabolism through reconstruction of microbial genomes from the environment. Nature 428:37–43

Valentine DL (2007) Adaptations to energy stress dictate the ecology and evolution of the Archaea. Nat Rev Microbiol 5:316–323

Wegener G, Krukenberg V, Riedel D, Tegetmeyer HE, Boetius A (2015) Intercellular wiring enables electron transfer between methanotrophic archaea and bacteria. Nature 526:587–590

Woese CR, Fox GE (1977) Phylogenetic structure of the prokaryotic domain: the primary kingdoms. Proc Natl Acad Sci U S A 74:5088–5090

Woese CR, Kandler O, Wheelis ML (1990) Towards a natural system of organisms: proposal for the domains Archaea, Bacteria, and Eucarya. Proc Natl Acad Sci U S A 87:4576–4579

Wuchter C, Abbas B, Coolen MJ, Herfort L, van Bleijswijk J, Timmers P et al (2006) Archaeal nitrification in the ocean. Proc Natl Acad Sci U S A 103:12317–12322

Xie W, Zhang CL, Wang J, Chen Y, Zhu Y, de la Torre JR et al (2014) Distribution of ether lipids and composition of the archaeal community in terrestrial geothermal springs: impact of environmental variables. Environ Microbiol 17:1600–1614

Zaremba-Niedzwiedzka K, Caceres EF, Saw JH, Backstrom D, Juzokaite L, Vancaester E et al (2017) Asgard archaea illuminate the origin of eukaryotic cellular complexity. Nature 541:353–358

Direct Interspecies Electron Transfer Between Archaea and Bacteria

Cuiping Zhao and Yuchen Liu

1 Introduction

In anoxic environment, bioavailable energy is often achieved by anaerobic degradation of large and complex compounds, e.g. cellulose, other polysaccharides, proteins, lipids and nucleic acids. As shown in Fig. 1, these large polymers are first degraded to monomers by hydrolytic reactions. Then these monomers are further degraded by primary fermenters to simple molecules, e.g. H_2, formate, acetate, and CO_2, which are substrates for methanogenic archaea to produce methane. These monomers can also be converted to short chain alcohols and fatty acids, e.g. ethanol, propionate, butyrate, and succinate, which can be further degraded to H_2, CO_2, and acetate by syntrophic microorganisms and then consumed by methanogens. Without methanogens, the degradation reactions by syntrophs under standard conditions are energetically unfavorable. Methanogens drive these reactions by consuming their products (acetate, $H_2 + CO_2$, and formate) quickly to very low concentrations. On the other hand, syntrophs provide the substrates for methanogens. This mutually beneficial process is syntrophy, which is interdependent lifestyle that the metabolism of a compound occurs only when multiple organisms coexist (Stams et al. 2006; McInerney et al. 2009; Purwantini et al. 2014).

As another example of syntrophy, anaerobic methane-oxidizing (or methanotrophic) archaea can form consortia with sulfate-reducing bacteria (SRB), oxidizing methane to CO_2 while reducing sulfate to H_2S (Joye 2012). The anaerobic oxidization of methane (AOM) process usually occurs when methane is the only available electron donor and the concentrations of suitable electron acceptors are

C. Zhao · Y. Liu (✉)
Department of Biological Sciences, Louisiana State University,
Baton Rouge, LA 70803, USA
e-mail: yuchenliu@lsu.edu

C. Zhao
e-mail: czhao14@lsu.edu

G. Witzany (ed.), *Biocommunication of Archaea*,
DOI 10.1007/978-3-319-65536-9_3

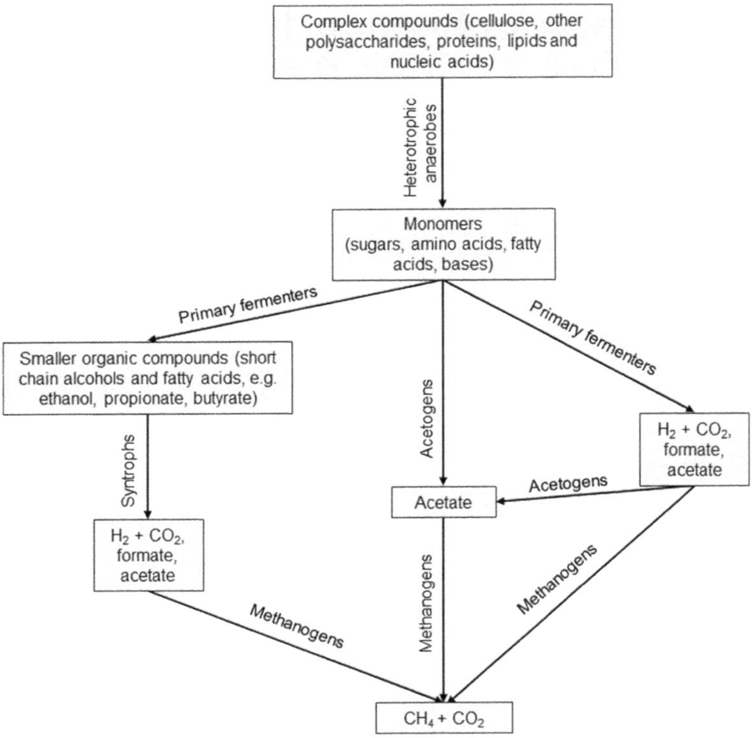

Fig. 1 The process of anaerobic degradation of complex compounds. Large polymers are first degraded to monomers by hydrolytic reactions. Monomers are further degraded to simple molecules, e.g. $H_2 + CO_2$, formate, and acetate, which are substrates for methanogenic archaea to produce methane; or to short chain alcohols and fatty acids, e.g. ethanol, propionate, and butyrate, which can be converted to $H_2 + CO_2$, formate, and acetate by syntrophic microorganisms and further consumed by methanogens

high, e.g. in marine sediments above methane hydrates with high sulfate concentrations or in organic-rich freshwater sediments with high nitrate concentrations (Thauer et al. 2008).

Both methanogenesis and AOM play important roles in global climate because methane, as a potent greenhouse effect gas, has 25 times higher of the global warming potential than CO_2 (Forster et al. 2007). The concentration of methane in the atmosphere has increased over the past 100 years from 0.9 to 1.8 parts per million (ppm) (Lelieveld et al. 1998). Currently, from the 2 Gt (1 Gt = 10^{15} g) of methane produced per year, ~1 Gt is formed globally from acetate, CO_2, formate by methanogenic archaea through methanogenesis in anoxic environments. Annually, ~1 Gt of methane is oxidized by anaerobic microbes through AOM, ~0.6 Gt is oxidized to CO_2 by aerobic microbes, and ~0.4 Gt escapes into the atmosphere (Thauer et al. 2008).

2 Interspecies Electron Transfer

In syntrophic processes, energy is conserved and transferred between different species by electron movement through chemical bonds and across biological membranes. Two different extracellular interspecies electron transfer mechanisms are possible: (i) electrons are transferred indirectly with electron shuttles, e.g. H_2, formate, acetate, or sulfur compounds, from one microbe to another (Fig. 2a); or (ii) electrons are transferred directly by electro-conductive cellular materials or minerals, known as direct interspecies electron transfer (DIET) (Fig. 2b) (Morris et al. 2013).

Interspecies hydrogen transfer is a well-documented strategy of extracellular interspecies electron transfer, in which H_2 is continuously produced by syntrophic bacteria and consumed by methanogens. Interspecies hydrogen transfer was first recognized in a coculture of the "S organism", which converts ethanol to acetate and H_2, and *Methanobacterium ruminantium*, which consumes H_2 for methanogenesis (Bryant et al. 1967). The generation of H_2 is energetically unfavorable at H_2 partial pressures (pH_2) > 100 Pa (Schink and Stams 2013). Methanogens can consume H_2 and maintain pH_2 < 10 Pa, allowing H_2-producing microbes to bypass the energetic barrier (Stams and Plugge 2009). H_2 plays an important role in many syntrophic process because H_2 is small, highly diffusible, and can be metabolized by diverse groups of microbes, such as methanogens, SRB, and denitrifiers (Stams et al. 2006).

As H_2 is poorly soluble in water, formate is a preferred interspecies electron carrier in aqueous environments (Thiele and Zeikus 1988; Boone et al. 1989; Hattori et al. 2001; de Bok et al. 2004; Stams et al. 2006; Stams and Plugge 2009),

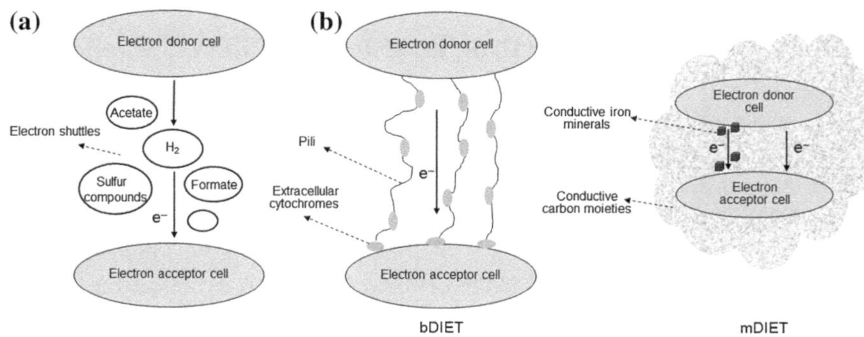

Fig. 2 Schemes of extracellular interspecies electron transfer. **a** Electron transfer via electron shuttles, e.g. H_2, formate, acetate, or sulfur compounds. **b** Direct interspecies electron transfer (DIET). *Left* During biological DIET (bDIET), electrons are transferred by electro-conductive cellular materials, e.g. pili and/or extracellular cytochromes either localized on the cell surface or along pili. *Right* During mineral DIET (mDIET), electrons are transferred via conductive iron minerals or conductive carbon moieties, e.g. magnetite, granulated activated carbon (GAC), biochar, graphite, or carbon cloth

especially in cocultures feeding on proteins (Zindel et al. 1988) or fatty acids like propionate and butyrate (de Bok et al. 2004; Sousa et al. 2007). Formate has ~ three times higher diffusion coefficient than H_2 in solutions and allows lager mass transfer to methanogens, so many syntrophic communities favor formate over H_2 transfer (Boone et al. 1989). Additionally, some syntrophic interactions can use both H_2 and formate to transfer electrons (Boone et al. 1989; Dong and Stams 1995; Stams et al. 2006; Rotaru et al. 2012). In a coculture with the dual H_2 and formate transfer mechanism, when the hydrogenase gene (*hybL*) of *Geobacter sulfurreducens* was deleted, the formate dehydrogenase gene (*fdnG*) was overexpressed (Rotaru et al. 2012).

In addition to H_2 and formate, other molecules can also function as electron shuttles. Acetate, the substrate for acetoclastic methanogenesis, can carry electrons among syntrophic partners (Platen and Schink 1987; Platen et al. 1994). Moreover, sulfur compounds (Biebl and Pfennig 1978; Boetius et al. 2000; Kaden et al. 2002; Milucka et al. 2012), humics and humics equivalents (Lovley et al. 1996, 1998, 1999; Newman and Kolter 2000; Liu et al. 2012), and flavins (Marsili et al. 2008; von Canstein et al. 2008; Brutinel and Gralnick 2012) can all work as interspecies electron carriers. For over 40 years, the indirect interspecies electron transfer with electron carriers had been considered as a main mechanism of extracellular interspecies electron transfer (McInerney et al. 2009).

DIET is an alternative interspecies electron transfer strategy discovered about two decades ago. In syntrophic cocultures, cells often aggregate in close physical contact. This phenomenon promotes effective electron transfer and makes DIET feasible. DIET has been found in both methanogenesis (Summers et al. 2010; Rotaru et al. 2014a, b) and AOM (McGlynn et al. 2015; Wegener et al. 2015) processes.

3 Direct Interspecies Electron Transfer (DIET)

DIET is achieved by electrical contacts between the electron donor and acceptor cells (Fig. 2b). Conductive pili, iron minerals, and carbon moieties can all function as electrical contacts during DIET. Two types of DIET have been proposed: the biological DIET (bDIET) and mineral DIET (mDIET), which are conducted by biological materials (e.g. pili and extracellular cytochromes) and conductive minerals, respectively (Shrestha and Rotaru 2014).

3.1 Biological DIET (bDIET)

Biological DIET was first documented in a coculture of two bacterial species, *Geobacter metallireducens* and *Geobacter sulfurreducens*, grown with ethanol as the electron donor and fumarate as the electron acceptor (Summers et al. 2010;

Shrestha et al. 2013). *G. metallireducens* can oxidize ethanol to CO_2 but cannot use fumarate as an electron acceptor, whereas *G. sulfurreducens* can reduce fumarate to succinate but cannot metabolize ethanol. This *Geobacter* coculture formed large (1–2 mm in diameter), electrically conductive aggregates that promote interspecies electron exchange using conductive pili and cytochromes for electrical connections. Deletion of *pilA* (encoding the structural pilin protein) or *omcS* (encoding a multiheme *c*-type cytochrome mainly associated with pili) eliminated syntrophic metabolism (Summers et al. 2010). Furthermore, deletion of the hydrogenase gene (*hyb*) in *G. sulfurreducens* resulted faster formation of the aggregates, suggesting that bDIET instead of interspecies hydrogen transfer is the primary interspecies electron transfer mechanism in this coculture (Summers et al. 2010).

Biological DIET has also been reported between bacteria and methanogenic archaea. The coculture of *G. metallireducens* and *Methanosaeta harundinacea* stoichiometrically converted ethanol to methane (Rotaru et al. 2014a) (Fig. 3). *M. harundinacea* is the first methanogen found to have the DIET ability, although it is still unclear how this archaeon accept electrons during DIET. In the aggregates, *M. harundinacea* highly expressed methanogenic genes, and *G. metallireducens* highly expressed *pilA* and the genes for ethanol metabolism. Furthermore, a *pilA*-deficient *G. metallireducens* strain was not able to metabolize ethanol or produce methane in the coculture with *M. harundinacea*, indicating that pili were important for DIET between *M. harundinacea* and *G. metallireducens*. The amount of methane produced in this coculture was consistent with a complete conversion of the added ethanol to methane based on the following reactions:

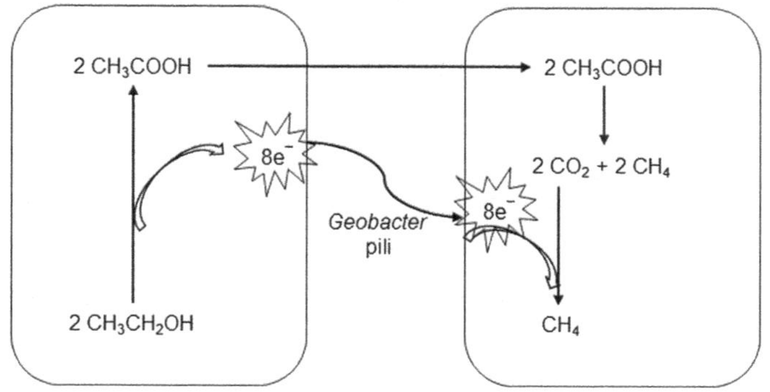

Fig. 3 Mechanism of bDIET between *Geobacter metallireducens* and *Methanosaeta harundinacea*. *G. metallireducens* oxidizes ethanol and transfers eight electrons to *M. harundinacea* via pili. *M. harundinacea* accepts electrons and produces methane from both acetate and CO_2

$$2\ CH_3\text{-}CH_2OH + 2\ H_2O \rightarrow 2\ CH_3\text{-}COOH + 8\ H^+ + 8\,e^- \quad (G.metallireducens)$$
$$2\ CH_3\text{-}COOH \rightarrow 2\ CH_4 + 2\ CO_2 \quad (M.harundinacea)$$
$$CO_2 + 8\ H^+ + 8\,e^- \rightarrow CH_4 + 2\ H_2O \quad (M.harundinacea)$$
$$\text{Sum reaction: } 2\ CH_3\text{-}CH_2OH \rightarrow 3\ CH_4 + CO_2$$

As another methanogen found to have the bDIET ability, *Methanosarcina barkeri* formed aggregates (~0.1 to 0.2 mm in diameter) with *G. metallireducens* and stoichiometrically converted ethanol to methane (Rotaru et al. 2014b). A *pilA*-deficient *G. metallireducens* strain was not able to initiate the coculture, suggesting that conductive pili are important for DIET. Unlike *M. harundinacea*, *M. barkeri* can also utilize H_2 as an electron carrier in the coculture with *Desulfovibrio* (McInerney and Bryant 1981) and *Pelobacter* (Rotaru et al. 2014b). However, other known H_2-utilizing methanogens, e.g. *Methanospirillum hungatei* and *Methanobacterium formicicum*, do not have the DIET ability as discovered so far (Rotaru et al. 2014a).

3.2 Mineral DIET (mDIET)

DIET has been shown to be possible without biological conductive materials (such as pili and cytochromes) in the presence of conductive minerals. Several examples of mDIET have been reported. (i) The addition of granulated activated carbon (GAC) can restore syntrophic metabolism in the coculture of *G. metallireducens* and *G. sulfurreducens* deficient in pili or cytochromes (Liu et al. 2012). GAC also accelerated methane production from ethanol in the coculture of *G. metallireducens* and *M. barkeri* (Liu et al. 2012). (ii) The addition of biochar or carbon cloth accelerated syntrophic metabolism in the coculture of *G. metallireducens* with *G. sulfurreducens* (Chen et al. 2014b) or *M. barkeri* (Chen et al. 2014a) with ethanol as the electron donor. Mutant strains lacking pili or pili-associated cytochromes restored DIET only in the presence of carbon cloth (Chen et al. 2014a). (iii) Magnetite (Fe_3O_4) nanoparticles attached to pili was able to compensate for the absence of OmcS in a *Geobacter* coculture with ethanol and fumarate as the substrates (Liu et al. 2015). (iv) Magnetite nanoparticles or carbon nanotubes stimulated methane production coupled to synthrophic butyrate oxidation in lake sediments (Zhang and Lu 2016). (v) A crystalline form of neutral red (2-amino-8-dimethylamino-3-methylphenazine) enhanced methane production in coal and food waste fed microbial communities and delivered electrons to *Methanosarcina mazei* (Beckmann et al. 2016).

4 Anaerobic Oxidation of Methane (AOM)

AOM often involves syntrophic associations between anaerobic methanotrophic archaea (ANME) and sulfate-, iron-, manganese-, or nitrate-reducing bacteria (Cui et al. 2015). ANME are phylogenetically closely related to methanogenic archaea, and three distinct ANME groups have been identified: ANME-1, ANME-2, and ANME-3. The ANME-1 cluster is phylogenetically related to the *Methanomicrobiales* and *Methanosarcinales* orders but forms a separate cluster, ANME-2 is related to cultivated members of *Methanosarcinales* (Hinrichs and Boetius 2003), and ANME-3 are more related to *Methanococcoides* spp. (Knittel et al. 2005). ANME-1 and ANME-2 are mostly found in marine environments, while ANME-3 is mainly found in mud volcanoes and seep sediments (Niemann et al. 2006; Lösekann et al. 2007; Knittel and Boetius 2009).

AOM was first found to be coupled with sulfate reduction (Fig. 4a), which is mediated by ANME (mostly belonging to the ANME-1 and ANME-2 clusters) and SRB (mostly belonging to the *Desulfosarcina*/*Desulfococcus* cluster of *Deltaproteobacteria*) (Kai-Uwe Hinrichs et al. 1999; Boetius et al. 2000). Sulfate-dependent anaerobic methane oxidation (S-DAOM) is mainly distributed in

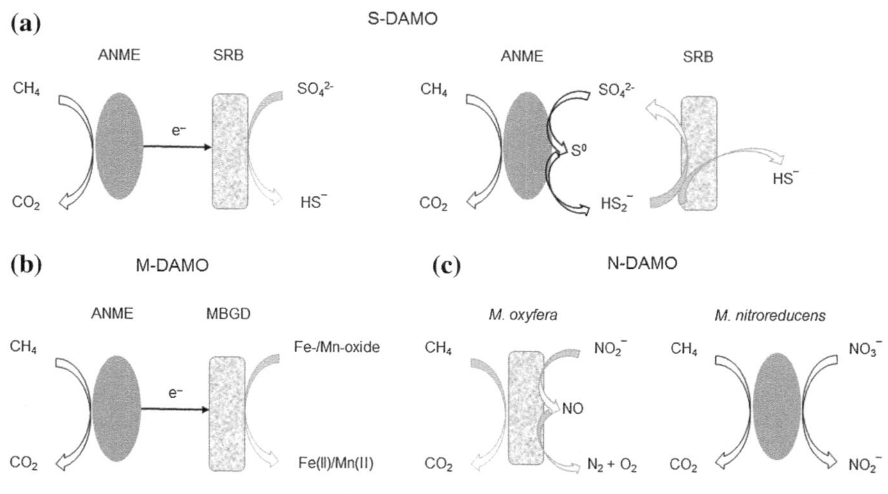

Fig. 4 Three different models of anaerobic methane oxidation (AOM) with different electron acceptors. **a** Sulfate-dependent anaerobic methane oxidation (S-DAOM). Anaerobic methanotrophic archaea (ANME) form aggregates with sulfate-reducing bacteria (SRB), coupling AOM with sulfate reduction. Alternative, ANME can autonomously perform AOM by reducing sulfate to zero-valent sulfur, which further reacts with environmental sulfide to form disulfide that is used by SRB. **b** Metal ion-dependent anaerobic methane oxidation (M-DAOM). ANME form aggregates with marine benthic group D (MBGD), coupling AOM with metal reduction. **c** Nitrite/nitrate-dependent anaerobic methane oxidation (N-DAOM). Both the bacterium *Candidatus* 'Methylomirabilis oxyfera' and the archaeon *Candidatus* 'Methanoperedens nitroreducens' can perform N-DAOM without partners

marine (Reeburgh 1976; Rusanov et al. 2002; Orphan et al. 2004; Durisch-Kaiser et al. 2005; Treude et al. 2007; Cui et al. 2015) and freshwater (Murase and Kimura 1996; Grossman et al. 2002; Eller et al. 2005; Alain et al. 2006; Smemo and Yavitt 2007; Miyashita et al. 2009) environments. It plays an important role in the biogeochemical cycling of carbon and sulfur. Besides S-DAOM, metal-ion dependent anaerobic methane oxidation (M-DAOM) (Fig. 4b) and nitrite/nitrate-dependent anaerobic methane oxidation (N-DAOM) (Fig. 4c) have also been discovered (Cui et al. 2015).

4.1 Reaction Mechanism of AOM

The AOM process has been proposed to follow a reverse methanogenesis pathway (Fig. 5) according to the following studies. First, methyl-coenzyme M reductase (MCR), the terminal enzyme of methanogenesis, likely plays an important role in AOM. In microbial mats from anoxic methane seeps in the Black Sea, ANME cells were abundant (Michaelis et al. 2002), and the rate of S-DAOM was 10 times higher than that of methanogenesis (Ye et al. 2003). In this environment,

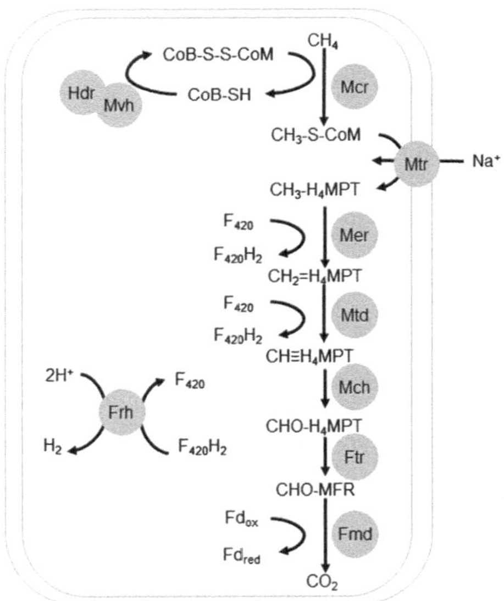

Fig. 5 Proposed mechanism of AOM through the reverse methanogenesis pathway. Enzymes: Hdr, coenzyme B-coenzyme M heterodisulfide (CoB-S-S-CoM) reductase; Mvh, F_{420}-nonreducing hydrogenase; Mcr, methyl-coenzyme M (CH_3-S-CoM) reductase; Mtr, N^5-methyl-tetrahydromethanopterin (H_4MPT):coenzyme M (CoM) methyltransferase; Mer, N^5, N^{10}-methylene-H_4MPT reductase; Mtd, methylene-H_4MPT dehydrogenase; Mch, N^5, N^{10}-methenyl-H_4MPT cyclohydrolase; Ftr, formyl-methanofuran (MFR)-H_4MPT formyltransferase; Fmd, formyl-MFR dehydrogenase; Frh, F_{420}-reducing hydrogenase

MCR accounted for 7% of extracted proteins (Krüger et al. 2003). Furthermore, the *mcrA* gene (encoding the á-subunit of MCR) was identified in ANME-1 and ANME-2 genomes (Hallam et al. 2003). A crystal structure of ANME-1 MCR was highly similar to methanogenic MCR, suggesting that this enzyme is capable to catalyze the reverse methane-forming step (Shima et al. 2012). Second, most other methanogenic genes besides *mcr* are also present in ANME genomes. Enzymes for all methanogenic steps are encoded in an ANME-1 genome, except *mer* encoding a N^5, N^{10}-methylene-tetrahydromethanopterin (H_4MPT) reductase (Meyerdierks et al. 2010). Also, all genes required for methanogenesis from CO_2 are present and actively expressed in an ANME-2a organism except putative hydrogenase genes (Wang et al. 2014). Furthermore, the complete methanogenic pathway genes, including all *mcr* subunits and *mer*, were identified in the genome of an ANME-2d organism (*Candidatus* 'Methanoperedens nitroreducens') that is capable of independent N-DAOM (Haroon et al. 2013).

4.2 Interspecies Electron Transfer in S-DAOM

In S-DAOM, electrons derived from methane oxidization by ANME presumably need to be transferred to SRB for sulfate reduction. However, the detailed mechanism of the electron transfer process is still not clear. Several hypotheses have been proposed. (i) Electrons are transferred from ANME to SRB through the production and consumption of a diffusible metabolite, such as H_2, formate or methanethiol (Moran et al. 2008; Meyerdierks et al. 2010) (Fig. 6a). This scenario is similar to the strategies for interspecies electron transfer during methanogenesis, but experimental evidence is lacking. (ii) An ANME-2 culture autonomously performed AOM by reducing sulfate to zero-valent sulfur (in the form of disulfide), which was disproportionated by SRB (Milucka et al. 2012) (Fig. 6b). However, this mechanism cannot be applied to ANME species without the ability of sulfate reduction. (iii) Biological DIET using cytochromes and pili has been proposed for electron transfer from ANME to SRB (Fig. 6c). The genes encoding secreted multiheme *c*-type cytochromes were expressed in ANME-1 (Meyerdierks et al. 2010). Multiheme cytochromes are also present in ANME-2 genomes, and redox active transition metal ions (present in cytochromes) were detected in the space between cells in the ANME-SRB consortia (McGlynn et al. 2015). Furthermore, bDIET via nanowires (composed of pili and extracellular cytochromes) has been proposed to be a principal electron transfer mechanism in thermophilic AOM (TAOM) between ANME-1 and their SRB HotSeep-1 partner (Krukenberg et al. 2016). The genome of HotSeep-1 (*Candidatus* 'Desulfofervidus auxilii') encodes genes for pili and cytochromes productions. During consortial growth, HotSeep-1 highly expressed pili genes, and both ANME and HotSeep-1 overexpressed cytochrome genes (Krukenberg et al. 2016). (iv) For ANME species without cytochromes and pili, mDIET is possible. For example, ANME-2a and ANME-2c can decouple AOM from SRB by using artificial electron acceptors, such as humic

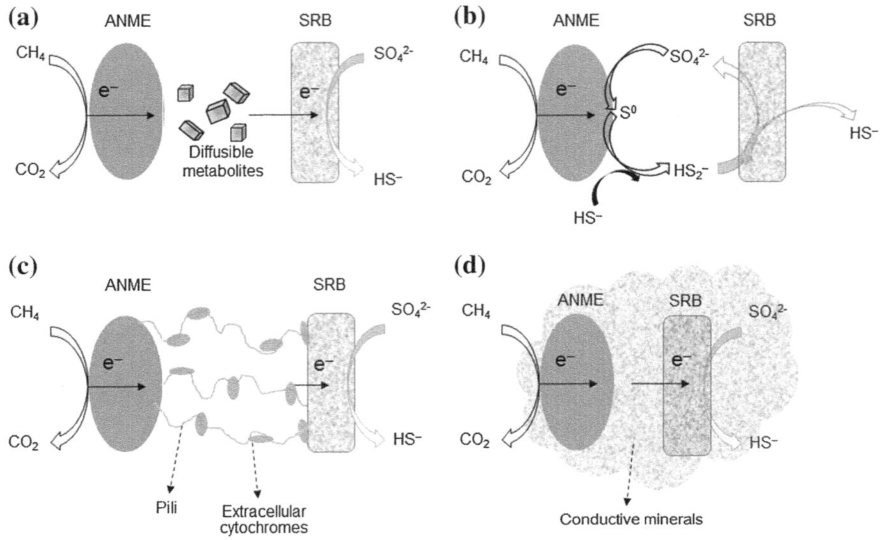

Fig. 6 Proposed interspecies electron transfer mechanisms during S-DAOM. **a** Electrons are transferred from ANME to SRB through diffusible metabolite, such as H_2, formate, or methanethiol. **b** ANME can autonomously perform AOM by reducing sulfate to zero-valent sulfur, which is disproportionated by SRB. **c** Electrons are transferred by bDIET with pili and extracellular cytochromes. **d** Electrons are transferred by mDIET with conductive minerals

acids and soluble iron (Scheller et al. 2016). This suggests that conductive materials can transfer electrons during AOM and that ANME-2 may couple AOM to metal-oxide reduction or other suitable electron acceptors. However, whether ANME-2/SRB can form aggregates without pili and whether the mechanism of DIET is fundamentally different between different ANME groups still need to be proved (Timmers et al. 2017). Moreover, how widespread DIET is in various known ANME/SRB consortia and whether DIET enables AOM to be coupled with electron acceptors other than sulfate await further studies.

5 Conclusion

DIET is potentially an important mechanism for electron transfer in syntrophic anaerobic consortia, especially in methanogenesis and AOM processes. DIET has so far been reported in the cocultures of *G. metallireducens* and *G. sulfurreducens*, *G. metallireducens* and *M. harundinacea*, *G. metallireducens* and *M. barkeri*, ANME-2 and sulfate-reducing *Deltaproteobacteria*, and ANME-1 and sulfate reducing HotSeep-1. However, there is still much work needed to understand the mechanism of DIET and its contribution to the global cycling of carbon and other nutrients.

References

Alain K, Holler T, Musat F, Elvert M, Treude T, Kruger M (2006) Microbiological investigation of methane- and hydrocarbon-discharging mud volcanoes in the Carpathian Mountains, Romania. Environ Microbiol 8:574–590

Beckmann S et al (2016) Novel phenazine crystals enable direct electron transfer to methanogens in anaerobic digestion by redox potential modulation. Energy Environ Sci 9:644–655

Biebl H, Pfennig N (1978) Growth yields of green sulfur bacteria in mixed cultures with sulfur and sulfate reducing bacteria. Arch Microbiol 117:9–16

Boetius A et al (2000) A marine microbial consortium apparently mediating anaerobic oxidation of methane. Nature 407:623–626

Boone DR, Johnson RL, Liu Y (1989) Diffusion of the interspecies electron carriers H_2 and formate in methanogenic ecosystems and its implications in the measurement of K_m for H_2 or formate uptake. Appl Environ Microbiol 55:1735–1741

Brutinel ED, Gralnick JA (2012) Shuttling happens: soluble flavin mediators of extracellular electron transfer in *Shewanella*. Appl Microbiol Biotechnol 93:41–48

Bryant MP, Wolin EA, Wolin MJ, Wolfe RS (1967) *Methanobacillus omelianskii,* a symbiotic association of two species of bacteria. Arch Microbiol 59:20–31

Chen S et al (2014a) Carbon cloth stimulates direct interspecies electron transfer in syntrophic co-cultures. Bioresour Technol 173:82–86

Chen S et al (2014b) Promoting interspecies electron transfer with biochar. Sci Rep 4:5019

Cui M, Ma A, Qi H, Zhuang X, Zhuang G (2015) Anaerobic oxidation of methane: an "active" microbial process. Microbiologyopen 4:1–11

de Bok FA, Plugge CM, Stams AJ (2004) Interspecies electron transfer in methanogenic propionate degrading consortia. Water Res 38:1368–1375

Dong X, Stams AJ (1995) Evidence for H_2 and formate formation during syntrophic butyrate and propionate degradation. Anaerobe 1:35–39

Durisch-Kaiser E, Klauser L, Wehrli B, Schubert C (2005) Evidence of intense archaeal and bacterial methanotrophic activity in the Black Sea water column. Appl Environ Microbiol 71:8099–8106

Eller G, Känel L, Krüger M (2005) Cooccurrence of aerobic and anaerobic methane oxidation in the water column of Lake Plusssee. Appl Environ Microbiol 71:8925–8928

Forster P et al (2007) Changes in atmospheric constituents and in radiative forcing. In: Solomon S et al (eds) Cambridge University Press, Cambridge, pp 129–234

Grossman EL, Cifuentes LA, Cozzarelli IM (2002) Anaerobic methane oxidation in a landfill-leachate plume. Environ Sci Technol 36:2436–2442

Hallam SJ, Girguis PR, Preston CM, Richardson PM, DeLong EF (2003) Identification of methyl coenzyme M reductase A (*mcrA*) genes associated with methane-oxidizing archaea. Appl Environ Microbiol 69:5483–5491

Haroon MF et al (2013) Anaerobic oxidation of methane coupled to nitrate reduction in a novel archaeal lineage. Nature 500:567–570

Hattori S, Luo H, Shoun H, Kamagata Y (2001) Involvement of formate as an interspecies electron carrier in a syntrophic acetate-oxidizing anaerobic microorganism in coculture with methanogens. J Biosci Bioeng 91:294–298

Hinrichs K-U, Boetius A (2003) The anaerobic oxidation of methane: new insights in microbial ecology and biogeochemistry. In: Wefer G, Billett D, Hebbeln D, Jørgensen BB, Schlüter M, van Weering TCE (eds) Ocean margin systems. Springer, Berlin, pp 457–477

Hinrichs K-U, Hayes JM, Sylva SP, Brewer PG, DeLong EF (1999) Methane-consuming archaebacteria in marine sediments. Nature 398:802–805

Joye SB (2012) Microbiology: a piece of the methane puzzle. Nature 491:538–539

Kaden J, S Galushko A, Schink B (2002) Cysteine-mediated electron transfer in syntrophic acetate oxidation by cocultures of *Geobacter sulfurreducens* and *Wolinella succinogenes*. Arch Microbiol 178:53–58

Knittel K, Boetius A (2009) Anaerobic oxidation of methane: progress with an unknown process. Annu Rev Microbiol 63:311–334

Knittel K, Losekann T, Boetius A, Kort R, Amann R (2005) Diversity and distribution of methanotrophic archaea at cold seeps. Appl Environ Microbiol 71:467–479

Krüger M et al (2003) A conspicuous nickel protein in microbial mats that oxidize methane anaerobically. Nature 426:878–881

Krukenberg V et al (2016) *Candidatus* Desulfofervidus auxilii, a hydrogenotrophic sulfate-reducing bacterium involved in the thermophilic anaerobic oxidation of methane. Environ Microbiol 18:3073–3091

Lelieveld JOS, Crutzen PJ, Dentener FJ (1998) Changing concentration, lifetime and climate forcing of atmospheric methane. Tellus B 50:128–150

Liu FH, Rotaru AE, Shrestha PM, Malvankar NS, Nevin KP, Lovley DR (2012) Promoting direct interspecies electron transfer with activated carbon. Energy Environ Sci 5:8982–8989

Liu F, Rotaru AE, Shrestha PM, Malvankar NS, Nevin KP, Lovley DR (2015) Magnetite compensates for the lack of a pilin-associated *c*-type cytochrome in extracellular electron exchange. Environ Microbiol 17:648–655

Lösekann T et al (2007) Diversity and abundance of aerobic and anaerobic methane oxidizers at the Haakon Mosby Mud Volcano, Barents Sea. Appl Environ Microbiol 73:3348–3362

Lovley DR, Coates JD, Blunt-Harris EL, Phillips EJP, Woodward JC (1996) Humic substances as electron acceptors for microbial respiration. Nature 382:445–448

Lovley DR, Fraga JL, Blunt-Harris EL, Hayes LA, Phillips EJP, Coates JD (1998) Humic substances as a mediator for microbially catalyzed metal reduction. Acta Hydroch Hydrob 26:152–157

Lovley DR, Fraga JL, Coates JD, Blunt-Harris EL (1999) Humics as an electron donor for anaerobic respiration. Environ Microbiol 1:89–98

Marsili E, Baron DB, Shikhare ID, Coursolle D, Gralnick JA, Bond DR (2008) *Shewanella* secretes flavins that mediate extracellular electron transfer. Proc Natl Acad Sci U S A 105:3968–3973

McGlynn SE, Chadwick GL, Kempes CP, Orphan VJ (2015) Single cell activity reveals direct electron transfer in methanotrophic consortia. Nature 526:531–535

McInerney MJ, Bryant MP (1981) Anaerobic degradation of lactate by syntrophic associations of *Methanosarcina barkeri* and *Desulfovibrio* species and effect of H_2 on acetate degradation. Appl Environ Microbiol 41:346–354

McInerney MJ, Sieber JR, Gunsalus RP (2009) Syntrophy in anaerobic global carbon cycles. Curr Opin Biotechnol 20:623–632

Meyerdierks A et al (2010) Metagenome and mRNA expression analyses of anaerobic methanotrophic archaea of the ANME-1 group. Environ Microbiol 12:422–439

Michaelis W et al (2002) Microbial reefs in the Black Sea fueled by anaerobic oxidation of methane. Science 297:1013–1015

Milucka J et al (2012) Zero-valent sulphur is a key intermediate in marine methane oxidation. Nature 491:541–546

Miyashita A et al (2009) Development of 16S rRNA gene-targeted primers for detection of archaeal anaerobic methanotrophs (ANMEs). FEMS Microbiol Lett 297:31–37

Moran JJ, Beal EJ, Vrentas JM, Orphan VJ, Freeman KH, House CH (2008) Methyl sulfides as intermediates in the anaerobic oxidation of methane. Environ Microbiol 10:162–173

Morris BE, Henneberger R, Huber H, Moissl-Eichinger C (2013) Microbial syntrophy: interaction for the common good. FEMS Microbiol Rev 37:384–406

Murase J, Kimura M (1996) Methane production and its fate in paddy fields. Soil Sci Plant Nutr 42:187–190

Newman DK, Kolter R (2000) A role for excreted quinones in extracellular electron transfer. Nature 405:94–97

Niemann H et al (2006) Novel microbial communities of the Haakon Mosby mud volcano and their role as a methane sink. Nature 443:854–858

Orphan VJ, Ussler Iii W, Naehr TH, House CH, Hinrichs KU, Paull CK (2004) Geological, geochemical, and microbiological heterogeneity of the seafloor around methane vents in the Eel River Basin, offshore California. Chem Geol 205:265–289

Platen H, Schink B (1987) Methanogenic degradation of acetone by an enrichment culture. Arch Microbiol 149:136–141

Platen H, Janssen PH, Schink B (1994) Fermentative degradation of acetone by an enrichment culture in membrane-separated culture devices and in cell suspensions. FEMS Microbiol Lett 122:27–32

Purwantini E, Torto-Alalibo T, Lomax J, Setubal JC, Tyler BM, Mukhopadhyay B (2014) Genetic resources for methane production from biomass described with the Gene Ontology. Front Microbiol 5:634

Reeburgh WS (1976) Methane consumption in Cariaco Trench waters and sediments. Earth Planet Sci Lett 28:337–344

Rotaru AE et al (2012) Interspecies electron transfer via hydrogen and formate rather than direct electrical connections in cocultures of *Pelobacter carbinolicus* and *Geobacter sulfurreducens*. Appl Environ Microbiol 78:7645–7651

Rotaru AE et al (2014a) A new model for electron flow during anaerobic digestion: direct interspecies electron transfer to *Methanosaeta* for the reduction of carbon dioxide to methane. Energy Environ Sci 7:408–415

Rotaru AE et al (2014b) Direct interspecies electron transfer between *Geobacter metallireducens* and *Methanosarcina barkeri*. Appl Environ Microbiol 80:4599–4605

Rusanov I, Levi A, Pimenov NV, Iusupov SK, Ivanov MV (2002) The biogeochemical cycle of methane in the northwestern shelf of the Black Sea. Microbiology 71:558–566

Scheller S, Yu H, Chadwick GL, McGlynn SE, Orphan VJ (2016) Artificial electron acceptors decouple archaeal methane oxidation from sulfate reduction. Science 351:703–707

Schink B, Stams AJM (2013) Syntrophism among prokaryotes. In: Rosenberg E, DeLong EF, Lory S, Stackebrandt E, Thompson F (eds) The prokaryotes: prokaryotic communities and ecophysiology. Springer, Berlin, pp 471–493

Shima S et al (2012) Structure of a methyl-coenzyme M reductase from Black Sea mats that oxidize methane anaerobically. Nature 481:98–101

Shrestha PM, Rotaru AE (2014) Plugging in or going wireless: strategies for interspecies electron transfer. Front Microbiol 5:237

Shrestha PM et al (2013) Syntrophic growth with direct interspecies electron transfer as the primary mechanism for energy exchange. Environ Microbiol Rep 5:904–910

Smemo KA, Yavitt JB (2007) Evidence for anaerobic CH_4 oxidation in freshwater peatlands. Geomicrobiology 24:583–597

Sousa DZ, Smidt H, Alves MM, Stams AJ (2007) Syntrophomonas zehnderi sp. nov., an anaerobe that degrades long-chain fatty acids in coculture with *Methanobacterium formicicum*. Int J Syst Evol Microbiol 57:609–615

Stams AJ, Plugge CM (2009) Electron transfer in syntrophic communities of anaerobic bacteria and archaea. Nat Rev Microbiol 7:568–577

Stams AJ, de Bok FA, Plugge CM, van Eekert MH, Dolfing J, Schraa G (2006) Exocellular electron transfer in anaerobic microbial communities. Environ Microbiol 8:371–382

Summers ZM, Fogarty HE, Leang C, Franks AE, Malvankar NS, Lovley DR (2010) Direct exchange of electrons within aggregates of an evolved syntrophic coculture of anaerobic bacteria. Science 330:1413–1415

Thauer RK, Kaster AK, Seedorf H, Buckel W, Hedderich R (2008) Methanogenic archaea: ecologically relevant differences in energy conservation. Nat Rev Microbiol 6:579–591

Thiele JH, Zeikus JG (1988) Control of interspecies electron flow during anaerobic digestion: significance of formate transfer versus hydrogen transfer during syntrophic methanogenesis in flocs. Appl Environ Microbiol 54:20–29

Timmers PHA, Welte CU, Koehorst JJ, Plugge CM, Jetten MSM, Stams AJM (2017) Reverse methanogenesis and respiration in methanotrophic archaea. Archaea 2017:1–22

Treude T, Orphan V, Knittel K, Gieseke A, House CH, Boetius A (2007) Consumption of methane and CO_2 by methanotrophic microbial mats from gas seeps of the anoxic black sea. Appl Environ Microbiol 73:2271–2283

von Canstein H, Ogawa J, Shimizu S, Lloyd JR (2008) Secretion of flavins by *Shewanella* species and their role in extracellular electron transfer. Appl Environ Microbiol 74:615–623

Wang FP et al (2014) Methanotrophic archaea possessing diverging methane-oxidizing and electron-transporting pathways. ISME J 8:1069–1078

Wegener G, Krukenberg V, Riedel D, Tegetmeyer HE, Boetius A (2015) Intercellular wiring enables electron transfer between methanotrophic archaea and bacteria. Nature 526:587–590

Ye K, Malinina L, Patel DJ (2003) Recognition of small interfering RNA by a viral suppressor of RNA silencing. Nature 426:874–878

Zhang J, Lu Y (2016) Conductive Fe_3O_4 nanoparticles accelerate syntrophic methane production from butyrate oxidation in two different lake sediments. Front Microbiol 7:1316

Zindel U, Freudenberg W, Rieth M, Andreesen JR, Schnell J, Widdel F (1988) Eubacterium acidaminophilum sp. nov., a versatile amino acid-degrading anaerobe producing or utilizing H_2 or formate. Arch Microbiol 150:254–266

Why Archaea Are Limited in Their Exploitation of Other, Living Organisms

Stephen T. Abedon

1 Introduction

Numerous direct interactions exist between organisms including where one individual, an exploiter, gains at the expense of another, a victim (Seger 1992; Calcagno et al. 2010). These interactions often are deliberate on the part of the gaining individual and often involve victim organisms themselves serving as an exploitable resource. Examples of such exploitive interactions include predation, herbivory, parasitoidism, infections by pathogens, or infestations by parasitic eukaryotes. Exploiter species can kill victim species (e.g., predation) or instead can non-mortally damage victim species (such as herbivory but the typical impact of many parasites and pathogens as well). A common theme is that victim organisms are alive, at least at the start of their interactions with exploiter organisms. Alternative trophic strategies, not involving the acquisition of living organisms for consumption, are to serve as saprophytes, detritivores, filter feeders (particular in terms of acquisition of already dead materials), or consumers of already solubilized organic carbon (dissolved organic matter or DOM). In addition, in terms of inter-organism interactions, are mutualisms, commensalisms, and competition, though of these only commensalism, as a potential gateway to parasitism, is considered here.

Not all organisms are well equipped to exploit other, living organisms, but instead these strategies typically will require specific exploitation-facilitating adaptations. For pathogens, these adaptations include "adhesins", "invasins", and/or "cytotoxins", following the perspective reviewed by Moxon and Tang (2000). In addition are adaptations involved in overcoming whatever defenses against exploitation would-be victims might display (such as "evasins") and also those that

S.T. Abedon (✉)
Department of Microbiology, The Ohio State University, Mansfield, OH 44906, USA
e-mail: abedon.1@osu.edu

G. Witzany (ed.), *Biocommunication of Archaea*,
DOI 10.1007/978-3-319-65536-9_4

result in an ability by the exploiter to benefit from having harmed the other organism, i.e., "pabulins". As a complementary perspective, it is fairly clear that pathogens (i) must possess some means by which they can harm another organism; (ii) often harm follows some degree of invasiveness while means of maintaining that invasiveness must be reasonably robust (i.e., immune system evasion); and then, for harm to be exploitive, (iii) there usually must be some gain that results from the exploitation effort, where gains generally involve increased access to resources. Pathogens (iv) also must encounter and then acquire would-be victims with sufficient likelihood, and for obligate pathogens—all else held constant—this would be with even greater likelihood. Eckburg et al. (2003) would describe this latter property as "opportunity", which they describe as "ample access to a host" (p. 591). They note further (v) that pathogens typically *replicate* in association with their hosts. Thus, hallmarks of pathogenic organisms generally will include an effecting of harm (to the host), some degree of persistence (within the host), gain (from the host), transmission (to a new host), and, in most cases, reproduction that to at least some degree occurs in association with a host.

Among exploiter species are numerous eukaryotes (i.e., domain Eukarya), Bacteria (capital "B" here to denote the domain), and viruses as well as viroids. Included among exploitive eukaryotes are a substantial fraction of animals along with pathogenic, parasitic, or predatory protozoa and fungi. There also are parasitic as well as carnivorous plants. Among members of domain Bacteria can be found predators as well as numerous pathogens (Brown and Inglis 2009). Parasites and pathogens, as well as predators, however, appear to be relatively lacking among members of domain Archaea as at best only a handful of disease-associated, pathogenic, or parasitic archaeal relationships with other organisms have been suggested. For the latter, see: Reeve (1999), Cavicchioli et al. (2003), Eckburg et al. (2003), Conway de Macario and Macario (2009), Horz and Conrads (2010, 2011), Moissl-Eichinger and Huber (2011), Shiffman and Charalambous (2012), Wrede et al. (2012), Aminov (2013), Bang and Schmitz (2015), Lurie-Weinberger and Gophna (2015), Nkamga et al. (2016).

The concept of parasitism implies net reductions in the health and/or evolutionary fitness of affected host organisms. This is rather than simply evidence that one organism obtains resources from another, since even mutualisms often display the latter. Parasitic organisms also may be "conditionally helpful" to hosts in specific environments (Fellous and Salvaudon 2009), such as "in colonizing hydrothermal vents" (Podar et al. 2008). In addition, commensalistic interactions in one host can be parasitic or pathogenic in another (Lawrence and Hendrickson 2008). Parasites and pathogens as a consequence are not necessarily always easily identified as such, and indeed even at the level of genes and molecules it can be difficult to explain why one organism may display more commensalistic behaviors in comparison with a close relative that instead is more exploitive towards a given host organism (Wassenaar and Gaastra 2001; Moxon and Tang 2000).

With these caveats in mind, here I consider the question, in somewhat general terms, of why Archaea seem to be relatively lacking among pathogens, parasites, or predators. Cavicchioli et al. (2003), Eckburg et al. (2003), Gill and Brinkman

(2011), Shiffman and Charalambous (2012) and others as cited below provide complementary though less ecologically oriented considerations of this question. See also Abedon (2013) as well as Valentine (2007). I conclude with an assertion that there likely exist multiple criteria—criteria which one sees in some abundance in both domains Bacteria and Eukarya—that in combination will tend to support the evolution by at least some members of these groups towards more exploitive behavior of living organisms. Archaea, meanwhile, may be deficient in one or more of these criteria. The emphasis primarily is on comparison of Archaea to Bacteria, but eukaryotes as exploiters of other organisms are considered as well.

2 Potentially Parasitic or Pathogenic Archaea

Simple propagation of symbiotic organisms can lead to increased opportunities for their transmission to new hosts. In addition, microbes under certain circumstances, such as following host injury, can enter into host tissues even if a microbe does not possess virulence factors that actively promote such invasion. If through some means of immune system evasion an organism were then able to persist within those tissues for long enough such that their presence resulted in reductions in host health—and especially were they then to successfully disseminate to a new hosts as a consequence of this persistence—then that organism could be viewed as displaying some minimal level of action as a pathogen or parasite.

An organism that is present within a mixed infection, that has entered into that infection in association with injury including following the entrance of other, more invasive organisms, and which was then able to contribute to the survival or propagation of other, especially virulent microorganisms within the context of the injury, could be similarly viewed as displaying, at some minimal level, characteristics of a pathogen or parasite. The latter would be particularly so if these co-infecting organisms were also capable of replicating in these sites, transmitting as a consequence of this replication to new hosts with greater likelihood, and to some degree evade host immunity. That is, to serve as evolved pathogens versus something more a kin to accidental pathogens (Rivera-Perez et al. 2017). Indeed, at an absolute minimum such evolved pathogens must be able to replicate during, or otherwise at least benefit from such interactions with a host organism, and in so doing in some manner cause or at least facilitate host harm. Though some infectious diseases can occur without in situ pathogen replication, e.g., such as food poisonings, it is difficult to argue that the poison/toxin producing organisms technically are serving as *symbiotic* parasites in such instances and therefore, though pathogenic, they may not be strictly parasitic.

Among Archaea only a few examples of harmful symbioses have been suggested. Methanogens associated with periodontal mixed infections—particularly *Methanobrevibacter* spp.—most prominently may fit such minimum pathogen criteria, e.g., Aminov (2013), Nguyen-Hieu et al. (2013). There also are examples of presumptive Archaea-mediated parasitism of other Archaea (Jahn et al. 2008,

Moissl-Eichinger and Huber 2011, Wrede et al. 2012, Podar et al. 2013, Giannone et al. 2015). Other instances of Archaea facilitation of disease have been suggested, but none conclusively (Horz and Conrads 2010; Shiffman and Charalambous 2012; Aminov 2013). Given this dearth of Archaea linked to disease states, I address the question of why it may be that Archaea which behave more robustly as pathogens or parasites, or indeed as predators of other, living organisms have not—at least yet—been described.

3 Why Are Archaeal Pathogens Rare?

It is likely that Archaea exploit other organisms to at least some degree, but whether Archaea function as pathogens, parasites, or predators to any large extent is uncertain. Indeed, a large majority of the members of domain Bacteria also are *not* pathogenic (Casadevall and Pirofski 2000; Gill and Brinkman 2011) or, for that matter, predatory. To address why Archaea appear to be rarer among pathogens than Bacteria, this section considers a number of possible issues, as presented in separate subsections.

3.1 Low Densities or Diversity of Co-located Potential Victims

It is possible for prokaryotic organisms to parasitize or serve as predators of other prokaryotic organisms, e.g., especially bdellovibrios for the latter. Classically, however, the question of whether a prokaryotic organism acts as a pathogen, or indeed as a parasite, is associated with the host organism being eukaryotic rather than prokaryotic. For an organism to enter into a symbiosis with another organism, such as a parasitic archaean with a eukaryotic host, then both environmental ranges and habitats must overlap, that is, the would-be parasite or pathogen must be able to physically interact with a host organism.

A defining though neither unique nor universal characteristic of Archaea is their growth and presence within extreme environments. These include environments that are very hot and/or highly acidic, very salty, or—perhaps only arguably as "extreme"—anoxic (that is, anaerobic). Generally the expectation is that Archaea dominate such environments (though this is not always true) or at least are present in relatively high concentrations (though this also is not always true). A number of studies support especially this latter, non-parenthetical contention, that is, that in extreme environments archaeal cells often can be found at relatively high densities.

Sahm et al. (2013), for example, describe archaeal densities within a high-temperature hot spring (92 °C) of roughly 10^8 cells/ml, though of interest more Bacteria were present than Archaea. Miller-Coleman et al. (2012) reported Archaea densities (specifically of phylum *Korarchaeota*) in excess of 10^5 cells/g in

Yellowstone hot springs sediments, with a maximum of 6.6×10^6 16S rRNA gene copies present also per gram. As the volume of a gram of sediments contains more than just fluid, and assuming two 16S rRNA genes per cell (Lee et al. 2009; Klappenbach et al. 2001), then this suggests that densities of archaeal cells external to non-archaeal sediment materials (e.g., such as attached to grains of sand) perhaps approached those seen by Sahm et al. Keep in mind, however, that these counts likely are of more than one archaeal species.

Prieur (2002), based on data presented in a Ph.D. dissertation by P. Chevaldonné, reported "10^7–10^9 cells per gram of smoker wall (dry weight)" (a "smoker" is a kind of hydrothermal vent). Li et al. (2014), on the other hand, detected 10^5 to roughly 10^6 archaeal cells per gram of hydrothermal vent wall. Again, keep in mind that solid structures can consist of much more than the fluid volumes in which Archaea actually occupy.

In two hypersaline lakes in Romania, Andrei et al. (2015) detected Archaea at densities ranging around 10^6 cells/ml in one lake and $>10^5$ cells/ml (one sample) to $>10^7$ cells/ml (one sample) to 10^8 cells/ml (two samples) in another. In the latter lake, decreasing densities were seen with depth. In another Romanian hypersaline lake, Archaea densities were estimated to range from 6×10^6 to 10^8 cells/ml (Baricz et al. 2014). Overall, then, the data seem to indicate that Archaea concentrations within extreme environments often can be relatively high, though they do not necessarily exceed Bacteria in concentration and also in absolute terms are not *always* particularly high in concentration (e.g., $>10^6$ cells/ml).

What about eukaryote presence within extreme environments? Protists exist, for example, that are halophilic (in some cases growing above 300 parts per thousand salinity, that is, as extreme halophiles); that are at least mildly thermophilic (up to at least 55 °C); that are acidophilic (below pH 1.5); and that are microaerophilic or anaerobic (Park et al. 2007, 2012). Photosynthesis in a hypersaline Antarctic lake was found to be dominated by eukaryotic organisms, though interestingly in this same lake Archaea were largely lacking (Yau et al. 2013). With few exceptions, animals as well as plants tend to be much less tolerant of extreme conditions. Thermophilic animals nevertheless do exist, particularly the Pompeii worm, *Alvinella pompejana* (Holder et al. 2013). In addition, animals known as loriciferans—which potentially possesses endosymbiotic methanogenic Archaea that are similar to those present in anaerobic protists—have been described as living in low oxygen, high saline, deep-sea conditions (Levin 2010). For further discussion of eukaryotes that spend some or all of their time anaerobically metabolizing, see (Muller et al. 2012).

In summary, environmental interactions between Archaea and eukaryotes in at least some extreme environments can occur. Archaea densities can be relatively high in these environments, though densities of Bacteria in many instances are even higher. It is likely that eukaryote densities, however, are lower in many or most extreme environments than they are in non-extreme environments. Nevertheless, based solely on numbers, there certainly should be a potential for Archaea to form symbiotic relationships with eukaryotes in extreme environments, and certainly such relationships exist. At a minimum, these observations therefore are suggestive that one place in which archaeal pathogens might be sought is in association with eukaryotes that are found within more extreme environments.

3.2 *Lower Densities of Non-extremophilic Archaea*

What of Archaea in less extreme environments? Densities of Archaea in oligotrophic Pacific Ocean water columns have been measured at less than $10^{4.5}$ cells/ml down to about 200 m, which in the study was approximately 10% that of Bacteria. Below 200 m densities of Bacteria drop to approximately the same as Archaea, with roughly 10^4 cells/ml for each (Karner et al. 2001). In the Sargasso Sea, which also is oligotrophic, densities of what presumably were a combination of Bacteria and Archaea were present at approximately 10^5 cells/ml at 200 m and two to four times that density above 100 m, implying rough consistency with the Karner et al. study, at least in terms of total numbers of these microorganisms (Li et al. 1992). Murray et al. (1998) determined Archaea numbers in Antarctic coastal waters ranging from about 1 to 3×10^4/ml or roughly 10% of total cells. DeLong et al. (1999) found ocean water column Archaea densities were roughly in the range of 10^5 cells/ml, with a maximum of approximately 3×10^5/ml and often making up somewhat more than 10% of cells. In wastewater treatment Archaea were found to make up less than 2% of total cells present in activated sludge, though in other steps of treatment archaea numbers instead were found to exceed those of bacteria (Fredriksson et al. 2012). In another study, reported densities of Archaea were closer to 10% of cells in activated sludge (Hagman et al. 2008).

In sediments, and especially multiple centimeters deep and deeper, Archaea can (Lipp et al. 2008) but do not always (Borrel et al. 2012) dominate Bacteria in terms of numbers. While eukaryotes tend to be less prevalent deep within sediments (Sinclair and Ghiorse 1987), nevertheless substantial overlap between eukaryotes and Archaea within sediments likely still occurs. In soils, by contrast, abundance of Archaea has been shown to average 2% of all 16S rRNA sequences, ranging from zero to 15.6%, with only five soil samples out 146 consisting of over 5% Archaea (Bates et al. 2011) (there as based upon PCR primers which these authors suggest are not overly biased), or closer to 1% (Ochsenreiter et al. 2003) for *Crenarchaeota* alone. On the other hand, Uroz et al. (2016) detected Archaea in soils in densities potentially in the vicinity of 10^8 per gram, or roughly 10% densities of Bacteria. Potentially over 90% of these were members of archaeal phylum Thaumarchaeota, which are ammonia-oxidizing chemolithotrophs. Cao et al. (2012) observed archaeal densities as based on 16S gene prevalence making up less than 10% of bacterial densities, or roughly 10^6–10^8 cells/g dry weight assuming two 16S genes per cell.

Eukaryotes of course are particularly abundant in less extreme environments, though less so where nutrients are less available. These studies of archaeal densities in these environments tend to suggest that in many but not all less-extreme environments, members of domain Bacteria tend to exceed, often by an order of magnitude, numbers of domain Archaea that are also present. Nevertheless, as based on this relative dearth alone, it is difficult to argue that Archaea are so rare that it would be unlikely that at least some archaeal pathogens would have evolved in these environments. Indeed, based on numbers of environmental, presumably free-living Archaea alone we presumably should expect that Archaea would also make up approximately 10% of prokaryote symbioses with eukaryotes, and perhaps also 10% of pathogens. An alternative perspective, however, may be that these transitions are not linear in their proportionality. Thus, hypothetically, perhaps a fraction consisting of 10% of free-living prokaryotes might translate to a fraction of only 1% of non-pathogenic symbiotic prokaryotes which in turn might translate to a fraction of only 0.1% of pathogenic prokaryotes. In other words, it is possible that environmental numbers of Archaea as well as diversity are important to consider in addressing the question of why Archaea are rare among pathogens. From looking at such numbers alone, however, it is difficult, at least for this author, to formulate robust conclusions or hypotheses.

3.3 Relatively Low Number of Symbioses

Assuming that pathogens often evolve from otherwise commensalistic symbiotic Bacteria (Raskin et al. 2006; Barash and Manulis-Sasson 2009; Groisman and Ochman 1996)—such as *Staphylococcus epidermidis* which appears to have evolved to display conditional, opportunistic nosocomial infections (Ziebuhr et al. 2006)—then one may relevantly question whether already exiting commensalistic Archaea are sufficiently numerous or diverse that pathogenic lifestyles should have evolved. Serving as commensals of larger organisms Eckburg et al. (2003) would categorize (p. 591) as "Access and colonization". Certainly Archaea can form symbiotic relationships, suggesting that (p. 1121) "there are no underlying reasons that prevent archaea from actively interacting with eukaryotes" (Cavicchioli et al. 2003).

Archaeal symbioses with other organisms do exist, including with humans (Cavicchioli et al. 2003; Eckburg et al. 2003; Lange et al. 2005; Moissl-Eichinger and Huber 2011; Shiffman and Charalambous 2012; Bang and Schmitz 2015; Nkamga et al. 2016). These include particularly methanogens (Saengkerdsub and Ricke 2013), and potentially Archaea found in association with plant roots (Buée et al. 2009). Indeed, Wrede et al. (2012) provide a recent review of the diversity of known archaeal symbiotic relationships. Do such intimate interactions with other organisms, perhaps particularly with eukaryotes, occur with lower frequency or diversity than is true for Bacteria?

One means of addressing the latter issues is to compare published 16S rRNA gene libraries. Diversity of Archaea, as defined in this manner, can be approximately ten-fold lower than bacterial diversity in "digestive systems" (Aller and Kemp 2008), that is, as digestive system symbioses; see also Eckburg et al. (2005). That observation would appear to be moderately consistent with methanogens in gastrointestinal tracks ranging from less than 1% of total cells in the bovine rumen, and up to 10% of anaerobic cells in certain humans (Saengkerdsub and Ricke 2013). In fact, in most but not all environments, archaeal 16S rRNA diversity was been found to be lower than bacterial diversity, with the exception of two hot and one methane-rich environment where archaea were found to be more diverse and in biofilms where archaea were found to be similarly diverse (Aller and Kemp 2008). In certain sponges archaea also have been found to be more numerous than bacteria (Pape et al. 2006). Notwithstanding these exceptions, the pools of Archaea symbionts from which pathogens might be evolutionarily drawn appears to often be less diverse than equivalent pools of Bacteria. Perhaps comparably, the diversity of Archaea found within wastewater anaerobic reactors, which draw their associated organisms from local environments, appears to be low in comparison to Bacteria found in the same environments (Curtis and Sloan 2004). Archaea as found in association with eukaryotes thus tend to be less abundant than Bacteria, with the types of interactions seen, as well as the archaeal types themselves, also less diverse.

An important caveat with regard to such analyses is that enumeration biases often are an issue such as in terms of the impact of choice of PCR primers. Such biases can be more problematic when making genera- or species-level comparisons versus distinguishing among higher taxa, however (Fredriksson et al. 2013). Nonetheless, absolute certainty about relative bacterial and archaeal abundance cannot be claimed, with the above studies better viewed as supporting a trend of higher bacterial numbers and diversity found in environmental and symbiotic association with eukaryotes than appears to be the case for Archaea. Overall, these tendencies would not appear to be absolute blocks on Archaea taking on pathogenic lifestyles. Nevertheless, they could be consistent with Archaea doing so, all else held constant, at rates that are lower than as seen with Bacteria. Indeed, up to this point in our discussion there would not appear to be any obvious blocks on Archaea making up approximately 10% of the prokaryotic pathogens found in association with eukaryotes. This predication, however, has not been observed to be even close to correct. Therefore, to at least a first approximation, we should assume that the more limited pool from which Archaea pathogens may be drawn versus Bacteria is not alone, at least in a linear fashion, in giving rise to the dearth of discovered archaeal pathogens.

3.4 Lower Frequency of Nutritionally Sophisticated, Easily Cultured Archaeal Organotrophs

Known bacterial pathogens such as of birds and mammals by and large are mesophilic, non-acidophilic, and non-halophilic organisms that tend to derive nutrients from organic compounds that are more complex than acetate, e.g., compounds such as glucose. Most known bacterial pathogens also are relatively easily cultured. The Archaea, by contrast, consist most prominently of acidophiles, extreme halophiles, hyperthermophiles, the anaerobic methanogens, methane oxidizers, and nitrifiers. Though mesophilic, non-acidophilic, non-halophilic Archaea may very well exist that utilize relatively complex organic molecules as, especially, carbon and energy sources (again, e.g., such as glucose), they are not obviously abundant. In addition, Nkamga et al. (2016) indicate a lack of capacity of sequenced methanogens as found in the human microbiome to break glycosidic bonds. By contrast, organisms with such general characteristics are abundant among Bacteria.

Archaea that are known to obtain nourishment from relatively complex organic compounds do exist (Chaban et al. 2006) and include the thermophilic *Desulfurococcus* (Perevalova et al. 2016) and *Thermoproteus* (Selig and Schönheit 1994); the thermoacidophilic *Sulfolobus* (Chen et al. 2005) and *Metallosphaera* (Liu et al. 2011); the mesoacidophilic *Cuniculiplasma* (Golyshina et al. 2016); cold-adapted secondary calcite-associated species (Reitschuler et al. 2014); and various halophiles (Andrei et al. 2012) including the haloalkaliphilic *Halalkalicoccus* (Xue et al. 2005). Certain methanogens also can utilize simpler organic molecules such as the noted acetic acid as well as formic acid (Chaban et al. 2006; Welte and Deppenmeier 2011; Saengkerdsub and Ricke 2013), and among halophiles there are also a limited number of methanogens (Andrei et al. 2012). A likely important source of organic materials to more extreme environments can be due to flow from less-extreme environments, such as in the form feces (Urschel et al. 2015).

Archaea that are able to take up amino acids can be found throughout marine water columns (Herndl et al. 2005). These latter, non-thermophilic, non-acidophilic, and non-halophilic organisms are, however, difficult to culture. Indeed typically, though with notable exceptions, "Archaea are notoriously difficult to culture" (Shiffman and Charalambous 2012). Consistently, as of 2011 the cultivation of "over 50 different archaeal strains", in total, was touted as "progress" in the archaeal field (Bates et al. 2011). It is uncertain to what extent these less easily cultured Archaea can utilize relatively complex organic molecules as carbon and, perhaps especially, energy sources, i.e., as presumably would be useful for parasites living directly off of nutrients derived from eukaryotic hosts. Though an inability to culture a microorganism can be a consequence of missing cofactors or due to toxic substances present in culture media (Stewart 2012; Tanaka et al. 2014), it nevertheless is the case that most pathogenic Bacteria are readily cultured.

The suggestion that Archaea might be less suited to exploiting pathogen niches due to nutritive needs—e.g., from p. 49 of (Shiffman and Charalambous 2012), "Obtaining essential nutrients from the host may limit Archaea from becoming pathogens"—serves as an extension of the nutrient-access argument, "Pathogenic archaebacteria: do they not exist because archaebacteria use different vitamins?", as presented by Martin (2004), and as clarified by Cavicchioli and Curmi (2004) as well as Shiffman and Charalambous (2012). Rather than "vitamins" or "unique cofactors", however, more basic nutrients are posited here as being unexploitable by would-be pathogenic Archaea, i.e., again such as glucose as a carbon and especially energy source (since pathogens presumably generally require some sort of host-derived energy source and, unless they are autotrophic, then some source of organic carbon as well). Returning to Martin (2004), he notes as well that (p. 592), "Pathogens, like every other organism on earth, are looking for a meal." Regarding Martin's central thesis, one might expect that a requirement for otherwise unavailable cofactors might limit Archaea development into pathogens at the level of symbiosis formation, that is, rather than strictly at the point of pathogen formation. An inability to thrive on those macronutrients which are readily available in association with host organisms, however, presumably also could limit the potential for Archaea to serve as symbionts of eukaryotes.

Given these considerations, we can perform a pertinent thought experiment: To what degree do archaea exist which would be capable of thriving, especially as pure cultures, were we to artificially deposit them within the tissues of a host organism under circumstances in where host-mediated immunity is suppressed? If the answer to this question is that Archaea under such circumstances generally would not possess a high potential to thrive, then Archaea potentially are not well suited in terms of their nutrient needs to exploit perhaps a majority of especially mesophilic eukaryote-associated niches. In particular, there may not be substantial coincidence between organotrophic Archaea and those organisms we typically would tend to characterize in terms of associated pathogens. This postulated comparative lack of relatively sophisticated, non-extremophilic Archaea organotrophs in part could be a consequence of Archaea inherently having had difficulty competing with Bacteria within relatively nutrient-rich niches (Karlsson et al. 2012).

3.5 Low-Energy-Availability Specialists

There exist potential conflicts between Archaea metabolic properties versus the benefits that Bacteria can derive from pathogenesis. Valentine (2007), in particular, suggests that Archaea are especially well adapted to the exploitation of niches that supply chronically low densities of energy (p. 321): "…archaea have evolved to thrive with [low-] energy stress, whereas bacteria can adapt to maximize the availability of energy and other resources." Possible consequences of such adaptation to consistently low energy availability could include poor suitability to exploit abundances of energy should they become available. Thus, as Valentine further suggests (also p. 321) that, "Bacteria [contrasting Archaea] are inherently better suited to pathogenesis as it is consistent with their capacity to seek new resources and adapt to changing environments."

While Archaea may thus be viewed as energy-conserving specialists, eukaryotes at an opposite extreme appear to serve as energy-capturing specialists, as in fact is required of them (Lane and Martin 2010) given their relatively large sizes and morphological sophistication in comparison to prokaryotes. Bacteria, in turn, would appear to be found somewhere in between, that is, neither highly robust in their ability to capture energy especially from other organisms (particularly, that is, in terms of ingesting or engulfing relatively large 'chunks' of food) and also, at the other extreme, not necessarily existing as energy-conservation specialists (previous paragraph). Vast numbers of eukaryotes, and particularly chemoheterotrophic eukaryotes, by contrast obtain their energy (and carbon) by consuming other, living organisms, while so far as we can tell Archaea predominantly do *not* obtain their energy (or carbon) by consuming other, living organisms (i.e., Sect. 3.4).

Chemoheterotrophic bacteria appear to be intermediate between eukaryotes and Archaea also in terms of their carbon- and energy-capturing abilities. Though bacteria, as noted, can neither engulf nor ingest, they do appear to be quite adept in terms of exploiting a huge diversity of dissolved or dissolvable organic carbon. Chemoheterotrophic eukaryotes thus, by and large, concentrate on consuming a diversity of relatively big things (e.g., prokaryote-sized things and up) while Bacteria concentrate on consuming a diversity of relatively small things (i.e., specific, individual molecules) and chemoheterotrophic Archaea, particularly as found within less extreme environments, seem to concentrate on consuming a smaller diversity of even smaller things (such as acetate and smaller). Potentially consistently, the Bacteria-as-pathogen niche may be viewed as something of an intermediate situation where relatively small fractions of Bacteria, as compared with eukaryotes, make their living by obtaining energy (and carbon) by parasitically consuming other, *living* organisms, versus Archaea where apparently a somewhat even smaller fraction equivalently survive by parasitically consuming other, living organisms. Like Bacteria, eukaryotes also are somewhat prevalent among intra-organismally parasitic species. I return to consideration of eukaryotes below.

3.6 Coevolutionary ("Arms Races") Mismatches

The potential for a pathogen to remain a pathogen over evolutionary time is dependent on its ability to overcome host defense mechanisms, e.g., immune systems, even as those mechanisms evolve to resist the action of specific organisms. This so-called antagonistic coevolution (Pal et al. 2007; Paterson et al. 2010)—such as selection for hosts that resist pathogens, where pathogens in turn are selected to overcome host defenses—likely is stronger the more virulent a pathogen, or the greater a pathogen's rate of evolution (Calcagno et al. 2010). That is, given greater motivation on the part of the host to resist a pathogen, or would-be pathogenic organism, then the stronger the selection driving the host side of antagonistic coevolution.

Pathogen evolution can be limited by the availability of beneficial mutations (LeClerc et al. 1996), i.e., as required for overcoming new degrees of host resistance. Holding mutation rates constant—though Archaea possibly display even lower mutation (base-substitution) rates per nucleotide site per generation than bacteria (Lynch 2010)—then small pathogen population sizes, as resulting from lower rates of pathogen replication, should show reduced rates of pathogen evolution. Indeed, low rates of pathogen replication also can reduce levels of pathogen virulence, at least to the extent that greater pathogen numbers are expected to result in greater host harm (again, all else held constant). Slower pathogen replication should also put off host harm until further into the future on a per infection basis, i.e., perhaps not until after the pathogen has been brought under control by the host immune system (and thereby making a pathogen a *not* very "good" pathogen). Alternatively, long-term evasion of immunity is not necessarily easy for pathogens to achieve. It often is also in the interest of a pathogen to display lower virulence if it must infect its host for long periods of time, that is, towards successful transmission to new hosts (Ewald 2004).

Slow growing pathogens or slow growing would-be pathogens thus could be handicapped in their mutational potential to overcome host defenses as well as their ability to harm hosts sufficiently to be recognized as or even to serve as pathogens. If they are so virulent as to select for specific host resistance mechanisms, then they may not be able to mutationally respond coevolutionarily fast enough to survive as populations. If they are so *not* virulent as to not select for specific host resistance mechanisms, then they may not be harming the host sufficiently to be recognized as pathogens. To the extent that Archaea which are capable of replicating symbiotically with other organisms tend to display relatively low growth rates, then only exceptional Archaea (e.g., *Mycobacterium tuberculosis* equivalents) may display substantial pathogenesis. Furthermore, to the extent that Archaea succeed in becoming pathogens, that is, being virulent against a host organism, then if they display low rates of mutation-limited evolution—such as relative to as displayed by successful bacterial pathogens—they may tend to go extinct due to faster evolution of host resistance.

3.7 Possible Infectious Dose Deficiencies

A further, potential consequence of slower population growth, if this results in smaller population sizes, is a failure of would-be pathogen populations to routinely achieve infectious doses, i.e. (p. 591), "gaining access to the host in sufficient numbers" (Eckburg et al. 2003). An infectious dose is a measure of either the potential for individual infectious organisms to become established as an infection or instead is a measure of requirements for cooperation within a population of organisms also to achieve infections (Leggett et al. 2012). Some organisms, such as *Shigella*, require very low infectious doses—implying that these organisms are quite good individually at establishing infections—whereas less infectious pathogens, or would-be pathogens, may be unable to establish infections even given relatively high exposure of a host, particularly of a healthy, immunologically sound host to the pathogen. Pathogens which are highly effective at transmitting to new hosts thus either are individually highly effective at establishing infections, or instead tend to become exposed to potential new hosts as populations consisting of high numbers of individuals (or, of course, both). In either case, the potential to establish infections is thought to be dependent on the secretion or at least display of molecules, extracellularly to the pathogen, which aid in bypassing host mechanisms that otherwise could interfere with infection establishment.

To the extent that Archaea are not able to establish large population sizes, then they may be more dependent than pathogens generally on being highly individually infectious. In a way, this is saying that Archaea would need to be particularly effective pathogens in order just to be pathogens, which actually is not altogether untrue generally for especially obligate pathogens. If we assume that being highly infectious is an evolved property, that is, with evolutionary improvement in the ability of a pathogen population to display lower infectious doses over time, then a low potential to achieve infectious doses could serve as an early barrier to the development of an organism as a pathogen. Thus, if higher doses of would-be pathogens are required to get a start as a pathogen, with subsequent evolution of lower-dose abilities, then organisms which are inherently less able to grow populations to relatively high densities may also be inherently less able to become initially established as pathogens.

This argument is essentially epidemiological, though has as its basis the inherent abilities of microorganisms to interact with their hosts. Specifically, the issue is that for pathogens to survive as pathogens then they require a basic reproductive number, R_0, which is greater than 1.0. In other words, for an obligate pathogen, since infections cannot last forever, each infected host must be turned into at least one other, new infected host or else the number of infected hosts, and thereby the pathogen population, will be in decline. Transmission ability therefore must be adequate, and that in turn can be dependent on an ability to produce sufficient numbers of potentially transmittable pathogen progeny. Alternatively, for non-obligate pathogens, infection of a host will require that adequate numbers of the pathogen be present to achieve infectious doses at the point of environmental contact with potentially infectable individuals.

For Archaea which already serve as evolved symbionts then a reasonable potential for transmission to new hosts, i.e., $R_0 \geq 1.0$, must already exist. For a transition from being a commensalistic individual to being a pathogen to be worthwhile to an organism, then either the more virulent varieties must outcompete less virulent ones within hosts (ideally without substantially impacting transmission ability) or instead more virulent varieties must display improvements in transmission abilities (Ewald 2004). One advantage in the latter case could be a greater potential to reach infectious doses to achieve such improved transmission. Therefore with regard to infectious doses, an important question is whether in fact commensalistic Archaea possess such an evolutionary potential to increase their typical population sizes within colonized individuals.

3.8 Difficulties Associated with Niche Invasion

Horz and Conrads (2010) ask whether bacteria, as making up the human microbiome, may actively exclude newcomers such that, perhaps, what Archaea we possess as normal microbiota reflect that fraction of prokaryotes we are exposed to during our initial colonization. Thus they state (p. 5), "One could also speculate that some archaea simply do not have the opportunity to colonize the human body but once they get access colonization is possible." This would be a hypothesis concerning niche invasion. More generally, if Archaea are inherently capable of colonizing niches found within host organisms, such as ourselves, then might their relatively low prevalence among symbionts be a consequence of a poor potential to compete with Bacteria for exploitation of overlapping ecological niches? The contrasting hypotheses—that Archaea in fact are not equivalently capable of colonizing other organisms—Horz and Conrads suggest as well, citing Martin (2004), i.e., as concerning the availability to Archaea of required nutrients.

These ideas expressed by Horz and Conrads (2010) are ones of (1) organism difficulty in invading niches due to issues of competition with other organisms, i.e. (p. 593), they "should be able to compete successfully with other endogenous flora for survival and persistence" (Eckburg et al. 2003), versus (2) simply lack of ability to exploit a given niche at all, that is, even absent competition. These actually represent differences in what can be described as (2) fundamental versus (1) realized niches (i.e., as numbered equivalently to the first list). A fundamental niche is the totality of habitats a population of organisms theoretically may persist in whereas a realized niche is the extent of habitat that this same population may persist in given competition with other organisms. Thus, a fundamental niche generally is broader than a realized niche. An organism may be blocked from invading a given ecological niche either because its fundamental niche is insufficiently broad (i.e., it can't exploit a given niche even absent competition from other organisms) or instead because its realized niche is insufficiently broad (i.e., it can't exploit a given niche *because* of competition from other organisms).

The invasion by organisms of new or existing niches requires possession of appropriate adaptations (Sakai et al. 2001; Brown and Inglis 2009). For microorganisms, such adaptations may be acquired not just through mutation but also via horizontal gene transfer (HGT), and a standard view is that pathogen evolution often is dependent on the acquisition by microbes, via HGT, of new, useful genes or gene complexes (Lawrence and Hendrickson 2008). If we assume that Archaea have the same potential as Bacteria to acquire niche-invasion adaptations via mutation—though note the questioning of that assumption under the heading of coevolutionary mismatches—and continue to at least assume that were Archaea to possess such adaptations then Archaea would be more prevalent among symbiotic organisms, then a reasonable place to look for deficiencies in Archaea acquisition of adaptations useful for invasion of symbiosis niches would be in terms of HGT.

Especially within less extreme environments, Archaea might be relatively lacking in terms HGT with other Archaea, in comparison, that is, to Bacteria with other Bacteria. The idea here is that due to lower numbers of individuals, e.g., approximately 10% of the numbers of Bacteria in many cases, the opportunity for HGT among Archaea may be lower. The theory underlying this suggestion is based on mass action, that is, to the extent that donors of genetic material as well as recipients of genetic material are lower in prevalence then so too should be the rate of donation and reception of genetic material (Smets and Lardon 2009). This general idea López-García et al. (2015) appear to describe instead as a "statistical bias". Though mass action theory for the occurrence of HGT was developed for conjugation, to a first approximation it should be true for transduction and transformation as well, i.e., fewer potential donors should result in less DNA donated to a 'transfer pool' and, for a transfer pool of a given size, fewer potential recipients should result in fewer successful DNA transfer events. An organism-density dependence of levels of HGT could result in a feedback loop in which the most prevalent organisms within an environment will also be the ones with greater access to HGT and therefore should collectively evolve faster. These organisms thereby may be able to more effectively maintain their status as the most prevalent within an environment. This basically means that more prevalent related-groups of organisms should be able to more efficiently acquire those adaptations necessary to defend as well as invade niches while less prevalent related-groups of organisms may be both less able to defend and less able to invade niches. (As noted in Sect. 3.6, higher population numbers also should result in higher rates of acquisition of beneficial alleles via mutation.)

If in fact we can assume that Archaea, due to smaller community sizes, are limited relative to Bacteria in terms of their potential to evolve towards more effective niche invasion, then we may expect Archaea to be relegated primarily to those niches in which barriers to niche invasion which are arrayed against Bacteria are particularly high, e.g., perhaps particularly oligotrophic environments (re: Archaea as low-energy specialists). Equivalently, in Bacteria-dominated niches, barriers to niche invasion by Archaea may also be too high to overcome. It is important to note, however, that the assumption that Archaea may experience fewer intra-domain HGT events than Bacteria might simply not be the case (Choi and Kim 2007); see as well (Turkarslan et al. 2011). If so, then we might speculate that HGT with other Archaea might be qualitatively insufficient rather than quantitatively insufficient to result in Archaea acquisition of competitiveness-enhancing, symbiosis-facilitating adaptations. That is, necessary adaptations to either exploit Bacteria-occupied niches or compete with already-present Bacteria simply may not be present to share among Archaea. If so, then we might ask whether limitations on such Archaea evolution are a consequence of an inability to access necessary niche-invasion adaptations as found instead among Bacteria.

3.9 Low Sharing of Genes with Bacteria

Corroborating various ideas presented in Sect. 3.8, Ochman et al. (2000) have suggested (p. 299) that "it is difficult to account for the ability of bacteria to exploit new environments by the accumulation of point mutations alone." Evolution of pathogenic tendencies, for example, can be viewed as the evolution of the exploitation of new environments, including in terms of competing with already established commensal organisms as well as overcoming host innate immunity before they can establish infections (Rohmer et al. 2011). Indeed (p. 836), "gene gain as a result of horizontal gene transfer remains the most potent source of 'innovation' and variation" among bacterial pathogens (Pallen and Wren 2007). In Sect. 3.8, that the impact of differences in intra-domain HGT—Archaea-to-Archaea versus Bacteria-to-Bacteria—might negatively impact Archaea niche invasion potential thus was explored. In this subsection it instead is inter-domain HGT that is considered, specifically Bacteria-to-Archaea.

Consistent with this perspective of the importance of HGT in the evolution of bacterial pathogens, low efficiency of Bacteria-to-Archaea HGT—posited by Gill and Brinkman (2011) as due to a lack of substantial sharing of viruses between Archaea and Bacteria—thus might serve as a limitation on the ability of Archaea to evolve into pathogens. Gill and Brinkman's basic argument is that temperate bacteriophages (a.k.a., bacterial viruses or phages) often encode bacterial virulence factors, e.g., Christie et al. (2012), while phages also can serve as vectors for genomic islands. An inability of phages to also infect Archaea, that is, not also serve as archaeal viruses, results in a lack of transfer of evolved factors, as required for bacterial pathogenesis, to Archaea. If nothing else, this would be expected to slow the rate at which Archaea might evolve into recognizable pathogens.

I don't dispute Gill and Brinkman's (2011) core argument of a lack of phage- or archaeal virus-mediated gene exchange between Bacteria and Archaea. The more basic argument of there not being substantial gene exchange between Bacteria and Archaea, however, does not necessarily follow (Shiffman and Charalambous 2012). López-García et al. (2015) in particular review evidence that mesophilic archaea (p. 447/abstract) "were recipients of massive horizontal gene transfer from bacteria." See also arguments by Groussin et al. (2016) of "continuous acquisition of genes over long periods in the evolution of Archaea." See also Lurie-Weinberger et al. (2012a, b) for consideration explicitly of improved archaeal adaptation (by methanogens) to host-associated niches as a consequence of HGT from Bacteria. Among studies exploring the question of the rarity of archaeal pathogens, Cavicchioli et al. (2003) make similar arguments.

Unless there are biases specifically against transfer of virulence factor genes (which, it should be noted, at least is possible), then such high levels of transfer of genes from Bacteria to Archaea at some point during the latter's evolution argues against the relative importance of a lack of sharing of viruses between the two domains as a reason for the observed dearth of archaeal pathogens. It should be noted that mass action concerns, as indicated in Sect. 3.8, also may be of lowered relevance for Bacteria-to-Archaea HGT since at least the donor organisms, Bacteria, would be expected to be present at higher densities, e.g., roughly ten-fold higher, than were the donor organisms instead members of the Archaea.

Note that there have been attempts to argue in the opposite direction, that is, that transfer of genes from Archaea to Bacteria might be important in the evolution of *bacterial* pathogens (Faguy 2003). See, though, Gophna et al. (2004) for skepticism of this latter idea. See also Bokhari et al. (2011). Inter-domain transfers nonetheless may be biased towards transfer from Bacteria to Archaea rather than the other way around (López-García et al. 2015). A lack of sharing of viruses between Bacteria and Archaea thus may impact qualitatively or quantitatively what genetic material is transferred, but transfer overall does not appear to be blocked to an extraordinary extent. HGT-based arguments for why Archaea appear to be deficient among pathogens thus might be limited to suggestions that whatever amount of HGT does occur, either quantitatively or qualitatively, it is simply not sufficient to result in Archaea possessing those qualities which within Bacteria apparently are sufficient for relatively frequent evolution of pathogens.

3.10 Conceptual Limitations

A pathogen is a microorganism that harms especially larger eukaryotic organisms. The mechanisms by which that harm may be effected vary but at some level the concept of pathogen is intimately entwined with the concept of health. Thus, as usually recognized, a pathogen reduces a host organism's health in some measurable or at least noticeable way. A microorganism therefore must have at least three properties to be likely to be recognized as a pathogen: (1) It must affect a host that is of sufficient interest that reductions in its health will come to our attention. (2) The presence of the would-be pathogen must be detected in association with noticeable reductions in health, and detection is not necessarily easily achieved (Bang and Schmitz 2015). Lastly, (3) a lack of equivalent reductions in host health must be associated with lack of presence of the pathogen. In other words, some variation on Koch's Postulates must be demonstrated, but made more challenging if reductions in health, presence of potential pathogens, or indeed absence of potential pathogens are not readily or at least easily detected or facilitated.

An important issue regarding the potential of Archaea to serve as pathogens is that we are largely interested in whether they can serve as pathogens of humans. Particularly to the extent that potential Archaea-associated health reductions do not appear to be obvious, one therefore is reduced to using statistical inferences, i.e., as one sees with periodontal disease and the potential for methanogens to participate in mixed infections. It may be, however, that Archaea nevertheless can play parasitic roles in humans which are difficult to detect, though it is likely that these roles are constrained by a combination of the tendencies of mesophilic Archaea to be limited in their potential to utilize as energy as well as carbon sources any but the simplest of host molecules (i.e., simpler than even glucose). Within more extreme environments this limitation does not seem to similarly hold, suggesting that Archaea as pathogens, or as parasites, may be more readily recognized in host organisms able to occupy these more extreme niches. Alternatively, as our ability to identify organisms and correlate them more subtlety to host potential to thrive, i.e., as metabiomic studies increasing seek to achieve (Li et al. 2016), then it may be that Archaea will be associated perhaps especially with more subclinical variations in human health. Host species which demonstrably are colonized with more diverse types of Archaea, particularly more than just methanogens, may very well experience more negative outcomes of Archaea colonization than so far has been demonstrated in humans. Indeed, as Shiffman and Charalambous (2012) suggest

(p. 50), “With insufficiencies in methods of detection and few examples of cultured Archaea, the lack of a clear-cut pathogen is perhaps not unsurprising.”

4 Why Aren’t Eukaryotes Rare Among Pathogens?

Numerous eukaryotes serve as pathogens or, semantically nearly equivalently, instead as within-organism parasites. The discussion presented above provides clues as to why the eukaryotes seem to be more inclined to take up pathogen-like lifestyles than Archaea. In what manner, then, do eukaryotes differ from Archaea that could lead to a greater propensity to evolve lifestyles involving the taking of resources directly from the bodies of other organisms? See Fig. 1 for summary.

First, vast numbers of eukaryotes are organotrophic. Indeed, with the exception of many photosynthetic eukaryotes along with those eukaryotes that obtain nutrients from dead organisms, the eukaryote niche is one of exploitation of other, living organisms for nutrients (i.e., as seen among animals, phagotrophic protists, carnivorous plants, and predatory fungi). In fact, such living-organism exploiters are so prevalent among eukaryotes that it may be worth flipping the original question. That is, rather than asking why pathogens tend to be rare among Archaea relative to Bacteria, instead we might ask why eating living cells and bodies of other organisms is relatively rare among Bacteria in comparison to among eukaryotes.

Second, and potentially addressing the issue of a relative rarity of such exploitation among Bacteria versus among eukaryotes, much of the exploitation of other organisms for organic nutrients as mediated by eukaryotes involves engulfment or ingestion. These are nutrient-acquisition strategies that are much less accessible to Bacteria or Archaea. Indeed, it may simply require greater sophistication for a saprophyte to come to consume living organisms that are bigger than itself, i.e., as an absorber of nutrients which are located outside of an exploiter’s cell (with ‘cell’ in this case serving as the exploiter’s body), than for a saprophyte to come to consume living organisms figuratively one “bite-sized” piece at a time, i.e., as otherwise ingesters or engulfers of nutrients. Another way of stating this is that protection by relatively large organisms from attack by relatively small organisms may be a more long-standing adaptation (i.e., immune systems) than protection of organisms from attack by relatively large organisms, particularly since smaller organisms as attackers (i.e., pathogens or parasites) presumably pre-date larger organisms as attackers (i.e., predators) by relatively large spans of evolutionary times.

Third, large numbers of eukaryotes appear to be less well adapted to conditions of severe nutrient restriction, particularly given continued organism metabolism, than has been suggested as typical for many Archaea. Instead, eukaryotes are inherently relatively energy inefficient, supporting large, sophisticated cells and bodies. Chemoheterotrophic eukaryotes therefore may not only be more inclined to consume the cells and bodies of other organisms, they may have little choice but to do so in terms of access to sufficient nutrients. To the extent that there are only so many dead organisms around to consume, therefore, then selection presumably has

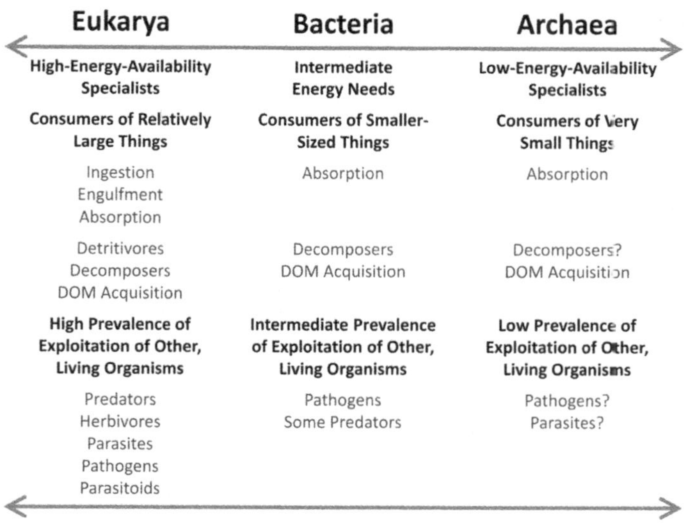

Fig. 1 Comparison of the three domains of cellular organisms in terms of general energy-utilization and trophic characteristics. *Bold, darker text* indicates general characteristics while *lighter text* indicates more specific characteristics. Not explicitly considered in the figure are phototrophs though eukaryote phototrophs (i.e., plants and algae) may also be viewed high-energy-availability specialists. Note that in terms of consumption what is meant by "Large" is roughly everything from clumps of relatively large macromolecules on up (though eukaryotes certainly can partake of smaller molecules as well), what is meant by "Small" is macromolecules on down, and what is mean by "Very Small" is roughly macromolecule subunits and down (though it is assumed that at least some Archaea must be capable of extracellular digestion). "DOM" refers to Dissolved Organic Matter while ingestion and engulfment (particularly phagocytosis for the latter) are mechanisms of acquisition of nutrients especially in forms which are large enough that those materials do not qualify as DOM. Parasitism of other, living organisms here primarily is towards nutrient acquisition directly from bodies. Bacteria that are predators include bdellovibrios but also at least arguably include some pathogens, such as members of genus *Clostridium*. Note that viruses presumably consist of even higher proportions of exploiters of other, living organisms than members of domain Eukarya, assuming that most endosymbioses are mutualistic rather than exploitive. Note too that the term "Pathogen" is used synonymously with that of a parasitic symbiont whose presence can result in reductions in a host's health

favored the de novo evolution of multiple strategies, among these otherwise voracious organisms, of seeking such nutrients directly from bodies and cells of living organisms.

Fourth, eukaryotes, including microscopic eukaryotes, tend to be prevalent within environments in which eukaryotes, including macroscopic eukaryotes, also are found. Thus, predation on other, living organisms—including Bacteria as well as Archaea—may represent a niche which is difficult, evolutionarily, for especially already relatively sophisticated organotrophic organisms to resist. So too, therefore, it may be that the most likely location within which Archaea are capable of exploiting other, living organisms for nutrients will be found within environments in which both more sophisticated organotrophic Archaea and sufficient numbers of potentially exploitable living organisms are simultaneously present, e.g., as within hot springs.

Fifth, eukaryotes as well as Bacteria tend to be both diverse and abundant within environments within which substantial trophic interactions are known to have developed. To the extent that exploiting living organisms as carbon and energy sources encourages the evolution of pathogenic and parasitic relationships, then three criteria likely underlie the evolution of most or all such strategies: (1) An ability to so exploit relatively large quantities of molecules making up these other organisms (i.e., relatively sophisticated organotrophic tendencies), (2) the presence within environments of so-exploitable organisms, and (3) sufficient time for such strategies of exploitation of other, living organisms to evolve. The latter presumably requires sufficient long-term environmental stability, or at least relatively long-term environment existence. More sophisticated eukaryotic and bacterial organotrophs living within less extreme environments presumably have had geologically longer spans of time to evolve such strategies than more sophisticated Archaea organotrophs, i.e., as tend to thrive within more ephemeral environments such as hot springs and geothermal vents. Though relevant, the question of exactly why sophisticated organotrophic Archaea are not conspicuous within less extreme environments is not addressed here.

5 Conclusion

We can consider the likelihood of organisms evolving to become pathogens to be a multifactorial process requiring a combination of perhaps prevalence (such as in association with potential eukaryote hosts), diversity (ditto), existing utility (such as already being relatively sophisticated organotrophs), an ability to respond relatively rapidly to high nutrient loads in terms of rates of population growth, and also requiring sufficient spans of evolutionary time. To varying degrees these factors may be depressed to a greater extent among Archaea than they are among Bacteria. They appear as well to be more robust among eukaryotes also than they are among Archaea. Indeed, eukaryotes in a number of ways appear to be better equipped to serve as exploiters of other organisms than Bacteria, while Bacteria appear to be better equipped to play the role of exploiter than Archaea.

At a minimum we might expect these factors to have additive effects, e.g., such that deficiencies in any one relevant factor may reduce the potential for Archaea to evolve into pathogens by, say, a hypothetical two-fold. Combine enough of these then the likelihood of Archaea evolving into pathogens may be reduced to the levels that we in fact observe among Archaea. Of course, the postulated reducing effect that each factor has could be greater (or smaller) than two-fold and/or the impacts may be synergistic against Archaea evolving into pathogens, etc. At this point in time, however, the impact of the different factors is essentially not known. Thus they represent nothing more certain than suggested hypotheses. It is possible too that greater surveillance for Archaea serving as pathogens, parasites, or even predators will yet turn up more definitive examples. A discovery of Archaea serving as significant pathogens turning out to be a norm, particularly within less extreme environments, does not, however, seem likely.

Acknowledgements I am grateful to John Reeve for both introducing me to the issue of rarity of pathogens among Archaea and also for helpful discussion of the ideas presented in this and a previously published essay on this subject.

References

Abedon ST (2013) Are archaeons incapable of being parasites or have we simply failed to notice? Bioessays: news and reviews in molecular. Cell Dev Biol 35:501

Aller JY, Kemp PF (2008) Are archaea inherently less diverse than bacteria in the same environments? FEMS Microbiol Ecol 65:74–87

Aminov RI (2013) Role of archaea in human disease. Front Cell Infect Microbiol 3:42

Andrei AS, Banciu HL, Oren A (2012) Living with salt: metabolic and phylogenetic diversity of archaea inhabiting saline ecosystems. FEMS Microbiol Lett 330:1–9

Andrei AS, Robeson MS, Baricz A, Coman C, Muntean V, Ionescu A, Etiope G, Alexe M, Sicora CI, Podar M, Banciu HL (2015) Contrasting taxonomic stratification of microbial communities in two hypersaline meromictic lakes. ISME J 9:2642–2656

Bang C, Schmitz RA (2015) Archaea associated with human surfaces: not to be underestimated. FEMS Microbiol Rev 39:631–648

Barash I, Manulis-Sasson S (2009) Recent evolution of bacterial pathogens: the gall-forming *Pantoea agglomerans* case. Annu Rev Phytopathol 47:133–152

Baricz A, Coman C, Andrei AS, Muntean V, Keresztes ZG, Pausan M, Alexe M, Banciu HL (2014) Spatial and temporal distribution of archaeal diversity in meromictic, hypersaline Ocnei Lake (Transylvanian Basin, Romania). Extremophiles 18:399–413

Bates ST, Berg-Lyons D, Caporaso JG, Waters WA, Knight R, Fierer N (2011) Examining the global distrbiution of dominant archaeal populations in soil. ISME J 5:908–917

Bokhari H, Anwar M, Mirza HB, Gillevet PM (2011) Evidences of lateral gene transfer between archaea and pathogenic bacteria. Bioinformation 6:293–296

Borrel G, Lehours AC, Crouzet O, Jezequel D, Rockne K, Kulczak A, Duffaud E, Joblin K, Fonty G (2012) Stratification of Archaea in the deep sediments of a freshwater meromictic lake: vertical shift from methanogenic to uncultured archaeal lineages. PLoS ONE 7:e43346

Brown SP, Inglis RF (2009) Evolutionary ecology of microbial wars: within-host competition and (incidental) virulence. Evol Appl 2:32–39

Buée M, De Boer W, Martin F, van Overbeek L, Jurkevitch E (2009) The rhizosphere zoo: an overview of plant-associated communities of microorganisms, including phages, bacteria, archaea, and fungi, and of some of their structuring factors. Plant Soil 321:189–212

Calcagno V, Dubosclard M, de MC (2010) Rapid exploiter-victim coevolution: the race is not always to the swift. Am Nat 176:198–211

Cao P, Zhang LM, Shen JP, Zheng YM, Di HJ, He JZ (2012) Distribution and diversity of archaeal communities in selected Chinese soils. FEMS Microbiol Ecol 80:146–158

Casadevall A, Pirofski LA (2000) Host-pathogen interactions: basic concepts of microbial commensalism, colonization, infection, and disease. Infect Immun 68:6511–6518

Cavicchioli R, Curmi PM (2004) Response to William Martin's letter. Bioessays 26:593

Cavicchioli R, Curmi PM, Saunders N, Thomas T (2003) Pathogenic archaea: do they exist? Bioessays (News and Reviews In Molecular Cellular and Developmental Biology) 25:1119–1128

Chaban B, Ng SY, Jarrell KF (2006) Archaeal habitats—from the extreme to the ordinary. Can J Microbiol 52:73–116

Chen L, Brugger K, Skovgaard M, Redder P, She Q, Torarinsson E, Greve B, Awayez M, Zibat A, Klenk HP, Garrett RA (2005) The genome of *Sulfolobus acidocaldarius*, a model organism of the *Crenarchaeota*. J Bacteriol 187:4992–4999

Choi IG, Kim SH (2007) Global extent of horizontal gene transfer. Proc Natl Acad Sci U S A 104:4489–4494

Christie GE, Allison HA, Kuzio J, McShan M, Waldor MK, Kropinski AM (2012) Prophage-induced changes in cellular cytochemistry and virulence. In: Hyman P, Abedon ST (eds) Bacteriophages in health and disease. CABI Press, Wallingford, pp 33–60

Conway de Macario E, Macario AJL (2009) Methanogenic archaea in health and disease: a novel paradigm of microbial pathogenesis. Int J Med Microbiol 299:99–108

Curtis TP, Sloan WT (2004) Prokaryotic diversity and its limits: microbial community structure in nature and implications for microbial ecology. Curr Opin Microbiol 7:221–226

DeLong EF, Taylor LT, Marsh TL, Preston CM (1999) Visualization and enumeration of marine planktonic archaea and bacteria by using polyribonucleotide probes and fluorescent in situ hybridization. Appl Environ Microbiol 65:5554–5563

Eckburg PB, Lepp PW, Relman DA (2003) Archaea and their potential role in human disease. Infect Immun 71:591–596

Eckburg PB, Bik EM, Bernstein CN, Purdom E, Dethlefsen L, Sargent M, Gill SR, Nelson KE, Relman DA (2005) Diversity of the human intestinal microbial flora. Science (New York, N Y) 308:1635–1638

Ewald PW (2004) Evolution of virulence. Infect Dis Clin North Am 18:1–15

Faguy DM (2003) Lateral gene transfer (LGT) between Archaea and *Escherichia coli* is a contributor to the emergence of novel infectious disease. BMC Infect Dis 3:13

Fellous S, Salvaudon L (2009) How can your parasites become your allies? Trends Parasitol 25:62–66

Fredriksson NJ, Hermansson M, Wilen BM (2012) Diversity and dynamics of archaea in an activated sludge wastewater treatment plant. BMC Microbiol 12:140

Fredriksson NJ, Hermansson M, Wilen BM (2013) The choice of PCR primers has great impact on assessments of bacterial community diversity and dynamics in a wastewater treatment plant. PLoS ONE 8:e76431

Giannone RJ, Wurch LL, Heimerl T, Martin S, Yang Z, Huber H, Rachel R, Hettich RL, Podar M (2015) Life on the edge: functional genomic response of *Ignicoccus hospitalis* to the presence of *Nanoarchaeum equitans*. ISME J 9:101–114

Gill EE, Brinkman FS (2011) The proportional lack of archaeal pathogens: Do viruses/phages hold the key? Bioessays 33:248–254

Golyshina OV, Lunsdorf H, Kublanov IV, Goldenstein NI, Hinrichs KU, Golyshin PN (2016) The novel extremely acidophilic, cell-wall-deficient archaeon *Cuniculiplasma divulgatum* gen. nov., sp. nov. represents a new family, *Cuniculiplasmataceae* fam. nov., of the order *Thermoplasmatales*. Int J Syst Evol Microbiol 66:332–340

Gophna U, Charlebois RL, Doolittle WF (2004) Have archaeal genes contributed to bacterial virulence? Trends Microbiol 12:213–219

Groisman EA, Ochman H (1996) Pathogenicity islands: bacterial evolution in quantum leaps. Cell 87:791–794

Groussin M, Boussau B, Szollosi G, Eme L, Gouy M, Brochier-Armanet C, Daubin V (2016) Gene acquisitions from bacteria at the origins of major archaeal clades are vastly overestimated. Mol Biol Evol 33:305–310

Hagman M, Nielsen JL, Nielsen PH, Jansen J (2008) Mixed carbon sources for nitrate reduction in activated sludge-identification of bacteria and process activity studies. Water Res 42:1539–1546

Herndl GJ, Reinthaler T, Teira E, van AH, Veth C, Pernthaler A, Pernthaler J (2005) Contribution of archaea to total prokaryotic production in the deep Atlantic Ocean. Appl Environ Microbiol 71:2303–2309

Holder T, Basquin C, Ebert J, Randel N, Jollivet D, Conti E, Jekely G, Bono F (2013) Deep transcriptome-sequencing and proteome analysis of the hydrothermal vent annelid *Alvinella pompejana* identifies the CvP-bias as a robust measure of eukaryotic thermostability. Biol Direct 8:2

Horz HP, Conrads G (2010) The discussion goes on: what is the role of Euryarchaeota in humans? Archaea 2010:967271

Jahn U, Gallenberger M, Paper W, Junglas B, Eisenreich W, Stetter KO, Rachel R, Huber H (2008) *Nanoarchaeum equitans* and *Ignicoccus hospitalis*: new insights into a unique, intimate association of two archaea. J Bacteriol 190:1743–1750

Karlsson AE, Johansson T, Bengtson P (2012) Archaeal abundance in relation to root and fungal exudation rates. FEMS Microbiol Ecol 80:305–311

Karner MB, DeLong EF, Karl DM (2001) Archaeal dominance in the mesopelagic zone of the Pacific Ocean. Nature (London) 409:507–510

Klappenbach JA, Saxman PR, Cole JR, Schmidt TM (2001) rrndb: the ribosomal RNA operon copy number database. Nucl Acids Res 29:181–184

Lane N, Martin W (2010) The energetics of genome complexity. Nature (London) 467:929–934

Lange M, Westermann P, Ahring BK (2005) Archaea in protozoa and metazoa. Appl Microbiol Biotechnol 66:465–474

Lawrence JG, Hendrickson H (2008) Genomes in motion: gene transfer as a catalyst for genome change. Horizontal gene transfer in the evolution of pathogenesis. Cambridge University Press, Cambridge, pp 3–22

LeClerc JE, Li B, Payne WL, Cebula TA (1996) High mutation frequencies among *Escherichia coli* and *Salmonella* pathogens. Science (New York, N Y) 274:1208–1211

Lee ZM, Bussema C III, Schmidt TM (2009) rrnDB: documenting the number of rRNA and tRNA genes in bacteria and archaea. Nucl Acids Res 37:D489–D493

Leggett HC, Cornwallis CK, West SA (2012) Mechanisms of pathogenesis, infective dose and virulence in human parasites. PLoS Pathog 8:e1002512

Levin LA (2010) Anaerobic metazoans: no longer an oxymoron. BMC Biol 8:31

Li WKW, Dickie PM, Irwin BD, Wood AM (1992) Biomass of bacteria, cyanobacteria, prochlorophytes and photosynthetic eukaryotes in the Sargasso Sea. Deep Sea Res Part A (Oceanographic Research Papers) 39:501–519

Li J, Zhou H, Fang J, Sun Y, Dasgupta S (2014) Microbial distribution in different spatial positions within the walls of a black sulfide hydrothermal chimney. Mar Ecol Prog Ser 508:67–85

Li D, Wang P, Wang P, Hu X, Chen F (2016) The gut microbiota: a treasure for human health. Biotechnol Adv 34:1210–1224

Lipp JS, Morono Y, Inagaki F, Hinrichs KU (2008) Significant contribution of Archaea to extant biomass in marine subsurface sediments. Nature (London) 454:991–994

Liu LJ, You XY, Guo X, Liu SJ, Jiang CY (2011) *Metallosphaera cuprina* sp. nov., an acidothermophilic, metal-mobilizing archaeon. Int J Syst Evol Microbiol 61:2395–2400

López-García P, Zivanovic Y, Deschamps P, Moreira D (2015) Bacterial gene import and mesophilic adaptation in archaea. Nat Rev Microbiol 13:447–456

Lurie-Weinberger MN, Gophna U (2015) Archaea in and on the human body: health implications and future directions. PLoS Pathog 11:e1004833

Lurie-Weinberger MN, Peeri M, Gophna U (2012a) Contribution of lateral gene transfer to the gene repertoire of a gut-adapted methanogen. Genomics 99:52–58

Lurie-Weinberger MN, Peeri M, Tuller T, Gophna U (2012b) Extensive inter-domain lateral gene transfer in the evolution of the human commensal methanosphaera stadtmanae. Front Genet 3:182

Lynch M (2010) Evolution of the mutation rate. Trends Genet 26:345–352

Martin W (2004) Pathogenic archaebacteria: do they not exist because archaebacteria use different vitamins? Bioessays 26:592–593

Miller-Coleman RL, Dodsworth JA, Ross CA, Shock EL, Williams AJ, Hartnett HE, McDonald AI, Havig JR, Hedlund BP (2012) Korarchaeota diversity, biogeography, and abundance in Yellowstone and Great Basin hot springs and ecological niche modeling based on machine learning. PLoS ONE 7:e35964

Moissl-Eichinger C, Huber H (2011) Archaeal symbionts and parasites. Curr Opin Mirobiol 14:364–370

Moxon R, Tang C (2000) Challenge of investigating biologically relevant functions of virulence factors in bacterial pathogens. Philos Trans R Soc Lond B Biol Sci 355:643–656

Muller M, Mentel M, van Hellemond JJ, Henze K, Woehle C, Gould SB, Yu RY, van der Giezen M, Tielens AG, Martin WF (2012) Biochemistry and evolution of anaerobic energy metabolism in eukaryotes. Microbiol Mol Biol Rev 76:444–495

Murray AE, Preston CM, Massana R, Taylor LT, Blakis A, Wu K, DeLong EF (1998) Seasonal and spatial variability of bacterial and archaeal assemblages in the coastal waters near Anvers Island, Antarctica. Appl Environ Microbiol 64:2585–2595

Nguyen-Hieu T, Khelaifia S, Aboudharam G, Drancourt M (2013) Methanogenic archaea in subgingival sites: a review. APMIS 121:467–477

Nkamga VD, Henrissat B, Drancourt M (2016) Archaea: essential inhabitants of the human digestive microbiota. Hum Microbiome J 3:1–8

Ochman H, Lawrence JG, Groisman EA (2000) Lateral gene transfer and the nature of bacterial innovation. Nature (London) 405:299–304

Ochsenreiter T, Selezi D, Quaiser A, Bonch-Osmolovskaya L, Schleper C (2003) Diversity and abundance of Crenarchaeota in terrestrial habitats studied by 16S RNA surveys and real time PCR. Environ Microbiol 5:787–797

Pal C, Macia MD, Oliver A, Schachar I, Buckling A (2007) Coevolution with viruses drives the evolution of bacterial mutation rates. Nature (London) 450:1079–1081

Pallen MJ, Wren BW (2007) Bacterial pathogenomics. Nature (London) 449:835–842

Pape T, Hoffmann F, Quéric N-V, von Juterzenka K, Reitner J, Michaelis W (2006) Dense populations of Archaea associated with demosponge *Tentorium semisuberites* Schmidt, 1870 from Arctic deep-waters. Polar Biol 29:662–667

Park JS, Simpson AG, Lee WJ, Cho BC (2007) Ultrastructure and phylogenetic placement within Heterolobosea of the previously unclassified, extremely halophilic heterotrophic flagellate *Pleurostomum flabellatum* (Ruinen 1938). Protist 158:397–413

Park JS, De Jonckheere JF, Simpson AG (2012) Characterization of *Selenaion koniopes* n. gen., n. sp., an amoeba that represents a new major lineage within heterolobosea, isolated from the Wieliczka salt mine. J Eukaryot Microbiol 59:601–613

Paterson S, Vogwill T, Buckling A, Benmayor R, Spiers AJ, Thomson NR, Quail M, Smith F, Walker D, Libberton B, Fenton A, Hall N, Brockhurst MA (2010) Antagonistic coevolution accelerates molecular evolution. Nature (London) 464:275–278

Perevalova AA, Kublanov IV, Bidzhieva SK, Mukhopadhyay B, Bonch-Osmolovskaya EA, Lebedinsky AV (2016) Reclassification of *Desulfurococcus mobilis* as a synonym of *Desulfurococcus mucosus*, *Desulfurococcus fermentans* and *Desulfurococcus kamchatkensis* as synonyms of *Desulfurococcus amylolyticus*, and emendation of the *D. mucosus* and *D. amylolyticus* species descriptions. Int J Syst Evol Microbiol 66:514–517

Podar M, Anderson I, Makarova KS, Elkins JG, Ivanova N, Wall MA, Lykidis A, Mavromatis K, Sun H, Hudson ME, Chen W, Deciu C, Hutchison D, Eads JR, Anderson A, Fernandes F, Szeto E, Lapidus A, Kyrpides NC, Saier MH Jr, Richardson PM, Rachel R, Huber H, Eisen JA, Koonin EV, Keller M, Stetter KO (2008) A genomic analysis of the archaeal system *Ignicoccus hospitalis-Nanoarchaeum equitans*. Genome Biol 9:R158

Podar M, Makarova KS, Graham DE, Wolf YI, Koonin EV, Reysenbach AL (2013) Insights into archaeal evolution and symbiosis from the genomes of a nanoarchaeon and its inferred crenarchaeal host from Obsidian Pool, Yellowstone National Park. Biol Direct 8:9

Prieur D (2002) Hydrothermal vents: prokaryotes in deep-sea hydrothermal vents. Encycl Environ Microbiol 1617–1628

Raskin DM, Seshadri R, Pukatzki SU, Mekalanos JJ (2006) Bacterial genomics and pathogen evolution. Cell 124:703–714

Reeve JN (1999) Archaebacteria then … Archaes now (are there really no archaeal pathogens?). J Bacteriol 181:3613–3617

Reitschuler C, Lins P, Wagner AO, Illmer P (2014) Cultivation of moonmilk-born non-extremophilic Thaum and Euryarchaeota in mixed culture. Anaerobe 29:73–79

Rivera-Perez JI, Gonzalez AA, Toranzos GA (2017) From evolutionary advantage to disease agents: forensic reevaluation of host-microbe interactions and pathogenicity. Microbiol Spectr 5:EMF-0009-2016

Rohmer L, Hocquet D, Miller SI (2011) Are pathogenic bacteria just looking for food? Metabolism and microbial pathogenesis. Trends Microbiol 19:341–348

Saengkerdsub S, Ricke SC (2013) Ecology and characteristics of methanogenic archaea in animals and humans. Crit Rev Microbiol

Sahm K, John P, Nacke H, Wemheuer B, Grote R, Daniel R, Antranikian G (2013) High abundance of heterotrophic prokaryotes in hydrothermal springs of the Azores as revealed by a network of 16S rRNA gene-based methods. Extremophiles 17:649–662

Sakai AK, Allendorf FW, Holt JS, Lodge DM, Molfsky J, With KA, Baughman S, Cabin RJ, Cohen JE, Ellstrand NC, McCauley DE, O'Neil P, Parker IM, Thompson JN, Weller SG (2001) The population biology of invasive species. Ann Rev Ecol Syst 32:305–332

Seger J (1992) Evolution of exploiter–victim relationships. In: Crawley MJ (ed) Natural enemies. The population biology of predators, parasites and diseases. Blackwell Scientific, Oxford, pp 3–25

Selig M, Schönheit P (1994) Oxidation of organic compounds to CO_2 with sulfur or thiosulfate as electron acceptor in the anaerobic hyperthermophilic archaea *Thermoproteus tenax* and *Pyrobaculum islandicum* proceeds via the citric acid cycle. Arch Microbiol 162:286–294

Shiffman ME, Charalambous BM (2012) The search for archaeal pathogens. Rev Med Microbiol 23:45–51

Sinclair JL, Ghiorse WC (1987) Distribution of protozoa in subsurface sediments of a pristine groundwater study site in Oklahoma. Appl Environ Microbiol 53:1157–1163

Smets BF, Lardon L (2009) Mass action models describing extant horizontal transfer of plasmids: inferences and parameter sensitivities. Meth Mol Biol 532:289–305

Stewart EJ (2012) Growing unculturable bacteria. J Bacteriol 194:4151–4160

Tanaka T, Kawasaki K, Daimon S, Kitagawa W, Yamamoto K, Tamaki H, Tanaka M, Nakatsu CH, Kamagata Y (2014) A hidden pitfall in the preparation of agar media undermines microorganism cultivability. Appl Environ Microbiol 80:7659–7666

Turkarslan S, Reiss DJ, Gibbins G, Su WL, Pan M, Bare JC, Plaisier CL, Baliga NS (2011) Niche adaptation by expansion and reprogramming of general transcription factors. Mol Syst Biol 7:554

Uroz S, Oger P, Tisserand E, Cebron A, Turpault MP, Buee M, De BW, Leveau JH, Frey-Klett P (2016) Specific impacts of beech and Norway spruce on the structure and diversity of the rhizosphere and soil microbial communities. Sci Rep 6:27756

Urschel MR, Kubo MD, Hoehler TM, Peters JW, Boyd ES (2015) Carbon source preference in chemosynthetic hot spring communities. Appl Environ Microbiol 81:3834–3847

Valentine DL (2007) Adaptations to energy stress dictate the ecology and evolution of the Archaea. Nat Rev Microbiol 5:316–323

Wassenaar TM, Gaastra W (2001) Bacterial virulence: can we draw the line? FEMS Microbiol Lett 201:1–7

Welte C, Deppenmeier U (2011) Proton translocation in methanogens. Methods Enzymol 494:257–280

Wrede C, Dreier A, Kokoschka S, Hoppert M (2012) Archaea in symbioses. Archaea 2012:596846

Xue Y, Fan H, Ventosa A, Grant WD, Jones BE, Cowan DA, Ma Y (2005) *Halalkalicoccus tibetensis* gen. nov., sp. nov., representing a novel genus of haloalkaliphilic archaea. Int J Syst Evol Microbiol 55:2501–2505

Yau S, Lauro FM, Williams TJ, Demaere MZ, Brown MV, Rich J, Gibson JA, Cavicchioli R (2013) Metagenomic insights into strategies of carbon conservation and unusual sulfur biogeochemistry in a hypersaline Antarctic lake. ISME J 7:1944–1961

Ziebuhr W, Hennig S, Eckart M, Kranzler H, Batzilla C, Kozitskaya S (2006) Nosocomial infections by Staphylococcus epidermidis: how a commensal bacterium turns into a pathogen. Int J Antimicrob Agents 28(Suppl 1):S14–S20

Archaeal Surface Structures and Their Role in Communication with the Extracellular Environment

Sonja-Verena Albers and Tessa E.F. Quax

1 Introduction

Cellular surface structures are instrumental for contact with the extracellular environment and can play important roles in communication with other organisms and viruses. They allow the formation of multicellular structures, such as biofilms or cellular aggregates, which enhance adaptation to a changing environment and increase survival (Battin et al. 2007; Palmer et al. 2007). And indeed, sessile microbial communities are the predominant mode of life in of microorganisms (Palmer and Stoodley 2007). Biofilms are characterized by complex multicellular architectures and the presence of EPS (extracellular polymeric substance). Cellular aggregates or biofilms allow the exchange of nutrients, DNA, and better protection to averse conditions, such as toxic compounds, temperature, UV and pH stress (Elasri and Miller 1999; Martinez and Casadevall 2007). Members of the three domains of life, Archaea, Bacteria and Eukarya, all have various surfaces structures at their disposal, of which some are shared between domains, and others are uniquely present within members of one domain. Archaea are without doubt champions in colonization of a huge range of habitats. Whereas the first archaea were mostly isolated from extreme environments, such as hot springs or salt lakes, metagenome sequencing allowed for the unexpected detection of archaea in a broad range of mesophilic habitats, including the world oceans and the human microbiome. Archaea were identified in biofilms in different ecosystems (VanDyke et al. 2008; Sanders et al. 1989; Boetius et al. 2000; Whitaker et al. 2005; Coman et al. 2013; Aoki et al. 2014). Archaea possess various surface structures which can be used for different purposes, such as attachment to (a)biotic surfaces, taxis, DNA exchange and biofilm formation. In some archaea, multiple surface structures were

S.-V. Albers (✉) · T.E.F. Quax
Molecular Biology of Archaea, Institute for Biology II, Faculty of Biology, University of Freiburg, Schänzlestrasse 1, 79104 Freiburg, Germany
e-mail: sonja.albers@biologie.uni-freiburg.de

G. Witzany (ed.), *Biocommunication of Archaea*,
DOI 10.1007/978-3-319-65536-9_5

shown to play a role in the same process. In this chapter we will give an overview of the currently known archaeal surface structures and their role in communication with the extracellular environment.

2 Type IV Pili like Structures

The large majority of described archaeal surface structures are build-up of proteins, which are assembled in a way that resembles the assembly process of bacterial type IV pili (Yu et al. 2012; Jarrell and Albers 2012; Henche et al. 2012a; Ng et al. 2011; Ajon et al. 2011). After recognition of a specific N-terminal signal peptide, the type IV pre-pilins are processed by signal peptidase (SP) III, and then transported across the cytoplasmic membrane via the Sec pathway. Archaea possess homologs of the bacterial SPIII, such as PibD in the crenarcheon *Sulfolobus solfataricus* (Albers et al. 2003) and *Haloferax volcanii* (Tripepi et al. 2010) and FlaK in the euryarchaeon *Methanococcus maripaludis* (Bardy and Jarrell 2002). Both PibD and FlaK share a conserved mechanism of processing class III signal peptides with the bacterial PilD, their recognition motifs are slightly different. After cleavage of the signal peptide, the mature pilin subunits are incorporated at the base of the growing surface structure. Even though the assembly process of bacterial and archaeal type IV pili is similar, these surface appendages display distinct structural arrangement and have only partly similar functions (Wang et al. 2008; Braun et al. 2016; Poweleit et al. 2016). Bacterial type IV pili employed for twitching motility, can assemble and disassemble, which has so far not been described for archaea. Moreover, the archaeal type IV pilins are usually N-glycosylated in contrast to bacterial type IV pilins (VanDyke et al. 2008; Tripepi et al. 2010). Well studied archaeal type IV pili like surface appendages are adhesive pili, required for attachment and biofilm formation (Pohlschroder and Esquivel 2015), and archaella (previously archaeal flagella), the archaeal motility structures, allowing swimming movement of cells (Albers and Jarrell 2015). These two structures are found in a large proportion of archaeal species. In addition, the UV-induced pili, essential for species specific DNA exchange, are found only in a crenarchaea (van Wolferen et al. 2016).

Recent genomic analysis has shown that type IV pili systems are very abundant and widespread amongst archaea (Makarova et al. 2016). There are many cases of horizontal gene transfer that can be detected, but the last common ancestor of all archaeal possessed at least one type IV pili system (Makarova et al. 2016). Strikingly, especially the hyperthermophiles encode a wide variety of type IV pili systems, which are thought to be responsible for interaction with other cells and abiotic surfaces (Makarova et al. 2016). The majority of these operons have not been experimentally studied and their exact function remains unclear.

2.1 Adhesive Pili

Several of the archaeal type IV like pili operons encode adhesive pili, which allow cells to attach to (a) biotic surfaces and to form biofilms.

2.1.1 Structure of Adhesive Pili

The adhesive pili of different species display quite distinct structures. The first structure of an adhesive pilus came from the euryarcheon *M. maripaludis*. It is 6 nm in diameter, possesses a hollow lumen of 20 Å, and displays two different helical symmetries in the same structure (Wang et al. 2008). The adhesive pilus structure of the crenarcheon *S. acidocaldarius* is very different. It is composed of three stranded helices with a rotation of $\sim 138°$, has a diameter of 11 nm, and, in contrast to the pili of *M. maripaludis,* has no hollow lumen (Henche et al. 2012b). In addition it is extremely stable, withstanding hot phenol and detergent treatment, which probably reflects the harsh growth conditions of *S. acidocaldarius* (70–90 °C and low pH) (Henche et al. 2012a). These findings demonstrate how small sequence changes might lead to large changes in quaternary structure.

2.1.2 Components of Adhesive Pili

Besides the structure, the operons encoding adhesive pili are also quite divergent. Adhesive pili of *M. maripaludis* are encoded by an operon of 11 genes, while the major pilin and the ATPase are not encoded in this locus (Szabo et al. 2007). Deletion analysis showed that the large majority of the genes in the operon are essential for piliation (Nair et al. 2014). In contrast, the operon encoding adhesive pili (Aap pili) in *S. acidocaldarius* consists of only five genes, including a pilB ATPase (AapE) and a pilC integral membrane protein (aapF). The operon encodes in addition the minor and major pilin, AapA and AapB, respectively. The *S. acidocaldarius* prepilins are processed by PibD, as is the case for components of Ups pili and archaella (see below) (Henche et al. 2012a). In contrast, *M. maripaludis* possesses a dedicated prepilin peptidase, named EppA, which can only cleave prepilins, while FlaK specifically processes archaellins (Bardy and Jarrell 2003a). Interestingly, most archaeal adhesive pili operons encode for multiple pilins. Usually there is one major pilin, while the minor pilins are still important for assembly such as in *M. maripaludis* and *S. acidocaldarius* (Henche et al. 2012a; Ng et al. 2011). But in *H. volcanii*, six different pilins are encoded which can all individually build pili, although these pili are shorter than those produced by the wild type (Tripepi et al. 2013). Expression of *pilA1* and *pilA2* can only partially rescue the adhesive properties of the wild type strain, while expression of only *pilA3* or *pilA4* leads to an even stronger adhesion phenotype (Tripepi et al. 2013). In *Halobacterium salinarium,* an operon homologous to the described *H. volcanii* operon encoding PilB and PilC proteins, was shown to be responsible for the production of adhesive surface structures (Losensky et al. 2015). Since the pilins are not encoded in the same operon, it remains unclear which of the 30 potential pilins in the *H. salinarium* genome are the potential major and minor pilins processed of these structures (Losensky et al. 2015). It has been suggested that different pilins maybe are expressed under different growth conditions, such as stress or are required for adhesion to different surfaces (Esquivel and Pohlschroder 2014). This would be in parallel to different archaellins that seem to be expressed and incorporated in the archaellum under varying growth conditions by the halophile *Haloarcula marismortui* (Syutkin et al. 2014).

2.1.3 Adhesive Pili and Biofilm Formation

Adhesive pili are not only important for adhesion to abiotic surfaces, but also required for biofilm formation in archaea. Biofilm formation in bacteria usually starts with the formation of microcolonies, which become encased by a polysaccharide matrix (Monds and O'Toole 2009). The biofilms of *Sulfolobus* and several haloarchaeal species seem to develop in the same way (Fröls et al. 2012; Orell et al. 2013; Chimileski et al. 2014). Deletion of the Aap pili of *S. acidocaldarius* leads to a very densely packed biofilm in comparison with wild type biofilms, suggesting that the Aap pili are capable of keeping the cells at a certain distance from each other, most likely to ensure optimal fluid and nutrient exchange (Henche et al. 2012b). For the euryarcheon *H. volcanii*, in wild type cells microcolony formation is observed after 24 h in static liquid medium. Exceptions are knock-outs strains of four of the six adhesive pilins, which suddenly do form microcolonies after 6 h. Overexpression of the two remaining pilins, *pilA5* and *pilA6*, in a wild type background never led to enhanced microcolony formation, suggesting that this process is inhibited by pilin 1–4 (Esquivel et al. 2013). During the maturation process of bacterial biofilms, EPS and extracellular DNA (eDNA) are secreted and form a matrix in which cells are entrapped. Archaeal biofilms were also reported to be constructed from these substances (Henche et al. 2012b; Orell et al. 2013; Tripepi et al. 2010; Fröls et al. 2012).

2.1.4 Regulation of Biofilm Formation

Little is known yet about regulation of biofilm formation. A proteome analysis of biofilms of *H. salinarium* showed that 34.5% of the predicted proteome was differentially regulated in sessile versus planktonic cells (Losensky et al. 2016). Importantly, some of the most profound downregulated proteins during biofilm formation were the archaellins and some components of the archaellum motor complex (Losensky et al. 2016). For the crenarchaeon *S. acidocaldarius* ~15% of genes were differentially transcribed in biofilms compared to planktonic cells (Koerdt et al. 2011). This transcriptome analysis also led to identification of the first regulator of archaeal biofilm formation. A group of Lrs14 proteins was differentially expressed. Mutational analysis of these Lrs14 proteins indicated that three of them influence biofilm formation. While two knock-outs led to reduced biofilm formation, deletion of the *abfr1* gene resulted in increased biofilm formation (Orell et al. 2013). Abfr1 was shown to be a repressor of biofilm formation by inhibiting EPS production and increasing archaellum expression (Orell et al. 2013).

S. solfataricus and *S. tokodaii* lack, in contrast to *S. acidocaldarius*, adhesive pili. Their apparent different dependence on Aap pili might be explained by their dissimilar growth environments. Whereas in nature *S. solfataricus* and *S. tokodaii* are mainly found inside the mud pool ecosystem of geothermal springs, is *S. acidocaldarius* isolated from the crusty edges of these springs, and is clearly preferring a biofilm lifestyle. This is also reflected in the respective biofilm morphologies of these strains, since *S. acidocaldarius* is the quickest of the three species to engage in biofilm formation and also builds the most elaborate biofilms with towering structures (Koerdt et al. 2010). Some homologues of most *aap* genes, except *aapA* and *aapB*, are found in other Sulfolobaceae (Henche et al. 2012a). Only *S. solfataricus* possesses an *aapB* gene, but a transposase in present in close vicinity in the genome, suggesting that the Aap pilus might previously have been present in all Sulfolobaceae (Henche et al. 2012a).

2.1.5 Pili Interaction with Viruses

Surface structures of archaea do not only mediate contact with other cells, but also represent an attachment platform for viruses. Several bacterial viruses are known to bind initially to surface structures, such as the Ff inoviruses or flagellotrophic phages that bind F-pili and flagella, respectively (Rakonjac et al. 2011; Guerrero-Ferreira et al. 2011). This first contact with the bacterial host is usually reversible and then in the second step, viruses attach irreversibly to a specific cell envelope receptor. Primary and secondary adsorptions can occur on the same receptor or usually involve different players (Poranen et al. 2002). Very little is known about the entry strategies of archaeal viruses. The first insights in their possible initial contact with the cell host came from electron microscopic analysis of purified virions from the order *Ligamenvirales*, such as *Acidianus* Filamentous Viruses (AFV). While all different AFV species are flexible filaments enveloped in a lipid layer, their terminal structures are very different, varying from several filaments to claws or brushes (Bettstetter et al. 2003; Haring et al. 2005; Vestergaard et al. 2008). This is thought to reflect the different host range they display. Several of these terminal virion structures were still bound to cellular surface structures after isolation, demonstrating that these viruses indeed bind filamentous surface structures (Bettstetter et al. 2003). The first thorough study of the entry process of an archaeal virus was performed using *Sulfolobus islandicus* Rod-shaped Virus 2 (SIRV2). This showed that SIRV2 initially binds to filamentous cellular surface structures, either at the tip or at the side of the filament archaella (Quemin and Quax 2015). Later the viruses are seen at the cell envelope, where they are shorter and partly distorted (Quemin and Quax 2015). How the viruses travel to the cell envelope is unknown, since currently retractable pili have not been identified in archaea yet. The structures to which the viruses bind have a diameter of 5–10 nm and are therefore probably not archaella (Quemin and Quax 2015). They could represent adhesive pili or yet unidentified surface structures, such as the enigmatic 'threads' that are displayed on the surface of *S. acidocaldarius* (Jarrell and Albers 2012). Intriguingly, *S. islandicus* and *S. solfataricus* are both reported not to possess adhesive pili (Henche et al. 2012a). However, mutational analysis of *S. solfataricus*, showed that AapE and AapF are important for the entry process of SIRV2, indicating a role for type IV like pili (Deng et al. 2014).

The identification of the surface structure involved in SIRV2 entry awaits deeper investigation. It is to be expected that further molecular exploration of archaeal virus-host interactions will lead to the description of more archaeal cellular surface structures important for virus attachment and entry.

2.2 *Archaella*

The archaeal motility structure, the archaellum, is an important tool to respond adequately to signals from the extracellular environment, since it allows for directed movement.

2.2.1 Structure and Composition of the Archaellum

The archaellum, is a rotating type IV pilus of ~10 to 15 nm in diameter (Alam and Oesterhelt 1984; Kalmokoff and Jarrell 1991; Bardy and Jarrell 2003b; Szabo et al. 2006, 2007; Shahapure et al. 2014). Functionally it resembles the bacterial flagellum that also is a filamentous structure capable of generating a rotational force allowing swimming motility (Berg and Anderson 1973). Whereas the flagellum consists of ~ 40 different proteins, the archaellum achieves motility with a minimal set of only ~8 proteins (FlaB-J), all without homology to the flagellum components (Chaban et al. 2007; Lassak et al. 2012). Operons encoding the archaellum are quite conserved, with the major differences between cren-and euryarchae being the presence of the FlaC-E proteins, and usually multiple archaellins, in the latter. The archaellum is hypothesized to have originated in the Euryarchaea and be transferred by horizontal gene transfer to anchestors of the Sulfolobales and Thaumarchaea (Makarova et al. 2016). New subunits are added to the base of the growing filament, in contrast to flagellum where subunits travel through the hollow interior to be added to the growing cap structure (Jarrell et al. 1996; Cohen-Krausz and Trachtenberg 2002, 2008). Rotation of the filamentous archaellum is driven by ATP hydrolysis, in contrast to the proton motive force required for rotation of the flagellum (Manson et al. 1977; Streif et al. 2008). Because of above mentioned differences, the archaellum and flagellum represent a clear example of convergent evolution.

2.2.2 Archaella Allow for Movement and Taxis

The archaellum allows motility and the movement towards environments with optimal growth conditions, named taxis. Members of all major archaeal clades, except the korarchaea, possess an archaellum. Tactic behavior of bacteria has been intensively studied and the chemotaxis system was found to be responsible for the transfer of signals from the extracellular environment to the base of the flagellum to induce a change in swimming direction (Porter et al. 2011). Interestingly, components of the bacterial chemotaxis system are present in Euryarchaea, Thaumarchaea and the recently discovered deep-branching Lokiarchaea, and were likely transferred from bacteria via horizontal gene transfer (Wuichet et al. 2010; Spang et al. 2015; Briegel et al. 2015). Studies on the euryarcheon *H. salinarium* demonstrated that archaea are capable of directed motility in response to various signals such as nutrients, oxygen and light and that the chemotaxis system is required for this behavior (Marwan and Oesterhelt 1987; Rudolph and Oesterhelt 1995; Lindbeck et al. 1995; Storch et al. 1999; Kokoeva and Oesterhelt 2000). Almost the complete set of bacterial chemotaxis genes are also encoded by archaea. Key player of the bacterial chemotaxis system is the response regulator CheY that, in phosphorylated state, binds to the switch complex at the base of the flagellum. In archaea, phosphorylation of CheY was also demonstrated to be important for taxis (Rudolph et al. 1995; Rudolph and Oesterhelt 1996). How CheY transfers signals to the base of the archaellum is not known yet. The first insights in this communication process came from a large protein interaction network established for *H. salinarium*. This demonstrated that archaeal CheY interacts with 1–2 adaptor proteins, named CheF (Schlesner et al. 2009, 2012). These proteins are unique to archaea, and in turn bind to components of the euryarchaeal archaellum (FlaC-E) (Schlesner et al. 2009, 2012). Structural organization of the base of the euryarchaeal archaellum, and details of the protein interaction of CheF with CheY and FlaC-E, await further investigation.

Above mentioned studies indicate that the chemotaxis system represents a flexible plug and play device that is capable of communication with both the bacterial and the archaeal motility structure. This finding is even more striking in the light of the biochemical and structural differences between the flagellum and archaellum. Surprisingly, the chemotaxis system has not been identified in crenarchaea, suggesting that they might possess an alternative system to respond to tactic signals from the environment.

2.2.3 Role of Archaella in Surface Adhesion

Besides the role of the archaellum in directed movement, it is in many archaea also important for surface attachment. It was shown to be indispensable for surface attachment of the crenarchaeon *S. solfataricus* (Zolghadr et al. 2010). In *S. acidocaldairus* indeed the Aap pili are the major players in attachment. However, also archaella were shown to have a minor role in this process (Henche et al. 2012a). Deletion of archaella does not influence the morphology of *S. acidocaldarius* biofilms, but affects the release of cells from the biofilm after maturation (Henche et al. 2012b). Aap pili and archaella are regulatory linked as deletion of the *aapF* gene leads to induction of the archaellum operon, hyperarchaellation and hypermotility (Henche et al. 2012b). Deletion of other Aap pili genes resulted in hypermotility without hyperarchaellation, possibly due to increased EPS secretion (Henche et al. 2012a).

Archaella were shown to be indispensable, together with pili, for the adhesion of *M. maripaludis* (Jarrell et al. 2011). In addition they promote attachment to different surfaces of *Methanocaldococcus villosus* and *Pyrococcus furiosus* cells (Nather et al. 2006), in which case the archaella wrap around each other and form large cable like structures of 100–200 nm diameter (Bellack et al. 2011). The only archaea for which was shown thus far that surface adhesion is archaellum independent was for the two haloarchaea *H. volcanii* and *H. salinarium* (Tripepi et al. 2010; Losensky et al. 2015).

2.3 *UV Induced Pili*

Cellular surface structures do not only assist in migration towards optimal growth environments, or adhesion to other cells and abiotic surfaces to form biofilms, but can also be important for repair of DNA damage. UV induced pili, Ups pili, are type IV like pili that are formed upon dsDNA breaks. They are important for species specific DNA exchange (Ajon et al. 2011).

2.3.1 Structure and Function of UV Induced Pili

The operon responsible for production of these pili was first identified during a transcriptome study of *S. solfataricus* cells after UV stress resulting in DNA damage (Fröls et al. 2008). The ups pili are $\sim$10 nm in diameter, composed of three evenly spaced helices and can be variable in length. Their main protein constituents UpsA and B are both produced as preproteins (Fröls et al. 2008). Deletion of *upsA* or *B* still leads to pili formation as long as the other pilin is still present (van Wolferen et al. 2013). The Ups pili mediate cellular aggregation after dsDNA breaks. This cellular aggregation happens after exposing cells to UV stress, and also after inducing dsDNA breaks with DNA damaging agents such as

Mitomycin C (Fröls et al. 2008). After a regeneration period the aggregates disappear again (Frols et al. 2007). dsDNA breaks are thought to be the intracellular signal inducing production of Ups pili, which in turn are required for the formation of cellular aggregates. This aggregation was shown to be species specific, as is the DNA exchange that occurs upon aggregation (Ajon et al. 2011). The mechanism by which species specificity is ensured is not elucidated yet, but it is speculated that the S-layer proteins might play a role in this. Only one of the mating strains needs to produce Ups pili in order to lead to successful DNA exchange (Ajon et al. 2011). Large parts of the chromosome can be transferred (but usually they are below 90 kbp) and DNA exchange is not limited to specific regions of the genome or plasmids (van Wolferen et al. 2016; Ajon et al. 2011). Recently it was demonstrated in *S. acidocaldarius* that the Ups operon is required for aggregation, but not responsible for the actual DNA exchange. The exchange is mediated by the Ced system, which is conserved amongst crenarchaea (van Wolferen et al. 2016). The Ced system mediates DNA uptake, most likely by forming a membrane pore after cellular aggregation (van Wolferen et al. 2016). An active DNA export system in archaea has not been identified yet.

2.3.2 Other Modes of DNA Exchange

Interestingly euryarchaea of the genus *Haloferax* are also known to exchange large parts of their genomic DNA (up to 500 kbp), presumably due to the formation of cytoplasmic bridges after aggregation (Rosenshine et al. 1989; Naor et al. 2012; Naor and Gophna 2013). Indeed, deletion of *pibD* in *H. volcanii* did not diminish efficiency of DNA exchange, demonstrating that type IV pili are likely not involved in the process (M. Tripepi et al. 2010). However, DNA exchange can be enhanced by biofilm formation, which in turn relies on adhesive pili (Chimileski et al. 2014). In biofilms of *H. saliniarium* the presence of extracellular DNA was demonstrated and seemed to be produced already after the first day of adhesion (Fröls et al. 2012). The molecular mechanism and the potential involvement of cellular surface structures in DNA exchange of *Haloferax* have not been determined.

2.3.3 Additional Function of UV Induced Pili

In addition to DNA exchange, the Ups pili of *S. solfataricus* were also reported to play a role in attachment to abiotic surfaces such as mica, pyrite and glass (Zolghadr et al. 2010). Biofilms formed by *S. solfataricus* and *S. acidocaldarius* strains lacking Ups pili, had a less dense morphology from wild type, with lower number of adhered cells in the bottom of the biofilms but increased cluster formation (Zolghadr et al. 2010; Henche et al. 2012b).

2.4 *Iho670 Fibers*

The last group of type IV like pili is exclusively observed at the surface of *Ignicoccus hospitalis*. The Iho670 fibers have a diameter of 14 nm and can be up to 20 μm in length. Their presence is not related with the co-culturing of the obligate symbiont *Nanoarchaeum equitans* (Muller et al. 2009). They are composed of a 33 kDa protein named Iho670, which contains a type IV pili-like signal sequence and is produced as preprotein (Muller et al. 2009). It has however no homology with archaellins, or any other reported surface structure. The only reported homology was with the *S. acidocaldarius* FlaF protein, which is not the actual archaellin, but thought to be the stator protein of the archaellum (Braun et al. 2016). The structure of the Iho670 fibers was elucidated revealing a ß-sandwich and an extra density, which suggested glycosylation of Iho670 (Braun et al. 2016). The Iho670 fibers cross the two membranes of *I. hospitalis* and are anchored in the cytoplasm under the inner membrane by a central knob surrounded by a 'cage-like' structure (Meyer et al. 2014). The function of the fibers is unknown, but they do not play a role in motility, since *I. hospitalis* is non-motile.

3 Non Type IV Pili Surface Structures

Besides the widespread type IV pili like structures, some archaea also possess other surface structures, which occurrence is usually restricted to one or a few archaeal species. Example are the cannulae of *Pyrodictium* and hami of *Candidatus Altiarchaeum hamiconexum*.

3.1 *Cannulae*

Three *Pyrodictium* species produce extracellular cannulae, which are hollow tubes of ~25 nm in diameter. They are essential for the production of large cellular aggregates. These tubes can enter the periplasmic space, between the cell membrane and the S-layer, but are not continuous with the cytoplasm (Nickell et al. 2003). Occasionally the tubes can appear filled with undefined matter. The cannulae link cells together and can form a large and dense network, which can even be observed with the naked eye as flakes of 1–10 mm in the growth medium (König et al. 1988). Several cannulae can together form large bundles (Rieger et al. 1995). In addition to the cannulae, *Pyrodictium abyssi* was also reported to produce structures resembling archaella (Rieger et al. 1995). What the role of these archaella in cells entrapped in the cannulae network is, remains unclear. Also the function or composition of cannulae were never elucidated. Since non cannulae producing mutants could never be obtained, it is assumed that they are essential for survival and might serve as protection against the harsh growth conditions (75–110 °C in strictly anaerobic environment) or for exchange of nutrients or other molecules (Nickell et al. 2003).

3.2 *Hami*

Hami are unique nano-grappling hook structures, which are produced on the surface of non-cultivated euryarcheon *Cd. A. hamiconexum*. Unusually for archaea *Cd. A. hamiconexum* has two membranes and does not have a protective S-layer surrounding the cell (Probst et al. 2014). Each hami is assembled from three protein sub-filaments wound around each other to a barbed-wire-like structure ending in a distal grappling hook (Moissl et al. 2005; Perras et al. 2014). The hami span both membrane of the cell, are ~8 nm in diameter and can be ~ 2 µm long (Moissl et al. 2005). They are likely formed by multiple copies of only one protein, which has a prominent ß-sandwich fold and an S-layer like N-terminal region (perras front micro 2015). It has no homology with type IV pili. The prominent ß sheet topology might prime the protein for self-assembly. In addition, the hamus protein contains several glycosylation sites. It is suggested that the hamus protein is also exported by the Sec-pathway, due to the presence of a Sec signal sequence (Perras et al. 2014). Due to its homology with S-layer proteins it might have evolved from ancestral cell surface proteins, thus accounting for the loss of the original S-layer (Perras et al. 2014).

Each cell can produce hundreds of these hami that allow it to form large biofilms, which are washed up from the subsurface of the cold sulfidic spring from which *Cd. A. hamiconexum* is isolated (Henneberger et al. 2006). They consist almost exclusively of *Cd. A. hamiconexum* cells, with only ~ 5% of sulfate reducing bacteria present in the biofilms (Probst et al. 2014). In the biofilms the coccoid archaeal cells are spaced rather regularly, and hami usually connect to 1–7 other cells (Probst et al. 2014). Besides the hami the EPS is formed by (sugar) polymers (Henneberger et al. 2006). The biofilms are thought to contribute to filtering of substrates from the subsurface waters or possibly to electron transfer (Probst et al. 2014).

3.3 *Mth60 Fimbriae*

The first archaeal surface structure described to play a role in adhesion are the fimbriae from *Methanothermobacter thermoautotrophicus*. These fimbriae have a diameter of ~5 nm and can reach lengths of 10 µm (Doddema et al. 1979). The fimbriae were shown to consist of the 15 kDa protein Mth60. This protein is thought to be processed on the N-terminus and has N-glycosylation sites (Thoma et al. 2008). It does not display any homology to other proteins in public databases from archaea or bacterial type IV pili. Since purified Mth60 formed fiber like structures in solution, it is likely the only component of the fimbriae of *M. thermoautotrophicus*. Planktonic cells were observed to display only 1–2 fimbriae on their surface, while sessile cells were covered in an abundance of fimbriae, demonstrating their importance for adhesion (Thoma et al. 2008). Indeed their role in adhesion to surfaces was confirmed using antibodies raised against Mth60, which, when mixed with *M. thermoautotrophicus* cells, abolished adhesion (Thoma et al. 2008).

4 Concluding Remarks

The development of adequate genetic systems for model archaea in the last decennia, has allowed for the first insights into the composition of archaeal surface structures and their role in communication with the environment. Moreover electron microscopy and especially cryo-tomography have proven techniques for the study of the diversity of archaeal surface structures, also in still uncultivated strains. Groundbreaking work of various research groups offers us a colorful and diverse picture of the whole scale of cell surface structures that archaea might possess. Some are very strain specific, such as the hami or Mth60 fimbriae, while others are widely dispersed amongst all archaea, such as the type IV pili like structures archaellum or adhesive pili.

Archaeal surface structures were shown essential for a range of processes, such as adhesion to (a) biotic surfaces, biofilm formation, taxis, motility and viral entry. In some cases there is one dedicated surface structure covering a certain function, like the archaellum that is responsible for motility, in other cases several surface structures work in concert, such as the archaellum and Ups pili that are both important for adhesion of *S. solfataricus*. A common strategy for adhesion has not been observed in archaea so far. *P. furiosus* seems to use its archaella both for motility as for adhesion (Nather et al. 2006). *M. maripaludus* and *S. acidocaldarius* have dedicated adhesive pili, but also rely on the presence of archaella for optimal adhesion (Jarrell et al. 2011; Henche et al. 2012b). And again, in haloarchaea such as *H. volcanii* and *H. salinarium*, only adhesive pili are required and the presence of archaella does not influence attachment to abiotic surfaces and biofilm formation (M. Tripepi et al. 2010; Losensky et al. 2015).

It is expected that the growing interest in archaea and their importance in a broad range of habitats, will result in more studies of their surface structures and their role in communication with the extracellular environment. Surely this will reveal new surface structures and maybe structures with unexpected functions. Moreover the role of surface structures in virus attachment and entry still represents a largely unexplored terrain (see Fig. 1).

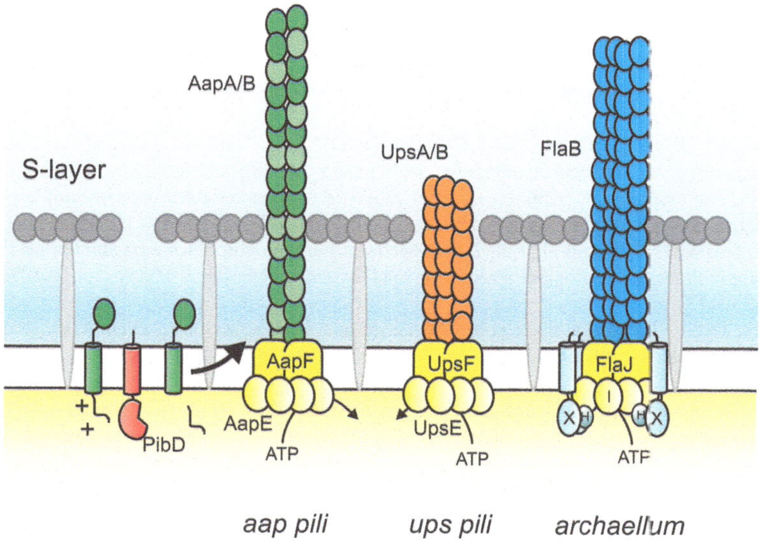

Fig. 1 Schematic representation of three different archaeal surface structures as found in *S. acidocaldarius*. All displayed structures have homology with type IV pili. The name of the major protein components of the three different filamentous structures are depicted at the *upper side* of the scheme. They are produced as preproteins, cleaved by the dedicated peptidase PibD. New subunits are added at the base of each of the growing structures in an ATP dependent manner. Aap pilus, adhesive pilus for attachment to surfaces. Ups pilus, UV inducible pilus for species specific DNA exchange. Archaellum, archaeal motility structure for directed movement

References

Ajon M, Fröls S, van Wolferen M et al (2011) UV-inducible DNA exchange in hyperthermophilic archaea mediated by type IV pili. Mol Microbiol 82:807–817. doi:10.1111/j.1365-2958.2011.07861.x

Alam M, Oesterhelt D (1984) Morphology, function and isolation of halobacterial flagella. J Mol Biol 176:459–475

Albers S-V, Jarrell KF (2015) The archaellum: how archaea swim. Front Microbiol 6:23. doi:10.3389/fmicb.2015.00023

Albers S-V, Szabó Z, Driessen AJM (2003) Archaeal homolog of bacterial type IV prepilin signal peptidases with broad substrate specificity. J Bacteriol 185:3918–3925

Aoki M, Ehara M, Saito Y et al (2014) A long-term cultivation of an anaerobic methane-oxidizing microbial community from deep-sea methane-seep sediment using a continuous-flow bioreactor. PLoS ONE 9:e105356. doi:10.1371/journal.pone.0105356

Bardy SL, Jarrell KF (2003a) Cleavage of preflagellins by an aspartic acid signal peptidase is essential for flagellation in the archaeon *Methanococcus voltae*. Mol Microbiol 50:1339–1347

Bardy SL, Jarrell KF (2002) FlaK of the archaeon *Methanococcus maripaludis* possesses preflagellin peptidase activity. FEMS Microbiol Lett 208:53–59

Bardy SL, Jarrell KF (2003b) Cleavage of preflagellins by an aspartic acid signal peptidase is essential for flagellation in the archaeon *Methanococcus voltae*. Mol Microbiol 50:1339–1347

Battin TJ, Sloan WT, Kjelleberg S et al (2007) Microbial landscapes: new paths to biofilm research. Nat Rev Microbiol 5:76–81. doi:10.1038/nrmicro1556

Bellack A, Huber H, Rachel R et al (2011) *Methanocaldococcus villosus* sp. nov., a heavily flagellated archaeon that adheres to surfaces and forms cell-cell contacts. Int J Syst Evol Microbiol 61:1239–1245. doi:10.1099/ijs.0.023663-0

Berg HC, Anderson RA (1973) Bacteria swim by rotating their flagellar filaments. Nature 245:380–382

Bettstetter M, Peng X, Garrett RA, Prangishvili D (2003) AFV1, a novel virus infecting hyperthermophilic archaea of the genus acidianus. Virology 315:68–79

Boetius A, Ravenschlag K, Schubert CJ et al (2000) A marine microbial consortium apparently mediating anaerobic oxidation of methane. Nature 407:623–626. doi:10.1038/35036572

Braun T, Vos MR, Kalisman N et al (2016) Archaeal flagellin combines a bacterial type IV pilin domain with an Ig-like domain. Proc Natl Acad Sci USA 113:10352–10357. doi:10.1073/pnas.1607756113

Briegel A, Ortega DR, Huang AN et al (2015) Structural conservation of chemotaxis machinery across Archaea and Bacteria. Environ Microbiol Rep 7:414–419. doi:10.1111/1758-2229.12265

Chaban B, Ng SY, Kanbe M et al (2007) Systematic deletion analyses of the fla genes in the flagella operon identify several genes essential for proper assembly and function of flagella in the archaeon, *Methanococcus maripaludis*. Mol Microbiol 66:596–609. doi:10.1111/j.1365-2958.2007.05913.x

Chimileski S, Franklin MJ, Papke RT (2014) Biofilms formed by the archaeon *Haloferax volcanii* exhibit cellular differentiation and social motility, and facilitate horizontal gene transfer. BMC Biol 12:65. doi:10.1186/s12915-014-0065-5

Cohen-Krausz S, Trachtenberg S (2002) The structure of the archeabacterial flagellar filament of the extreme halophile *Halobacterium salinarum* R1M1 and its relation to eubacterial flagellar filaments and type IV pili. J Mol Biol 321:383–395

Cohen-Krausz S, Trachtenberg S (2008) The flagellar filament structure of the extreme acidothermophile *Sulfolobus shibatae* B12 suggests that archaeabacterial flagella have a unique and common symmetry and design. J Mol Biol 375:1113–1124. doi:10.1016/j.jmb.2007.10.048

Coman C, Drugă B, Hegedus A et al (2013) Archaeal and bacterial diversity in two hot spring microbial mats from a geothermal region in Romania. Extremophiles 17:523–534. doi:10.1007/s00792-013-0537-5

Deng L, He F, Bhoobalan-Chitty Y et al (2014) Unveiling cell surface and type IV secretion proteins responsible for archaeal rudivirus entry. J Virol 88:10264–10268. doi:10.1128/JVI.01495-14

Doddema HJ, Derksen JWM, Vogels GD (1979) Fimbriae and flagella of methanogenic bacteria. FEMS Microbiol Lett 5:135–138. doi:10.1111/j.1574-6968.1979.tb03264.x

Elasri MO, Miller RV (1999) Study of the response of a biofilm bacterial community to UV radiation. Appl Environ Microbiol 65:2025–2031

Esquivel RN, Pohlschroder M (2014) A conserved type IV pilin signal peptide H-domain is critical for the post-translational regulation of flagella-dependent motility. Mol Microbiol 93:494–504. doi:10.1111/mmi.12673

Esquivel RN, Xu R, Pohlschroder M (2013) Novel archaeal adhesion pilins with a conserved N terminus. J Bacteriol 195:3808–3818. doi:10.1128/JB.00572-13

Fröls S, Ajon M, Wagner M et al (2008) UV-inducible cellular aggregation of the hyperthermophilic archaeon *Sulfolobus solfataricus* is mediated by pili formation. Mol Microbiol 70:938–952. doi:10.1111/j.1365-2958.2008.06459.x

Fröls S, Dyall-Smith M, Pfeifer F (2012) Biofilm formation by haloarchaea. Environ Microbiol 14:3159–3174. doi:10.1111/j.1462-2920.2012.02895.x

Frols S, Gordon PMK, Panlilio MA et al (2007) Response of the hyperthermophilic archaeon *Sulfolobus solfataricus* to UV damage. J Bacteriol 189:8708–8718. doi:10.1128/JB.01016-07

Guerrero-Ferreira RC, Viollier PH, Ely B et al (2011) Alternative mechanism for bacteriophage adsorption to the motile bacterium *Caulobacter crescentus*. Proc Natl Acad Sci 108:9963–9968. doi:10.1073/pnas.1012388108

Haring M, Vestergaard G, Brugger K et al (2005) Structure and genome organization of AFV2, a novel archaeal lipothrixvirus with unusual terminal and core structures. J Bacteriol 187:3855–3858. doi:10.1128/JB.187.11.3855-3858.2005

Henche A-L, Ghosh A, Yu X et al (2012a) Structure and function of the adhesive type IV pilus of *Sulfolobus acidocaldarius*. Environ Microbiol 14:3188–3202. doi:10.1111/j.1462-2920.2012.02898.x

Henche A-L, Koerdt A, Ghosh A, Albers S-V (2012b) Influence of cell surface structures on crenarchaeal biofilm formation using a thermostable green fluorescent protein. Environ Microbiol 14:779–793. doi:10.1111/j.1462-2920.2011.02638.x

Henneberger R, Moissl C, Amann T et al (2006) New insights into the lifestyle of the cold-loving SM1 euryarchaeon: natural growth as a monospecies biofilm in the subsurface. Appl Environ Microbiol 72:192–199. doi:10.1128/AEM.72.1.192-199.2006

Jarrell KF, Albers S-V (2012) The archaellum: an old motility structure with a new name. Trends Microbiol 20:307–312. doi:10.1016/j.tim.2012.04.007

Jarrell KF, Bayley DP, Kostyukova AS (1996) The archaeal flagellum: a unique motility structure. J Bacteriol 178:5057–5064

Jarrell KF, Stark M, Nair DB, Chong JPJ (2011) Flagella and pili are both necessary for efficient attachment of *Methanococcus maripaludis* to surfaces. FEMS Microbiol Lett 319:44–50. doi:10.1111/j.1574-6968.2011.02264.x

König H, Messner P, Stetter KO et al (1988) The fine structure of the fibers of *Pyrodictium occultum*. FEMS Microbiol Lett 49:207–212. doi:10.1111/j.1574-6968.1988.tb02717.x

Kalmokoff ML, Jarrell KF (1991) Cloning and sequencing of a multigene family encoding the flagellins of *Methanococcus voltae*. J Bacteriol 173:7113–7125

Koerdt A, Gödeke J, Berger J et al (2010) Crenarchaeal biofilm formation under extreme conditions. PLoS ONE 5:e14104. doi:10.1371/journal.pone.0014104

Koerdt A, Orell A, Pham TK et al (2011) Macromolecular fingerprinting of sulfolobus species in biofilm: a transcriptomic and proteomic approach combined with spectroscopic analysis. J Proteome Res 10:4105–4119. doi:10.1021/pr2003006

Kokoeva MV, Oesterhelt D (2000) BasT, a membrane-bound transducer protein for amino acid detection in *Halobacterium salinarum*. Mol Microbiol 35:647–656

Lassak K, Neiner T, Ghosh A et al (2012) Molecular analysis of the crenarchaeal flagellum. Mol Microbiol 83:110–124. doi:10.1111/j.1365-2958.2011.07916.x

Lindbeck JC, Goulbourne EA, Johnson MS, Taylor BL (1995) Aerotaxis in *Halobacterium salinarium* is methylation-dependent. Microbiology 2945–2953. doi:10.1099/13500872-141-11-2945

Losensky G, Jung K, Urlaub H et al (2016) Shedding light on biofilm formation of *Halobacterium salinarum* R1 by SWATH-LC/MS/MS analysis of planktonic and sessile cells. Proteomics. doi:10.1002/pmic.201600111

Losensky G, Vidakovic L, Klingl A et al (2015) Novel pili-like surface structures of *Halobacterium salinarum* strain R1 are crucial for surface adhesion. Front Microbiol 5:755. doi:10.3389/fmicb.2014.00755

Makarova KS, Koonin EV, Albers S-V (2016) Diversity and evolution of type IV pili systems in archaea. Front Microbiol 7:667. doi:10.3389/fmicb.2016.00667

Manson MD, Tedesco P, Berg HC et al (1977) A protonmotive force drives bacterial flagella. Proc Natl Acad Sci U S A 74:3060–3064

Martinez LR, Casadevall A (2007) Cryptococcus neoformans biofilm formation depends on surface support and carbon source and reduces fungal cell susceptibility to heat, cold, and UV light. Appl Environ Microbiol 73:4592–4601. doi:10.1128/AEM.02506-06

Marwan W, Oesterhelt D (1987) Signal formation in the halobacterial photophobic response mediated by a fourth retinal protein (P480). J Mol Biol 195:333–342

Meyer C, Heimerl T, Wirth R et al (2014) The Iho670 fibers of Ignicoccus hospitalis are anchored in the cell by a spherical structure located beneath the inner membrane. J Bacteriol 196:3807–3815. doi:10.1128/JB.01861-14

Moissl C, Rachel R, Briegel A et al (2005) The unique structure of archaeal "hami", highly complex cell appendages with nano-grappling hooks. Mol Microbiol 56:361–370. doi:10.1111/j.1365-2958.2005.04294.x

Monds RD, O'Toole GA (2009) The developmental model of microbial biofilms: ten years of a paradigm up for review. Trends Microbiol 17:73–87. doi:10.1016/j.tim.2008.11.001

Muller DW, Meyer C, Gurster S et al (2009) The Iho670 fibers of Ignicoccus hospitalis: a new type of archaeal cell surface appendage. J Bacteriol 191:6465–6468. doi:10.1128/JB.00858-09

Nair DB, Uchida K, Aizawa S-I, Jarrell KF (2014) Genetic analysis of a type IV pili-like locus in the archaeon *Methanococcus maripaludis*. Arch Microbiol 196:179–191. doi:10.1007/s00203-014-0956-4

Naor A, Gophna U (2013) Cell fusion and hybrids in archaea. Bioengineered 4:126–129. doi:10.4161/bioe.22649

Naor A, Lapierre P, Mevarech M et al (2012) Low species barriers in halophilic archaea and the formation of recombinant hybrids. Curr Biol 22:1444–1448. doi:10.1016/j.cub.2012.05.056

Nather DJ, Rachel R, Wanner G, Wirth R (2006) Flagella of *Pyrococcus furiosus*: multifunctional organelles, made for swimming, adhesion to various surfaces, and cell-cell contacts. J Bacteriol 188:6915–6923. doi:10.1128/JB.00527-06

Ng SYM, Wu J, Nair DB et al (2011) Genetic and mass spectrometry analyses of the unusual type IV-like pili of the archaeon *Methanococcus maripaludis*. J Bacteriol 193:804–814. doi:10.1128/JB.00822-10

Nickell S, Hegerl R, Baumeister W, Rachel R (2003) Pyrodictium cannulae enter the periplasmic space but do not enter the cytoplasm, as revealed by cryo-electron tomography. J Struct Biol 141:34–42

Orell A, Peeters E, Vassen V et al (2013) Lrs14 transcriptional regulators influence biofilm formation and cell motility of Crenarchaea. ISME J 7:1886–1898. doi:10.1038/ismej.2013.68

Palmer J, Flint S, Brooks J (2007) Bacterial cell attachment, the beginning of a biofilm. J Ind Microbiol Biotechnol 34:577–588. doi:10.1007/s10295-007-0234-4

Palmer RJ, Stoodley P (2007) Biofilms 2007: broadened horizons and new emphases. J Bacteriol 189:7948–7960. doi:10.1128/JB.00787-07

Perras AK, Wanner G, Klingl A et al (2014) Grappling archaea: ultrastructural analyses of an uncultivated, cold-loving archaeon, and its biofilm. Front Microbiol 5:397. doi:10.3389/fmicb.2014.00397

Pohlschroder M, Esquivel RN (2015) Archaeal type IV pili and their involvement in biofilm formation. Front Microbiol 6:190. doi:10.3389/fmicb.2015.00190

Poranen MM, Daugelavičius R, Bamford DH (2002) Common principles in viral entry. Annu Rev Microbiol 56:521–538. doi:10.1146/annurev.micro.56.012302.160643

Porter SL, Wadhams GH, Armitage JP (2011) Signal processing in complex chemotaxis pathways. Nat Rev Microbiol 9:153–165. doi:10.1038/nrmicro2505

Poweleit N, Ge P, Nguyen HH et al (2016) CryoEM structure of the *Methanospirillum hungatei* archaellum reveals structural features distinct from the bacterial flagellum and type IV pili. Nat Microbiol 2:16222. doi:10.1038/nmicrobiol.2016.222

Probst AJ, Weinmaier T, Raymann K et al (2014) Biology of a widespread uncultivated archaeon that contributes to carbon fixation in the subsurface. Nat Commun 5:5497. doi:10.1038/ncomms6497

Quemin ERJ, Quax TEF (2015) Archaeal viruses at the cell envelope: entry and egress. Front Microbiol 6:552. doi:10.3389/fmicb.2015.00552

Rakonjac J, Bennett NJ, Spagnuolo J et al (2011) Filamentous bacteriophage: biology, phage display and nanotechnology applications. Curr Issues Mol Biol 13:51–76

Rieger G, Rachel R, Hermann R, Stetter KO (1995) Ultrastructure of the hyperthermophilic archaeon *Pyrodictium abyssi*. J Struct Biol 115:78–87. doi:10.1006/jsbi.1995.1032

Rosenshine I, Tchelet R, Mevarech M (1989) The mechanism of DNA transfer in the mating system of an archaebacterium. Science 245:1387–1389

Rudolph J, Oesterhelt D (1995) Chemotaxis and phototaxis require a CheA histidine kinase in the archaeon *Halobacterium salinarium*. EMBO J 14:667–673

Rudolph J, Oesterhelt D (1996) Deletion analysis of the che operon in the archaeon *Halobacterium salinarium*. J Mol Biol 258:548–554. doi:10.1006/jmbi.1996.0267

Rudolph J, Tolliday N, Schmitt C et al (1995) Phosphorylation in halobacterial signal transduction. EMBO J 14:4249–4257

Sanders DA, Gillece-Castro BL, Stock AM et al (1989) Identification of the site of phosphorylation of the chemotaxis response regulator protein, CheY. J Biol Chem 264:21770–21778

Schlesner M, Miller A, Besir H et al (2012) The protein interaction network of a taxis signal transduction system in a halophilic archaeon. BMC Microbiol 12:272. doi:10.1186/1471-2180-12-272

Schlesner M, Miller A, Streif S et al (2009) Identification of Archaea-specific chemotaxis proteins which interact with the flagellar apparatus. BMC Microbiol 9:56. doi:10.1186/1471-2180-9-56

Shahapure R, Driessen RPC, Haurat MF et al (2014) The archaellum: a rotating type IV pilus. Mol Microbiol 91:716–723. doi:10.1111/mmi.12486

Spang A, Saw JH, Jørgensen SL et al (2015) Complex archaea that bridge the gap between prokaryotes and eukaryotes. Nature 521:173–179. doi:10.1038/nature14447

Storch K-F, Rudolph J, Oesterhelt D (1999) Car: a cytoplasmic sensor responsible for arginine chemotaxis in the archaeon *Halobacterium salinarum*. EMBO J 18:1146–1158. doi:10.1093/emboj/18.5.1146

Streif S, Staudinger WF, Marwan W, Oesterhelt D (2008) Flagellar rotation in the archaeon *Halobacterium salinarum* depends on ATP. J Mol Biol 384:1–8. doi:10.1016/j.jmb.2008.08.057

Syutkin AS, Pyatibratov MG, Galzitskaya OV et al (2014) *Haloarcula marismortui* archaellin genes as ecoparalogs. Extremophiles 18:341–349. doi:10.1007/s00792-013-0619-4

Szabo Z, Albers SV, Driessen AJ (2006) Active-site residues in the type IV prepilin peptidase homologue PibD from the archaeon *Sulfolobus solfataricus*. J Bacteriol 188:1437–1443. doi:10.1128/jb.188.4.1437-1443.2006

Szabo Z, Stahl AO, Albers SV et al (2007) Identification of diverse archaeal proteins with class III signal peptides cleaved by distinct archaeal prepilin peptidases. J Bacteriol 189:772–778. doi:10.1128/jb.01547-06

Thoma C, Frank M, Rachel R et al (2008) The Mth60 fimbriae of *Methanothermobacter thermoautotrophicus* are functional adhesins. Environ Microbiol 10:2785–2795. doi:10.1111/j.1462-2920.2008.01698.x

Tripepi M, Esquivel RN, Wirth R, Pohlschroder M (2013) *Haloferax volcanii* cells lacking the flagellin FlgA2 are hypermotile. Microbiology 159:2249–2258. doi:10.1099/mic.0.069617-0

Tripepi M, Imam S, Pohlschroder M (2010) *Haloferax volcanii* flagella are required for motility but are not involved in PibD-dependent surface adhesion. J Bacteriol 192:3093–3102. doi:10.1128/JB.00133-10

van Wolferen M, Ajon M, Driessen AJM, Albers S-V (2013) Molecular analysis of the UV-inducible pili operon from *Sulfolobus acidocaldarius*. Microbiologyopen 2:928–937. doi:10.1002/mbo3.128

van Wolferen M, Wagner A, van der Does C, Albers S-V (2016) The archaeal Ced system imports DNA. Proc Natl Acad Sci 113:2496–2501. doi:10.1073/pnas.1513740113

VanDyke DJ, Wu J, Ng SYM et al (2008) Identification of a putative acetyltransferase gene, MMP0350, which affects proper assembly of both flagella and pili in the archaeon *Methanococcus maripaludis*. J Bacteriol 190:5300–5307. doi:10.1128/JB.00474-08

Vestergaard G, Aramayo R, Basta T et al (2008) Structure of the *Acidianus filamentous* virus 3 and comparative genomics of related archaeal lipothrixviruses. J Virol 82:371–381. doi:10.1128/JVI.01410-07

Wang YA, Yu X, Ng SYM et al (2008) The structure of an archaeal pilus. J Mol Biol 381:456–466. doi:10.1016/j.jmb.2008.06.017

Whitaker RJ, Grogan DW, Taylor JW (2005) Recombination shapes the natural population structure of the hyperthermophilic archaeon *Sulfolobus islandicus*. Mol Biol Evol 22:2354–2361. doi:10.1093/molbev/msi233

Wuichet K, Cantwell BJ, Zhulin IB (2010) Evolution and phyletic distribution of two-component signal transduction systems. Curr Opin Microbiol 13:219–225. doi:10.1016/j.mib.2009.12.011

Yu X, Goforth C, Meyer C et al (2012) Filaments from Ignicoccus hospitalis show diversity of packing in proteins containing N-terminal type IV pilin helices. J Mol Biol 422:274–281. doi:10.1016/j.jmb.2012.05.031

Zolghadr B, Klingl A, Koerdt A et al (2010) Appendage-mediated surface adherence of *Sulfolobus solfataricus*. J Bacteriol 192:104–110. doi:10.1128/JB.01061-09

Archaeal Biocommunication in Hot Springs Revealed by Metagenomics

María-Eugenia DeCastro, Juan-José Escuder-Rodríguez, Manuel Becerra, Esther Rodríguez-Belmonte and María-Isabel González-Siso

1 Introduction

Archaea are organisms that can tolerate "extreme" conditions inhospitable to most life, such as the high salt concentration of the Great Salt Lake, Utah (Almeida-Dalmet et al. 2015) or the extreme pH of acid mine drainage emplacements (Méndez-García et al. 2015), colonizing habitats with a wide range of temperatures, from the Antarctic sediments (Purdy et al. 2003) to hot springs (Chan et al. 2015; López-López et al. 2015).

Hyperthermophilic microorganisms are considered to be the closest relatives to the earliest life forms on Earth (Schwartzman and Lineweaver 2004). As co-evolution of a microbial community could not take place without communication between cells, the study of cell signaling systems in hot springs archaea may increase our knowledge in the microbial interactions that have led to adaptation and survival in these extreme environments (Montgomery et al. 2013).

One of the main systems of interspecies communication, quorum sensing, has been recently identified in Archaea (Paggi et al. 2003; Zhang et al. 2012), suggesting that communication at multiple levels may be crucial in extreme environments. But many aspects of archaeal (inter-, intra- and trans-) communication remain to be explored, as cultivation and growth of these microorganisms under laboratory conditions is still a difficult task, hindering the characterization of their function and metabolic potential.

The development of cultivation-independent methods such as metagenomics has enabled scientists to directly access the genetic information of a given environment.

M.-E. DeCastro · J.-J. Escuder-Rodríguez · M. Becerra ·
E. Rodríguez-Belmonte · M.-I. González-Siso (✉)
Grupo EXPRELA, Centro de Investigacións Científicas Avanzadas (CICA),
Departamento de Bioloxía Celular e Molecular, Facultade de Ciencias,
Universidade da Coruña, A Coruña, Spain
e-mail: migs@udc.es

G. Witzany (ed.), *Biocommunication of Archaea*,
DOI 10.1007/978-3-319-65536-9_6

Therefore, metagenomics facilitates the study of the entire microbial community as a whole, where all microorganisms are interacting with the environment, taking part in the biochemical cycles and responding to abiotic and community structure changes in the ecosystem, for which biocommunication must take place.

Using this approach, a large number of archaeal populations (Song et al. 2010; Huang et al. 2011; Merkel et al. 2015) and archaeal viruses (Bolduc et al. 2012; Servín-Garcidueñas et al. 2013; Hochstein et al. 2016) have been studied in hot springs, leading to the assembly of novel archaeal genomes (Servín-Garcidueñas and Martínez-Romero 2014). Most of these studies are focused on the analysis of diversity (Eme et al. 2013; Coman et al. 2013; Kambura et al. 2016) or in functional analysis related to the carbon, sulfur, and nitrogen metabolism (Chan et al. 2015). There are only a few metagenomic studies of hot springs that investigate the interactions among microbes (with or within Archaeal populations); as an example, Lin et al. 2015 suggested a potential host-parasite interaction between *Nanoarchaea* and *Ignicoccus* in an acidic hot spring in Taiwan.

In the present chapter, we focus on the use of shotgun metagenomics to bring light to archaeal biocommunication in hot springs. For this purpose, we have analyzed nine previously published hot spring DNA metagenomes from distant locations, with different temperatures and pH values (Table 1). Eight of them were characterized and published by Menzel et al. (2015) and the one from Ourense, Galicia, was studied by our group and published by López-López et al. (2015). Metagenomic contigs, previously uploaded to the MG-RAST server (Meyer et al. 2008), were analyzed using this pipeline. After assessing the taxonomy of the nine hot springs in the MG-RAST against the GenBank database (Fig. 1a), all the sequences annotated as archaea (Fig. 1b), were selected for further functional analysis in the MG-RAST using the SEED Subsystems database (Fig. 2). From all

Table 1 Names and characteristics of the selected hot spring metagenomes

Sample name	T (°C)	pH	Site	MG-RAST accession no	Reference
Ch2-EY65S	65	7.0	Eryuan, China	4530144.3	Menzel et al. (2015)
Sun Spring	61–64	5.8–6.0	Uzon Caldera, Russia	4544453.3	
It6	76	3.0	Pozzuoli, Italy	4529716.3	
It3	86	5.5	Pisciarelli, Italy	4529718.3	
Is2-5S	85–90	5.0	Grensdalur, Iceland	4530143.3	
Is3-13	90	3.5–4.0	Krísuvík, Iceland	4583585.3	
CH1102	79	1.8	YNP, USA	4529720.3	
NL10	92	3.0–4.0	YNP, USA	4529719.3	
RCW	76	8.2	Ourense, Galicia, Spain	4570559.3	López-López et al. (2015)

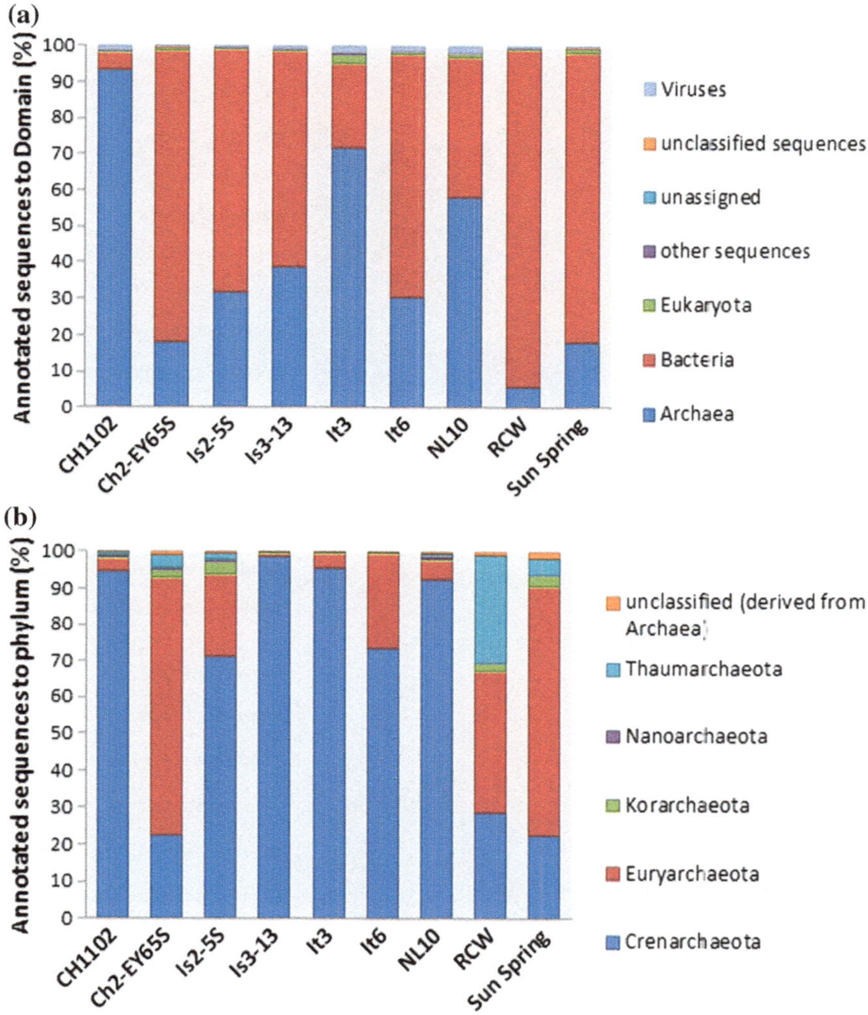

Fig. 1 Taxonomical analysis of the nine metagenomes described in Table 1. **a** Domain level. **b** Phylum level within the domain Archaea

the functions annotated, we have focused in those sequences related to the different aspects of biocommunication (interpretation of abiotic influences, trans-organismic communication, inter-organismic communication and intra-organismic communication) (Witzany 2016).

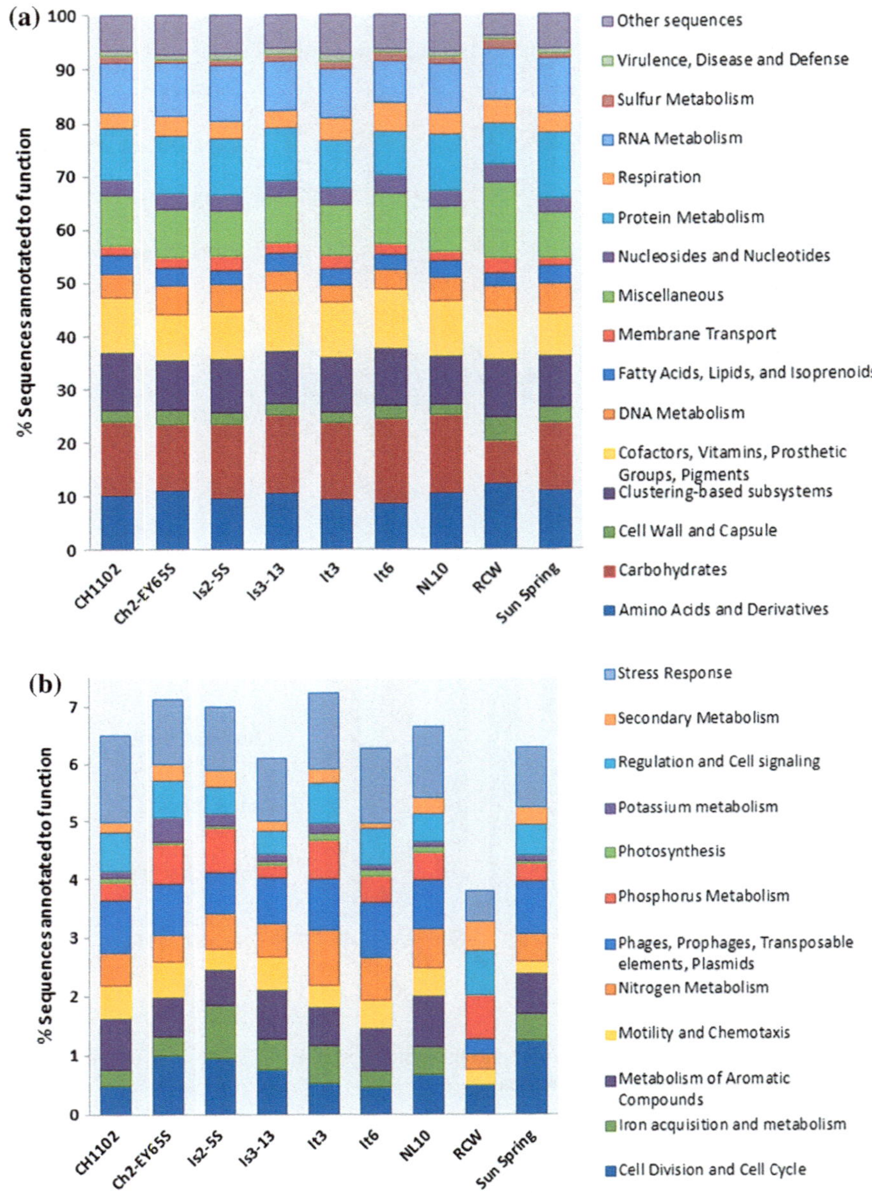

Fig. 2 Functional analysis of the nine metagenomes obtained through the MG-RAST pipeline using the SEED Subsystems database. **a** Main functions with more than 1% of sequences. **b** Detailed representation of those sequences represented as other sequences in (**a**)

2 Interpretation and Response to Abiotic Factors

Microorganisms depend on signal transduction pathways to sense and response to chemical (variations in pH, concentrations of nutrients, toxins and oxygen) and physical stimuli (intensity and wavelength of light, among others).

The signaling process in which single cells can detect and move toward favorable compounds (attractants) and away from unfavorable molecules (repellents) is termed chemotaxis. In this process, the transduction of the signal starts with the recognition of the attractants and repellents by the cell's signal sensing proteins, called chemoreceptors. Collins et al. (2014) analyzed 8384 chemoreceptors from bacterial and archaeal complete genomes, revealing the prevalence of cytoplasmic chemoreceptors in Archaea.

A bacterial-like chemotaxis system has been identified in *Euryarchaeota* (Rudolph and Oesterhelt 1996) even when the archaeal flagellum is not homologous to that of Bacteria (Jarrell and Albers 2012). From 0.2% in Sun Spring up to 0.6% in Ch2-EY65S sequences annotated as motility and chemotaxis-related genes were present in all the analyzed metagenomes (Fig. 2). In addition to sequences matching the archaellum-related proteins FlaH, FlaI and FlaJ, other important chemotaxis-related proteins are present, as the histidine kinase CheA, which starts the transduction, being autophosphorylated when a change is detected in the chemoreceptor. Other sequences detected in the metagenomes were assigned to chemotaxis regulators like CheW and CheY. These signaling proteins participate in the signal transduction from the chemoreceptors to the flagellar motor (Boyd 2015).

Besides the short-term fluctuations in environmental parameters that microorganisms need to rapidly detect and respond to, hot spring microbial communities experience a variety of continuous environmental stresses like permanent high temperature, extreme pH or low oxygen, among others. Thus, sequences matching genes involved in stress response were found in all the studied metagenomes (Fig. 2). The most abundant were those associated with oxidative stress and the adaptation to heat stress, such as heat shock protein GrpE and chaperones DnaJ and DnaK. The nucleotide exchange factor GrpE, DnaK and its co-chaperone DnaJ participate not only in heat shock, but also in heavy metal cellular stress responses, assisting other cellular proteins in folding correctly and maintaining their native structure (Rani et al. 2016). The presence of DnaK, DnaJ and GrpE has been previously reported in several archaeal genomes (Macario et al. 2006). More archaeal chaperone families can be found in the CrAgDb (Chaperone repertoire in Archaeal genomes) database (Rani et al. 2016).

Among all the sequences related to oxidative stress, those sequences annotated as alkyl hydroperoxide reductase subunit C-like protein were present in all the metagenomes studied here. Close homologs of this enzyme, responsible for alkyl peroxide metabolism, have been found in the genomes of several *Pyrococcus* species (*P.furiosus*, *P. abyssi* and *P. horikoshii*) and in the closely related archaea *Thermococcus kodakaraensis* (Strand et al. 2010). These results, along with the high abundance of this sequence in the studied metagenomes, suggest that alkyl hydroperoxide reductase could be a first-line constitutive and protective defense mechanism against oxidative stress.

3 Archaeal Symbiosis in Hot Springs

Beneficial relationships between organisms such as nutrient exchange or protection against environmental stresses are generally named symbioses. This cooperation represents an intimate, often long-term interaction of two or more organisms that can be considered trans-organismic communication, as it usually takes place between distantly related species.

Within the domain archaea, the *Nanoarchaeaota* constitutes a remarkable phylum in terms of symbiotic interactions, as all its members have been shown to be ectosymbionts, dependent on direct interaction with other Archaea. This association was firstly described between *Nanoarchaeum equitans* and the host *Ignicoccus hospitalis* in a submarine hot vent (Huber et al. 2002). Although the interaction was initially thought to be exclusive from the marine environments, Podar et al. 2013 sequenced the genome of the first hyperthermophilic *Nanoarchaeota* from a terrestrial environment (Obsidian Pool hot spring in Yellowstone National Park, YNP) and its archaeal host, a sulfolobales-like organism, evidencing that this symbiosis also occurs in hot springs and with a non-*Ignicoccus* host. Since then, the presence of the phylum *Nanoarchaeota* and its dependence on direct interaction with other Archaea has been described in several hot springs like the Shi-Huang-Ping acidic hot spring in Taiwan, where the co-existence of *Nanoarchaeota* and *Ignicoccus* was revealed by metagenomics (Lin et al. 2015). Members of this phylum were also present in the CH1102, CH2-EY65S, Is2-5S, NL10 and Sun Spring metagenomes, studied in this chapter, with a 0.72, 0.25, 0.32, 0.96 and 0.26% of all the archaeal sequences respectively, as shown in Fig. 1b.

A better comprehension on how *Nanoarchaeota* became dependent on other archaea and the way this relationship has influenced the genomes and physiology of the interacting species was given by Wurch et al. (2016), who isolated and characterized the first terrestrial *Nanoarchaeota*-host system, revealing that they lack almost all primary biosynthetic functions as well as respiration and ATP synthesis. This genome reduction is very frequent in all symbioses, as gene loss is compensated by the cooperative microenvironment leading to the final dependence between the implicated organisms and thus blocking the transient relationship in what is known as the Black Queen Hypothesis (Morris et al. 2012; Fullmer et al. 2015).

4 Inter-organismic Communication in Hot Springs

4.1 Quorum Sensing

In hot springs, as in most environments, microorganisms such as Archaea do not live alone, they co-exist with other species of microorganisms, interacting and coordinating with them by sign-mediated communication, as reflected in the analyzed metagenomes (Fig. 1a, b). One of the best studied cell-cell communication systems is quorum sensing. It is based on the production of intra- and interspecific signaling molecules, often called autoinducers, which are constitutively released by the cell, initiating a gene transcription response when they reach a critical concentration. This process enables the microbial community to regulate gene expression, synchronizing metabolism and behavior. At present, three different types of quorum sensing signaling molecules have been described: acyl-homoserine lactones (AHLs), oligopeptide autoinducers and autoinducer-2 (AI-2) (Hense and Schuster 2015). All of them are collected in the SigMol, a specialized repository of quorum sensing signaling molecules in prokaryotes, which includes archaeal molecules (Rajput et al. 2016).

Because of the inherent heat-labile nature of the AHLs lactone ring, an AHL-based signaling in thermophilic environments was not expected to be found (Montgomery et al. 2013). Despite these initial thoughts, the first archaeal AHLs were recently characterized in *Methanosaeta harundinacea*. These AHLs, responsible for filament formation, have a carboxylation modification previously unseen in bacteria (Zhang et al. 2012) but are still able to activate bacterial biosensors and are predicted to be more resistant to degrading conditions like high temperature and alkaline pH. The detection of similar carboxylated AHLs in cultures of *Methanosarcina mazei* and *Methanothermobacter thermautotrophicus* (Zhang et al. 2012), suggests that all AHL-like molecules in Archaea would possess longer chain or other characteristics that improve their stability.

Although no sequences homologous to the *lux* genes were found in the studied hot springs metagenomes, S-adenosyl methionine synthetase, responsible for the synthesis of S-adenosyl methionine, a precursor of AHLs (Schaefer et al. 1996), was present in all of them. Moreover, in two metagenomes (Is2-5S and NL10), sequences matching an AHL hydrolase have been found, confirming the potential of Archaea to produce quorum quenching molecules able to interfere with quorum sensing signaling. This interception of AHL signaling systems via enzymatic degradation of the signal was previously reported in hyperthermophilic archaea, as seen in *Sulfolobus solfataricus* which produces a lactonase, SsoPox, able to hydrolyze AHLs (Ng et al. 2011).

AI-2 has been reported to be a furan-like universal signaling molecule for interspecies communication in bacteria (Pereira et al. 2012). The process of quorum-sensing mediated by AI-2 involves the uptake of AI-2 via the Lsr transporter followed by the phosphorylation to AI-2-phosphate (P-AI-2) and subsequent sequestration of the signal on the cytoplasm. There, P-AI-2 can activate the transcription of the *lsr* operon by inhibiting the repressor LsrR and, consequently, incrementing the expression of the transporter and the kinase (Marques et al. 2014).

In Sun Spring metagenome (Russia), a sequence with a nucleotide homology of 77% with the uncultured *Thaumarchaeota* AD1000 and annotated as an autoinducer 2 (AI-2) aldolase LsrF has been found. The enzyme LsrF is a class I aldolase involved in the cleavage of P-AI-2 that also shows thiolase activity as it transfers an acetyl group from the P-AI-2 isomer P-HPD to the co-enzyme A (CoASH), producing dihydroxyacetone phosphate (DHAP) and acetyl-CoA (Marques et al. 2014).

Sequences matching *lsrF* have been previously found in other hot spring metagenomes and in thermophilic archaea like the *Thaumarchaeota* MY3 archaeon (GenBank accession number: ALI36583.1). Figure 3 shows the multiple alignments of the archaeal LsrF amino acid sequence found in the Sun Spring metagenome with other LsrF aminoacidic sequences found in the databases. Despite no *luxS* homologue genes have ever been found in archaeal genomes, the presence of *lsrF* sequences suggests that Archaea could at least be able to respond to AI-2 signaling, Furthermore, it has been described that an AI-2 type signal is produced when the hyperthermophilic archaeon *Pyrococcus furiosus* was cultured with *Thermotoga maritima*, although no phenotypic change was observed in response to this molecule (Nichols et al. 2009).

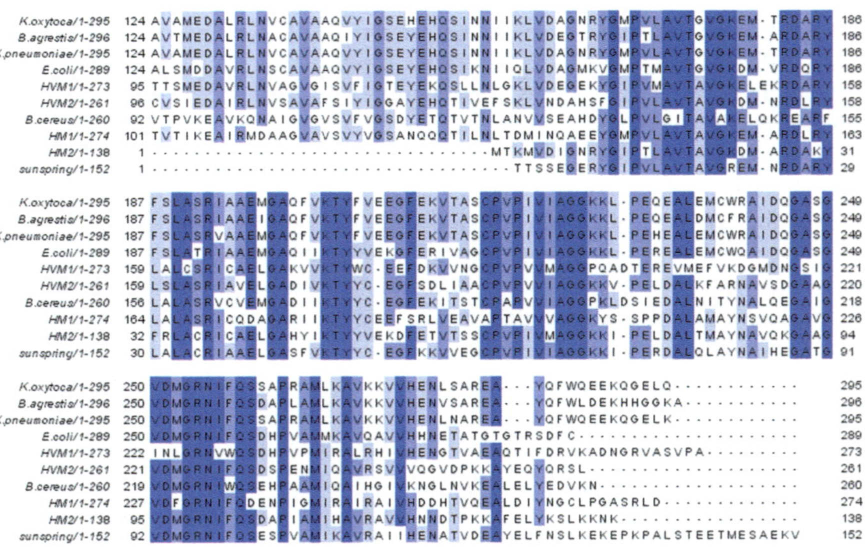

Fig. 3 Multiple alignments of the archaeal LsrF amino acid sequence found in the Sun Spring metagenome with other LsrF aminoacidic sequences found in the databases: *Klebisella oxyotica*, *Buttiauxella agrestis*, *Klebisella pneumoniae*, *Escherichia coli*, HVM1 (Hydrotermal vent metagenome A0A170Q9C1), HMV2 (Hydrotermal vent metagenome A0A160VIV8), *Bacillus cereus*, HM1 (Hydrocarbon metagenome A0A0W8FKB7), HM2 (Hydrocarbon metagenome A0A090S824), *Thaumarchaeota* MY3 archaeon and Sun Spring metagenome

Finally, it is possible that different compounds such as diketopiperazines act as quorum sensing molecules in archaea, as it has been described that the cyclo-(L-prolyl-valine) produced by *Halotherrigena hispanica* are able to activate AHL-biosensors (Tommonaro et al. 2012).

4.2 Archaeocins (Sulfolobicins and Halocins)

Archaea are widely represented in extreme environments, particularly in thermophilic habitats, as shown in the nine studied hot springs metagenomes, particularly in those with lowest pHs (CH1102, It3 and NL10) (Fig. 1a). In these conditions, antimicrobial compounds or other secondary metabolites can be produced in order to gain competitive advantage and establish new niches. Archaeocins are small ribosomally synthesized peptides able to inhibit the growth of microorganisms, mostly produced by haloarchaea (halocins) (Cheung et al. 1997; Platas et al. 2002) and by the thermophilic *Sulfolobus* genus (sulfolobicins) (Prangishvili et al. 2000).

Sulfolobicins were first identified in the thermophilic crenarchaeote *Sulfolobus islandicus* isolated from different solfataric fields around Iceland. These strains, growing at pH range of 2–4 and temperatures between 65 °C and 85 °C, were able to produce small antimicrobial peptides bound to the membrane of vesicles, termed sulfolobicins, inhibiting the growth of other closely related strains of *S. islandicus* and different *Sulfolobus* species such as *Sulfolobus solfataricus* P1 and *Sulfolobus shibatae* B12 (Prangishvili et al. 2000). The same membrane vesicle-associated condition was observed in the sulfolobicins released by *Sulfolobus acidocaldaricus* isolated from the Beppu Hot Springs in Kyushu Island, Japan. In the same study, the high stability of sulfolobicins was revealed, as they could resist SDS treatment and exposure to low and high pH (3–10.7), showing antimicrobial activity against other related *Sulfolobus* species (Ellen et al. 2011).

In six of the nine studied hot spring metagenomes (CH1102, Is2-5S, Is3-13, It3, It6 and NL10), the phylum *Crenarchaeota* is the most abundant within the domain Archaea (Fig. 1b), therefore the presence of sulfolobicins would be expected in all the studied hot springs, nevertheless no sequences matching sulfolobicins have been found in the metagenomic sequences. This fact reveals the main limitation of sequence metagenomics: the low number of sulfolobicin genes that have been annotated, characterized and deposited in the databases at present, makes difficult to find sulfolobicins using this approach. In this situation, functional metagenomics would be a better approach, as it does not rely on the sequence and could lead to the discovery of novel sulfolobicin genes.

No activity of sulfolobicins against organisms outside the *Sulfolobus* genus has been recorded, on the contrary, halocins, produced by the halophilic members of the phylum *Euryarchaeota*, less abundant in hot springs, have shown antimicrobial activity against members of the hyperthermophilic crenarchaeal genus *Sulfolobus*

(Haseltine et al. 2001). Thus, halocins can trespass the phylum barrier and inhibit the growth of distantly related archaea.

Recently, the production of novel compounds with an antimicrobial activity called cationic antimicrobial peptides (CAMPs), have been detected for the first time in the archaeon *Sulfolobus islandicus* (Notomista et al. 2015). These small positively charged peptides with amphipathic structure were able to induce cell death in *Escherichia coli*, three different *P. aeruginosa* strains (PAOI, PA14 and RP73), *S. aureus*, *B. subtilis* AZ54 and *C. albicans* cultures.

4.3 Vesicle and Cell Surface Communication

The release of membrane vesicles (MVs) is an evolutionarily conserved process in all domains of life, including Archaea. Recent studies suggest an important role of vesicles in cell-cell communication, as they can carry antimicrobial peptides, DNA or quorum sensing agents, among others. *Sulfolobus* species were shown to release MVs harbouring sulfolobicins (Prangishvili et al. 2000; Ellen et al. 2011), as mentioned before. Other archaeal species, like *Ignicoccus hospitalis*, produce and accumulate periplasmic MVs that could be involved in the transfer of metabolites and macromolecules between this microorganism and its symbiotic partner, *Nanoarchaeum equitans* (Marguet et al. 2013).

MVs can also mediate the transfer of DNA between cells, as has been observed in *Thermococcus onnurineus* (Choi et al. 2015) and *Thermococcus kodakaraensis* in which cells transformed with the shuttle plasmid pLC70 release MVs harbouring this plasmid (Gaudin et al. 2013). Moreover, a posterior study demonstrated that *Thermococcus nautilus can produce MVs harboring viral genomes, suggesting that* MVs could contribute to the intercellular transport of viral genomes, thus promoting recombination between viral, plasmid and/or cellular DNA in the absence of viral infection (Gaudin et al. 2014). Therefore, the production of MVs may have contributed to horizontal gene transfer (HGT). The high number of genes from hyperthermophilic bacteria that have an archaeal origin reveals an important role of HGT in the adaptation of bacteria to hot environments. As an example, the hyperthermophilic bacteria *Thermotoga maritima* has acquired 24% of its genes from archaea (van Wolferen et al. 2013).

In several groups of archaea, the release of MVs is facilitated by the cellular Endosomal Sorting Complex Required for Transport (ESCRT-III) machinery, conserved in a subset of Archaea, like the Sulfolobales (Deatherage and Cookson 2012). In fact, this apparatus can be used by archaeal viruses to complete their replication cycle and egress from the infected cells, as seen in the *Sulfolobus* turreted icosahedral virus isolated from Yellowstone National Park's acidic hot springs (Snyder et al. 2013). In other archaeal species that do not contain ESCRT-III homologs, such as *Thermococcus*, the mechanism for vesicle formation remains unknown, although a possible role of stomatin and Vps4/FtsH-like proteins has been suggested (Gaudin et al. 2013). Nevertheless, it is well documented that

Thermococcus species such as *Thermococcus kodakaraensis* and *Thermococcus gammatolerans* use structures called nanopods to project MVs several micrometers away from the cell, thus increasing the efficiency of vesicle transfer (Marguet et al. 2013). The production of tubular networks connecting neighbor cells, known as nanotubes, have also been recently described in the hyperthermophilic *Thermococcus* sp. 5-4, these structures would allow the transfer of several compounds, including DNA (Marguet et al. 2013).

Other archaeal surface structures involved in cell communication are the pili. Fröls et al. (2008) showed that, in response to UV stress, the hyperthermophilic archaeon *Sulfolobus solfataricus* induces pili formation with subsequent cellular aggregation, suggesting that this mechanism could enhance DNA transfer between cells to mediate repair of damaged DNA via homologous recombination.

Conjugational exchange of plasmid DNA has been observed both in hyperthermophilic bacteria and archaea; however, it seems that conjugation at high temperatures requires different mechanisms to take place, as the classical conjugation proteins have been scarcely found in these microorganisms (van Wolferen et al. 2013). Recently, two *Sulfolobus acidocaldaricus* proteins, Saci_0568 (named CedA) and Saci_0748 (named CedB), were identified as part of a previously unknown DNA importer, in what has been called the Crenarchaeal system for exchange of DNA (Ced), as homologous of these *Ced* genes are widely spread among *Crenarchaeota* (van Wolferen et al. 2016).

5 Intra-organismic Communication

The ability of individual cells to produce, detect and respond to cytoplasmic and endosomal signs is crucial to regulate gene expression and thus, to generate an appropriate response. But intra-cellular communication not only includes sign-mediated interactions within the organism, it also comprises all the events of HGT and the incorporation of persistent viral sequences that contribute to natural genome editing (She et al. 2001; Witzany 2016). Archaeal integrative viruses and mobile genetic elements require an integrase for the process of integrating into the host genome (Wang et al. 2015). Sequences matching a phage integrase associated with archaea were annotated in CH1102, Is3-13, It6 and NL10 hot spring metagenomes, revealing that viral integration and thus, genome editing, is a common process between archaea in hot spring environments. Moreover, other sequences related to transposable elements, phages, and prophages sequences, are present in all the studied archaeal hot spring metagenomes with percentages between 0.25 and 0.92% (Fig. 2).

The coexistence of viruses and cells has led to the development of a broad number of attack/defense mechanisms, such as the small non-coding RNAs (sRNAs), the clustered regularly interspaced short palindromic repeats (CRISPR) or toxin–antitoxin systems (TA) (Forterre and Prangishvili 2013).

sRNAs have been identified in Eukarya, Bacteria, and Archaea, playing an important role in many cellular and regulatory processes. In addition to the regulation of gene expression, these non-coding RNAs participate in the establishment of immunity against invading mobile genetic elements. Archaeal sRNAs have been studied in some thermophilic members such as *Sulfolobus solfataricus* (Tang et al. 2004) and the symbiotic *N. equitans* (Randau 2012). Most of them belong to the class of C/D-box or H/ACA-box sRNAs, which contribute to ribosomal RNA biogenesis and tRNA maturation (Prasse et al. 2013). In thermophiles, RNA methylation is an important sRNA guided RNA modification in which two proteins, Nop5 and fibrillarin, are recruited to generate a ribonucleoprotein complex. There, fibrillarin subunits catalyze the transfer of the methyl group of *S*-adenosyl methionine onto the target ribose. This modification increases the melting temperature of the RNA and reduces its degradation (Randau 2015). With this mechanism, Archaea can increase rRNA stability at elevated temperatures. Although no archaeal fibrillarin sequences were found in the metagenomic sequences from Ourense, some sequences matching archaeal fibrillarin were present in the sequences of the eight hot spring metagenomes studied by Mentzel et al. (2015). On the other hand, sequences annotated by subsystems as Nop5 were found in all the studied hot spring archaeal sequences.

The CRISPR adaptive immune system constitutes an important cell protection mechanism against invading nucleic acids, including viruses, through a genetic interference pathway (Prasse et al. 2013). It was shown that cell resistance against a virus can be acquired by integrating a genome fragment of this specific virus into the CRISPR cluster (Barrangou et al. 2007). Consequently, the CRISPR system contains a set of short spacer sequences that determine specificity and more novel spacer sequences are constantly being added to a growing CRISPR locus, as new resistances are acquired. Therefore, the CRISPR loci serve as 'memories' of a cell's interaction with viruses and spacer content present in an environmental sample can be used to link the viruses living in that environment with their hosts (Marchfelder et al. 2012). This method was used to assign the putative RNA viral genomes directly to a specific archaeal host in 28 hot springs of Yellowstone National Park (YNP), USA (Bolduc et al. 2012) and to determine the host of the archaeal *Acidianus* Tailed Spindle Virus in the acidic hot spring Crater Hills area of YNP, Alice Spring (CHAS) (Hochstein et al. 2016).

Toxin-antitoxin (TA) systems are autoregulated operons composed of tandem genes encoding small (approximately 10 kb) antitoxin and toxin proteins. In normally growing cells, toxins are co-expressed and neutralized by their correspondent antitoxins. However, as antitoxins are more labile than toxins, they are easily degraded by cell proteases under stress conditions. The free toxins can then interact with their cellular targets inhibiting DNA replication, protein and cell wall biosynthesis thus inhibiting cell growth and even leading to cell death (Yamaguchi et al. 2011).

TA systems are abundant in archaea as reported in 2011, where at least 86 from the 94 characterized archaeal genomes were shown to contain TA systems homologous to bacterial systems (Yamaguchi et al. 2011). In the metagenomes studied here, sequences matching the Doc toxin and the VapC and VapB TA systems were found. All of them are members of the Type II TA system.

Doc toxin is present in the archaeal sequences of CH1102, It3 and NL10 metagenomes. This toxin has an important role in the control of cell growth and death as it is able to block translation elongation through phosphorylation of Thr-382 of the essential elongation factor EF-Tu. This modification inhibits protein synthesis and arrests cell growth (Cruz et al. 2014).

VapC toxin protein and its antitoxin Vap B are annotated in the archaeal sequences of CH1102, Ch2-EY65S, It3, It6 and NL10 metagenomes. It has been demonstrated that this TA system is controlled by changes in temperature, as the knockout of vapBC locus in *S. solfataricus* considerably changed its transcriptomic profile and made these archaea heat shock labile (Cooper et al. 2009). Therefore, these two proteins might play an important role in the adaptation of microorganisms to high temperatures.

6 Concluding Remarks and Perspectives

Biocommunication comprises a complex network of signaling pathways that are essential to the survival and adaptation of the microbial communities, especially in harsh environments, such as hot springs, where high temperatures are a limiting factor. In these processes, microorganisms are able to respond to their environment, interacting and coordinating with others. Archaea are not an exception, as has been revealed in this study by the significant number of sequences related with archaeal biocommunication recovered from nine hot springs metagenomes. This finding reveals the potential of metagenomics to assess the interactions between microorganisms in a given ecosystem. Nevertheless, using this approach has some limitations such as the fact that the presence of a gene on a metagenome does not mean that it is expressed and, therefore, other techniques like functional metagenomics or metatranscriptomics could be used to provide a better understanding about biocommunication pathways. Moreover, the low number of annotated biocommunication-related sequences deposited in the databases at present makes difficult to find some genes using this approach, as shown with sulfolobicins in this study.

Apart from these drawbacks, the improvement of sequencing technologies together with their continuous drop of prize and the relatively fast generation of the sequences is resulting in a growing number of metagenomes that are being deposited in the databases and an increasing proportion of new annotated and characterized genes. Altogether, these factors contribute to the continuous improvement of the process of functional annotation of metagenomes, making metagenomics a more robust and affordable tool to study biocommunication processes.

Acknowledgements Funding both from the European Union Seventh Framework Programme (FP7/2007–2013) under grant agreement n° 324439, and from the Xunta de Galicia (Consolidación D.O.G. 20-12-2016, Contract Number: ED431C2016-012) co-financed by FEDER. The work of María-Eugenia DeCastro was supported by a FPU fellowship (Ministerio de Educación Cultura y Deporte) FPU12/05050. The authors acknowledge CESGA (www.cesga.es) in Santiago de Compostela, Spain, for providing access to computing facilities.

Bibliography

Almeida-Dalmet S, Sikaroodi M, Gillevet P, Litchfield C, Baxter B (2015) Temporal study of the microbial diversity of the North Arm of Great Salt Lake, Utah, U.S. Microorganisms 3:310–326. doi:10.3390/microorganisms3030310

Barrangou R, Fremaux C, Deveau H, Richards M, Boyaval P, Moineau S, Romero DA, Horvath P (2007) CRISPR provides acquired resistance against viruses in prokaryotes. Science 315:1709–1712. doi:10.1126/science.1138140

Bolduc B, Shaughnessy DP, Wolf YI, Koonin EV, Roberto FF, Young M (2012) Identification of novel positive-strand RNA viruses by metagenomic analysis of archaea-dominated Yellowstone hot springs. J Virol 86:5562–5573. doi:10.1128/JVI.07196-11

Boyd ES (2015) Archaea on the move. Environ Microbiol Rep 7:385–387. doi:10.1111/1758-2229.12281

Chan CS, Chan K-G, Tay Y-L, Chua Y-H, Goh KM (2015) Diversity of thermophiles in a Malaysian hot spring determined using 16S rRNA and shotgun metagenome sequencing. Front Microbiol 6:177. doi:10.3389/fmicb.2015.00177

Cheung J, Danna KJ, O'Connor EM, Price LB, Shand RF (1997) Isolation, sequence, and expression of the gene encoding halocin H4, a bacteriocin from the halophilic archaeon *Haloferax mediterranei* R4. J Bacteriol 179:548–551. doi:10.1128/JB.179.2.548-551.1997

Choi DH, Kwon YM, Chiura HX, Yang EC, Bae SS, Kang SG, Lee J-H, Yoon HS, Kim S-J (2015) Extracellular vesicles of the hyperthermophilic archaeon *Thermococcus onnurineus* NA1T. Appl Environ Microbiol 81:4591–4599. doi:10.1128/AEM.00428-15

Collins KD, Lacal J, Ottemann KM (2014) Internal sense of direction: sensing and signaling from cytoplasmic chemoreceptors. Microbiol Mol Biol Rev 78:672–684. doi:10.1128/MMBR.00033-14

Coman C, Drugă B, Hegedus A, Sicora C, Dragoş N (2013) Archaeal and bacterial diversity in two hot spring microbial mats from a geothermal region in Romania. Extremophiles 17:523–534. doi:10.1007/s00792-013-0537-5

Cooper CR, Daugherty AJ, Tachdjian S, Blum PH, Kelly RM (2009) Role of vapBC toxin–antitoxin loci in the thermal stress response of *Sulfolobus solfataricus*. Biochem Soc T 37:123–126

Cruz JW, Rothenbacher FP, Maehigashi T, Lane WS, Dunham CM, Woychik NA (2014) Doc toxin is a kinase that inactivates elongation factor Tu. J Biol Chem 289:7788–7798. doi:10.1074/jbc.M113.544429

Deatherage BL, Cookson BT (2012) Membrane vesicle release in bacteria, eukaryotes, and archaea: a conserved yet underappreciated aspect of microbial life. Infect Immun 80:1948–1957. doi:10.1128/IAI.06014-11

Ellen AF, Rohulya OV, Fusetti F, Wagner M, Albers S-V, Driessen AJM (2011) The sulfolobicin genes of *Sulfolobus acidocaldarius* encode novel antimicrobial proteins. J Bacteriol 193:4380–4387. doi:10.1128/JB.05028-11

Eme L, Reigstad LJ, Spang A, Lanzén A, Weinmaier T, Rattei T, Schleper C, Brochier-Armanet C (2013) Metagenomics of Kamchatkan hot spring filaments reveal two new major (hyper) thermophilic lineages related to *Thaumarchaeota*. Res Microbiol 164:425–438. doi:10.1016/j.resmic.2013.02.006

Fröls S, Ajon M, Wagner M, Teichmann D, Zolghadr B, Folea M, Boekema EJ, Driessen AJM, Schleper C, Albers SV (2008) UV-inducible cellular aggregation of the hyperthermophilic archaeon Sulfolobus solfataricus is mediated by pili formation. Molecular Microbiology 70(4): 938–952 doi: 10.1111/j.1365-2958.2008.06459.x.

Forterre P, Prangishvili D (2013) The major role of viruses in cellular evolution: facts and hypotheses. Curr Opin Virol 3:558–565. doi:10.1016/j.coviro.2013.06.013

Fullmer MS, Soucy SM, Gogarten JP (2015) The pan-genome as a shared genomic resource: mutual cheating, cooperation and the black queen hypothesis. Front Microbiol 6:728. doi:10.3389/fmicb.2015.00728

Gaudin M, Gauliard E, Schouten S, Houel-Renault L, Lenormand P, Marguet E, Forterre P (2013) Hyperthermophilic archaea produce membrane vesicles that can transfer DNA. Environ Microbiol Rep 5:109–116. doi:10.1111/j.1758-2229.2012.00348.x

Gaudin M, Krupovic M, Marguet E, Gauliard E, Cvirkaite-Krupovic V, Le Cam E, Oberto J, Forterre P (2014) Extracellular membrane vesicles harbouring viral genomes. Environ Microbiol 16:1167–1175. doi:10.1111/1462-2920.12235

Haseltine C, Hill T, Montalvo-Rodriguez R, Kemper SK, Shand RF, Blum P (2001) Secreted euryarchaeal microhalocins kill hyperthermophilic *Crenarchaea*. J Bacteriol 183:287–291. doi:10.1128/JB.183.1.287-291.2001

Hense BA, Schuster M (2015) Core principles of bacterial autoinducer systems. Microbiol Mol Biol Rev 79:153–169. doi:10.1128/MMBR.00024-14

Hochstein RA, Amenabar MJ, Munson-McGee JH, Boyd ES, Young MJ (2016) *Acidianus* tailed spindle virus: a new archaeal large tailed spindle virus discovered by culture-independent methods. J Virol 90:3458–3468. doi:10.1128/JVI.03098-15

Huang Q, Dong CZ, Dong RM, Jiang H, Wang S, Wang G, Fang B, Ding X, Niu L, Li X, Zhang C, Dong H (2011) Archaeal and bacterial diversity in hot springs on the Tibetan Plateau, China. Extremophiles 15:549–563. doi:10.1007/s00792-011-0386-z

Huber H, Hohn MJ, Rachel R, Fuchs T, Wimmer VC, Stetter KO (2002) A new phylum of archaea represented by a nanosized hyperthermophilic symbiont. Nature 417:63–67. doi:10.1038/417063a

Jarrell KF, Albers S-V (2012) The archaellum: an old motility structure with a new name. Trends Microbiol 20:307–312. doi:10.1016/j.tim.2012.04.007

Kambura AK, Mwirichia RK, Kasili RW, Karanja EN, Makonde HM, Boga HI (2016) Bacteria and archaea diversity within the hot springs of Lake Magadi and Little Magadi in Kenya. BMC Microbiol 16:136. doi:10.1186/s12866-016-0748-x

Lin KH, Liao BY, Chang HW, Huang SW, Chang TY, Yang CY, Wang YB, Lin YTK, Wu YW, Tang SL, Yu HT (2015) Metabolic characteristics of dominant microbes and key rare species from an acidic hot spring in Taiwan revealed by metagenomics. BMC Genom 16:1029. doi:10.1186/s12864-015-2230-9

López-López O, Knapik K, Cerdán ME, González-Siso MI (2015) Metagenomics of an alkaline hot spring in Galicia (Spain): microbial diversity analysis and screening for novel lipolytic enzymes. Front Microbiol 6:1291. doi:10.3389/fmicb.2015.01291

Macario AJL, Brocchieri L, Shenoy AR, Conway De Macario E, Freeland S (2006) Evolution of a protein-folding machine: genomic and evolutionary analyses reveal three lineages of the archaeal hsp70(dnaK) gene. J Mol Evol 63:74–86. doi:10.1007/s00239-005-6207-1

Marchfelder A, Fischer S, Brendel J, Stoll B, Maier L-K, Jäger D, Prasse D, Plagens A, Schmitz RA, Randau L (2012) Small RNAs for defence and regulation in archaea. Extremophiles 16:685–696. doi:10.1007/s00792-012-0469-5

Marguet E, Gaudin M, Gauliard E, Fourquaux I, Le Blond Du, Plouy S, Matsui I, Forterre P (2013) Membrane vesicles, nanopods and/or nanotubes produced by hyperthermophilic archaea of the genus *Thermococcus*. Biochem Soc Trans 41:436–442. doi:10.1042/BST20120293

Marques JC, Oh IK, Ly DC, Lamosa P, Ventura MR, Miller ST, Xavier KB (2014) LsrF, a coenzyme A-dependent thiolase, catalyzes the terminal step in processing the quorum sensing signal autoinducer-2. PNAS 111:14235–14240. doi:10.1073/pnas.1408691111

Méndez-García C, Peláez AI, Mesa V, Sánchez J, Golyshina OV, Ferrer M (2015) Microbial diversity and metabolic networks in acid mine drainage habitats. Front Microbiol 6:475. doi:10.3389/fmicb.2015.00475

Menzel P, Gudbergsdóttir SR, Rike AG, Lin L, Zhang Q, Contursi P, Moracci M, Kristjansson JK, Bolduc B, Gavrilov S, Ravin N, Mardanov A, Bonch-Osmolovskaya E, Young M, Krogh A, Peng X (2015) Comparative metagenomics of eight geographically remote terrestrial hot springs. Microb Ecol 70:411–424. doi:10.1007/s00248-015-0576-9

Merkel AY, Podosokorskaya OA, Chernyh NA, Bonch-Osmolovskaya EA (2015) Occurrence, diversity, and abundance of methanogenic archaea in terrestrial hot springs of Kamchatka and Saõ Miguel Island. Microbiology 84:577–583. doi:10.1134/S002626171504013X

Meyer F, Paarmann D, D'Souza M, Olson R, Glass E, Kubal M, Paczian T, Rodriguez A, Stevens R, Wilke A, Wilkening J, Edwards R (2008) The metagenomics RAST server—a public resource for the automatic phylogenetic and functional analysis of metagenomes. BMC Bioinform 9:386. doi:10.1186/1471-2105-9-386

Montgomery K, Charlesworth J, LeBard R, Visscher P, Burns B (2013) Quorum sensing in extreme environments. Life 3:131–148. doi:10.3390/life3010131

Morris JJ, Lenski RE, Zinser ER (2012) The Black Queen Hypothesis: evolution of dependencies through adaptive gene loss. MBio 3:e00036–12. doi:10.1128/mBio.00036-12

Ng FSW, Wright DM, Seah SYK (2011) Characterization of a phosphotriesterase-like lactonase from *Sulfolobus solfataricus* and its immobilization for disruption of quorum sensing. Appl Environ Microbiol 77:1181–1186. doi:10.1128/AEM.01642-10

Nichols JD, Johnson MR, Chou CJ, Kelly RM (2009) Temperature, not LuxS, mediates AI-2 formation in hydrothermal habitats. FEMS Microbiol Ecol 68:173–181. doi:10.1111/j.1574-6941.2009.00662.x

Notomista E, Falanga A, Fusco S, Pirone L, Zanfardino A, Galdiero S, Varcamonti M, Pedone E, Contursi P (2015) The identification of a novel *Sulfolobus islandicus* CAMP-like peptide points to archaeal microorganisms as cell factories for the production of antimicrobial molecules. Microb Cell Fact 14:126. doi:10.1186/s12934-015-0302-9

Paggi RA, Martone CB, Fuqua C, De Castro RE (2003) Detection of quorum sensing signals in the haloalkaliphilic archaeon *Natronococcus occultus*. FEMS Microbiol Lett 221:49–52. doi:10.1016/s0378-1097(03)00174-5

Pereira CS, Thompson JA, Xavier KB (2012) AI-2-mediated signalling in bacteria. FEMS Microbiol Rev 37:156–181. doi:10.1111/j.1574-6976.2012.00345.x

Platas G, Meseguer I, Amils R (2002) Purification and biological characterization of halocin H1 from *Haloferax mediterranei* M2a. Int Microbiol 5:15–19. doi:10.1007/s10123-002-0053-4

Podar M, Makarova KS, Graham DE, Wolf YI, Koonin EV, Reysenbach AL (2013) Insights into archaeal evolution and symbiosis from the genomes of a nanoarchaeon and its inferred crenarchaeal host from Obsidian Pool, Yellowstone National Park. Biol Direct 8:9. doi:10.1186/1745-6150-8-9

Prangishvili D, Holz I, Stieger E, Nickell S, Kristjansson JK, Zillig W (2000) Sulfolobicins, specific proteinaceous toxins produced by strains of the extremely thermophilic archaeal genus *Sulfolobus*. J Bacteriol 182:2985–2988. doi:10.1128/JB.182.10.2985-2988.2000

Prasse D, Ehlers C, Backofen R, Schmitz RA (2013) Regulatory RNAs in archaea: first target identification in Methanoarchaea. Biochem Soc Trans 41:344–349. doi:10.1042/BST20120280

Purdy KJ, Nedwell DB, Embley TM (2003) Analysis of the sulfate-reducing bacterial and methanogenic archaeal populations in contrasting Antarctic sediments. Appl Environ Microbiol 69:3181–3191. doi:10.1128/AEM.69.6.3181-3191.2003

Rajput A, Kaur K, Kumar M (2016) SigMol: repertoire of quorum sensing signaling molecules in prokaryotes. Nucleic Acids Res 44:D634–D639. doi:10.1093/nar/gkv1076

Randau L (2012) RNA processing in the minimal organism Nanoarchaeum equitans. Genome Biol 13:R63. doi:10.1186/gb-2012-13-7-r63

Randau L (2015) Evolution of small guide RNA genes in hyperthermophilic archaea. Ann NY Acad Sci 1341:188–193. doi:10.1111/nyas.12643

Rani S, Srivastava A, Kumar M, Goel M, Lund P (2016) CrAgDb—a database of annotated chaperone repertoire in archaeal genomes. FEMS Microbiol Lett 363:1–6. doi:10.1093/femsle/fnw030

Rudolph J, Oesterhelt D (1996) Deletion analysis of the che operon in the Archaeon *Halobacterium salinarium*. J Mol Biol 258:548–554

Schaefer AL, Valt DL, Hanzelka BL, Cronan JE, Greenberg EP, Wolfe RS (1996) Generation of cell-to-cell signals in quorum sensing: acyl homoserine lactone synthase activity of a purified *Vibrio fischeri* LuxI protein (autoinduction/lux genes/cell density-dependent gene expression). Biochemistry 93:9505–9509

Schwartzman DW, Lineweaver CH (2004) The hyperthermophilic origin of life revisited. Biochem Soc T 32:168–171

Servín-Garcidueñas LE, Martínez-Romero E (2014) Draft genome sequence of the *Sulfolobales* Archaeon AZ1, Obtained through Metagenomic Analysis of a Mexican Hot Spring. Genome Announc 2:e00164–14. doi:10.1128/genomeA.00164-14

Servín-Garcidueñas LE, Peng X, Garrett RA, Martínez-Romero E (2013) Genome sequence of a novel archaeal rudivirus recovered from a mexican hot spring. Genome Announc 1:e00040–12. doi:10.1128/genomeA.00040-12

She Q, Peng X, Zillig W, Garrett RA (2001) Genome evolution: gene capture in archaeal chromosomes. Nature 409:478. doi:10.1038/35054138

Snyder JC, Samson RY, Brumfield SK, Bell SD, Young MJ, Van Etten JL (2013) Functional interplay between a virus and the ESCRT machinery in archaea. PNAS 110:10783–10787. doi:10.1073/pnas.1301605110

Song ZQ, Chen JQ, Jiang HC, Zhou EM, Tang SK, Zhi XY, Zhang LX, Zhang CLL, Li WJ (2010) Diversity of Crenarchaeota in terrestrial hot springs in Tengchong, China. Extremophiles 14:287–296. doi:10.1007/s00792-010-0307-6

Strand KR, Sun C, Li T, Jenney FE, Schut GJ, Adams MWW (2010) Oxidative stress protection and the repair response to hydrogen peroxide in the hyperthermophilic archaeon *Pyrococcus furiosus* and in related species. Arch Microbiol 192:447–459. doi:10.1007/s00203-010-0570-z

Tang T-H, Polacek N, Zywicki M, Huber H, Brugger K, Garrett R, Bachellerie JP, Hüttenhofer A (2004) Identification of novel non-coding RNAs as potential antisense regulators in the archaeon *Sulfolobus solfataricus*. Mol Microbiol 55:469–481. doi:10.1111/j.1365-2958.2004.04428.x

Tommonaro G, Abbamondi GR, Iodice C, Tait K, De Rosa S (2012) Diketopiperazines produced by the halophilic archaeon, *Haloterrigena hispanica*, activate AHL bioreporters. Microb Ecol 63:490–495. doi:10.1007/s00248-011-9980-y

van Wolferen M, Ajon M, Driessen AJM, Albers SV (2013) How hyperthermophiles adapt to change their lives: DNA exchange in extreme conditions. Extremophiles 17:545–563. doi:10.1007/s00792-013-0552-6

van Wolferen M, Wagner A, van der Does C, Albers SV (2016) The archaeal Ced system imports DNA. Proc Natl Acad Sci U S A 113:2496–2501. doi:10.1073/pnas.1513740113

Wang Q, Cen Z, Zhao (2015) The survival mechanisms of thermophiles at high temperatures: an angle of omics. Physiology 30:97–106. doi:10.1152/physiol.00066.2013

Witzany G (2016) The biocommunication method: on the road to an integrative biology. Commun Integr Biol 9:e1164374. doi:10.1080/19420889.2016.1164374

Wurch L, Giannone RJ, Belisle BS, Swift C, Utturkar S, Hettich RL, Reysenbach A-L, Podar M (2016) Genomics-informed isolation and characterization of a symbiotic Nanoarchaeota system from a terrestrial geothermal environment. Nat Commun 7:12115. doi:10.1038/ncomms12115

Yamaguchi Y, Park J-H, Inouye M (2011) Toxin-Antitoxin systems in bacteria and archaea. Ann Rev Genet 45:61–79. doi:10.1146/annurev-genet-110410-132412

Zhang G, Zhang F, Ding G, Li J, Guo X, Zhu J, Zhou L, Cai S, Liu X, Luo Y, Zhang G, Shi W, Dong X (2012) Acyl homoserine lactone-based quorum sensing in a methanogenic archaeon. ISME J 6:1336–1344. doi:10.1038/ismej.2011.203

Sexual Communication in Archaea, the Precursor to Eukaryotic Meiosis

Harris Bernstein and Carol Bernstein

1 Introduction

Recent evidence indicates that, under stressful environmental conditions that cause DNA damage, archaeal cells of specific species aggregate and transfer DNA from one cell to another through direct contact (van Wolferen et al. 2016). DNA transfer, a form of information transfer, can be regarded as a kind of communication between individual cells. The function of this communication appears to be the replacement of damaged DNA sequence information in one cell by undamaged sequence information from another cell. DNA damage is potentially lethal to a cell if unrepaired. Thus DNA transfer between neighboring cells appears to be a mechanism for promoting survival of nearby (and likely genetically related) archaeal cells containing DNA damage.

Accumulating evidence indicates that numerous eukaryotic species that were once regarded as reflecting ancient asexual lineages actually are either capable of meiosis, or were recently capable of meiosis. These findings have led to the conclusion that meiosis was likely present in the ancestor of all eukaryotes. This suggests that a precursor of meiosis may have been already present in the prokaryotic ancestor of the eukaryotes.

Evidence indicates that the eukaryotic cell evolved as a consequence of the establishment of a symbiotic relationship between a host archaeal cell and an internalized aerobic eubacterium. The nuclear genome of eukaryotic cells appears to have evolved primarily from the ancestral archaea genome (Cotton and McInerney 2010).

The central step characteristic of meiosis is the intimate alignment of homologous chromosomes originating from different parents followed by genetic

H. Bernstein · C. Bernstein (✉)
Department of Cellular and Molecular Medicine, College of Medicine, University of Arizona, Tucson, AZ, USA
e-mail: bernstein324@yahoo.com

G. Witzany (ed.), *Biocommunication of Archaea*,
DOI 10.1007/978-3-319-65536-9_7

recombination (transfer of sequence information between homologs). We suggest here that the capability for DNA transfer and recombination between DNA of different individuals that occurs in the archaea was the precursor to eukaryotic meiosis, and that both of these processes function to repair DNA damage.

2 DNA Transfer in Archaea

2.1 *Sulfolobus solfataricus*

UV irradiation of the crenarchaeon *Sulfolobus solfataricus,* strongly induces type IV pili formation which, in turn, leads to cellular aggregation (Frols et al. 2008; Allers 2011). Exposure of *S. solfataricus* to bleomycin or mitomycin C, agents that cause double-strand breaks and other DNA damage, also induces cellular aggregation (Frols et al. 2008). Other physical stressors, such as pH or temperature shift, do not induce aggregation, suggesting that induction of aggregation is caused specifically by DNA damage.

Ajon et al. (2011) showed that UV-induced cellular aggregation mediates intercellular chromosomal marker exchange with high frequency. UV-induced cultures have recombination rates exceeding those of uninduced cultures by up to three orders of magnitude. The UV-induced pili of three different *Sulfolobus* species had distinct morphologies, and individual cells of these three species, including *S. solfataricus,* were able to aggregate only with other members of their own species (Ajon et al. 2011). Frols et al. (2008, 2009) and Ajon et al. (2011) hypothesized that the UV-inducible DNA transfer process and subsequent homologous recombinational repair represents an important mechanism to maintain chromosome integrity. This response appears to be a primitive form of sexual interaction.

In eubacteria, the RecA protein plays a central role in recombinational repair of DNA during bacterial transformation (Bernstein and Bernstein 2013). In eukaryotes, Rad51 and Dmc1 proteins, which are homologs of RecA, catalyze DNA pairing and strand exchange, central steps in recombinational repair during meiosis (Bernstein and Bernstein 2013). In the archaea *S. solfataricus*, another homolog of RecA, termed RadA, similarly catalyzes DNA pairing and strand exchange (Seitz et al. 1998). The RadA protein is more similar to the eukaryotic Rad51protein than to the eubacterial RecA protein (Seitz et al. 1998).

In eukaryotes, the Rad50 and Mrell proteins function in double-strand break repair by homologous recombination. In thermophilic archaea, *rad50* and *mre11* homologs are clustered in the genome with two genes *nurA* and *herA*. Quaiser et al. (2008) reported that, in *S. solfataricus*, the Mre11 protein interacts with both Rad50 and HerA, and is recruited to DNA following γ-irradiation. RadA recombinase is also recruited to DNA following γ-irradiation. The HerA ATPase cooperates with the Nur nuclease and the Mre11-Rad50 complex for the recombinational repair of double-strand breaks in thermophilic archaea (White 2011). The HerA protein of *S. solfataricus* catalyzes translocation-coupled DNA end processing, a key step in recombination (Rzechorzek et al. 2014). These findings indicate that in *S. solfataricus* the Rad50-Mre11-HerA complex, as well as the RadA recombinase, are employed in DNA repair via homologous recombination. This process is likely active during DNA exchange between *S. solfataricus* cells, an apparent primitive form of sexual interaction.

2.2 *Sulfolobus acidocaldarius*

In the thermophilic species, *Sulfolobus acidocaldarius*, the ability to exchange and recombine chromosomal markers appears to be an intrinsic property, and marker exchange is observed at temperatures up to 84 °C (Grogan 1996). Exposure to UV induces pili formation and cellular aggregation in *S. acidocaldarius* (Ajon et al. 2011). Cells with the ability to aggregate also have higher survival than mutants lacking pili and unable to aggregate. DNA damage induced by UV-irradiation increases the frequency of recombination in *S. acidocaldarius* (Wood et al. 1997). DNA strand-breaking agents (γ-radiation or bleomycin), as well as DNA cross-linking agents (mechlorethamine, butadiene epoxide, or cisplatin) stimulate formation of recombinants via conjugation in *S. acidocaldarius* (Reilly and Grogan 2002).

The *ups* operon (UV-inducible pili operon of *Sulfolobus*), containing 5 genes, is highly induced by UV irradiation. The proteins encoded by the *ups* operon are involved in UV-induced pili assembly and cellular aggregation (van Wolferen et al. 2013). The *ups* pili mediate cellular aggregation leading to intercellular DNA exchange and homologous recombination. Since this system increases the fitness of *S. acidocaldarius* cells after UV exposure, van Wolferen et al. (2013) thought that transfer of DNA takes place in order to repair UV-induced DNA damages by homologous recombination.

A further study of the *S. acidocaldarius ups* operon showed that one of the genes of the operon, *saci-1497*, encodes an endonuclease III that nicks UV-damaged DNA; and another gene of the operon, *saci-1500*, encodes a RecQ-like helicase that is able to unwind homologous recombination intermediates such as Holliday junctions (van Wolferen et al. 2015). It was proposed that the Saci-1497 and Saci-1500 proteins function in a homologous recombination-based DNA repair mechanism that uses transferred DNA as a template (van Wolferen et al. 2015). Thus it appears that the *ups* system in combination with homologous recombination provides a DNA damage mating response that rescues *Sulfolobales* from DNA damaging threats (van Wolferen et al. 2015).

Four genes, *cedA*, *cedA1*, *cedA2*, and *cedB*, are clustered in the *S. acidocaldarius* genome. The expression of two of these genes, *cedA* and *cedB*, is highly induced by UV light (Van Wolferen et al. 2016). CedA and CedB proteins are both essential for DNA transfer between *S. acidocaldarius* cells. These proteins act downstream of the Ups–pili system, and are employed in the import of DNA rather than its export (van Wolferen et al. 2016). The Ups system, described above, appears to function in mating pair formation, whereas the Ced system functions actively in DNA import (van Wolferen et al. 2016). These functions appear to be components of a primitive sexual interaction system that facilitates repair of DNA damage.

2.3 Halobacterium volcanii

Halobacterium volcanii has a distinctive mating system employing cytoplasmic bridges between cells (Rosenshine et al. 1989). Following the formation of the cytoplasmic bridges, two *H. volcani* cells can undergo fusion to form a heterodiploid cell (Naor and Grophna 2013). Heterodiploid cells contain two different chromosome originating from two different cells. Recombination may then occur between the chromosomes. Upon further cell division two haploid cells are formed.

2.4 Halobacterium sp. *Strain NRC-1*

UV irradiation of *Halobacterium* sp. strain NRC-1 induces several gene products employed in homologous recombination (McCready et al. 2005; Boubriak et al. 2008). For example, *radA1*, a homolog of the *rad51*and *recA* genes that have a central role in recombination, is induced 7-fold by UV. Homologous recombination facilitates recombinational repair of DNA damage. In the natural habitat of this halobacterium, homologous recombination is likely induced by the UV in sunlight (McCready et al. 2005).

2.5 Halorubrum

In wild populations of *Halorubrum*, genetic exchange and recombination occur frequently (Papke et al. 2004).

The findings reviewed, above, in this section show that, in response to DNA damaging conditions, some archaeal species can transfer DNA between cells or undergo cell fusion and/or upregulate genes that facilitate recombinational repair of DNA damage.

3 The Common Ancestor of all Eukaryotes was Likely Capable of Meiosis

Eukaryotes evolved from prokaryotic ancestors more than 1.5 billion years ago (Javaux et al. 2001). *Bangiomorph pubescens*, a red algae, is the oldest taxonomically resolved eukaryote in the fossil record. It existed more than 1.2 billion years ago and was sexually reproducing (Butterfield 2000).

Among present day eukaryotes, sexual reproduction is widespread. On the basis of a phylogenetic analysis, Dacks and Roger (1999) proposed that facultative sex was present in the common ancestor of all eukaryotes. In support of the idea that sex is a primitive characteristic of eukaryotes, many species that were once thought to be "ancient asexuals" have proven to have, or to have once had, the capability for sex. Examples are given below.

A core set of genes that function specifically in meiosis was found to be present in both *Trichomonas vaginalis* and *Giardia intestinalis*, two species once thought to be asexual (Ramesh et al. 2005; Malik et al. 2008). Since these two extant species are descendants of lineages that diverged early from the eukaryotic evolutionary tree, it was proposed that the common ancestor of all eukaryotes possessed the core genes necessary for meiosis (Malik et al. 2008).

Parasitic protozoa of the genus *Leishmania*, once thought to be asexual, are now considered to have a sexual cycle (Akopyants et al. 2009).

Amoeba have been traditionally regarded as asexual. Evidence now indicates that most amoeboid are anciently sexual, and that the majority of present day asexual groups likely arose recently and independently (Lahr et al. 2011).

The arbuscular mycorrhizal fungi, were considered to have propagated colonally for over 500 million years. However, recently, several members of this group (*Glomus* spp.) were found to contain in their genome homologs of 51 meiotic genes including seven meiosis-specific genes (Halary et al. 2011). These findings suggest that the meiotic machinery of these fungi has been conserved and that they may have a cryptic sexual cycle (Halary et al. 2011).

The pathogenic yeast *Candida albicans* had long been considered to be an "asexual" eukaryote. However, it is now evident that this organism has maintained an elaborate—but largely hidden—mating apparatus (Johnson 2003).

Peacock et al. (2014) showed that *Trypanosoma brucei* undergoes meiosis in its tsetse fly vector, and that meiosis is a normal part of their developmental cycle. Thus *T. brucei* is essentially a sexual organism. Trypanosomes belong to the supergroup Excavata, one of the earliest diverging lineages among eukaryotes (Hampl et al. 2009). Peacock et al. (2014) noted that their finding of a sexual stage in *T. brucei* supports the hypothesis that meiosis and sexual reproduction are ancestral and ubiquitous features of eukaryotes (Ramesh et al. 2005; Malik et al. 2008; Bernstein and Bernstein 2013).

Speijer et al. (2015) have also reviewed evidence that sex is a ubiquitous, ancient and inherent attribute of eukaryotic life. It now appears likely, considering all the recent evidence, that the common ancestor of eukaryotes was capable of sex and meiosis. The earliest eukaryotic organisms were probably single celled protists. Thus it is reasonable to hypothesize that meiosis evolved from sexual processes that were already present in prokaryotes. One such process is the type of gene transfer and cell fusion that occurs in the archaea.

4 The Prokaryotic to Eukaryotic Transition

Microbes were the sole forms of life on earth from about 3.4 billion to 570 million years ago. A major event in microbial evolution was the emergence of the eukaryotic cell at least 1.5 billion years ago (Javaux et al. 2001). The eukaryotic cell likely arose when an anaerobic host archaeal cell developed a stable relationship with an internalized aerobic bacterium (Speijer 2015). The internalized aerobe provided the capability for respiration and eventually evolved into the mitochondrion.

The internalized aerobic bacterium may initially have been an intracellular parasite that eventually became tamed (Guerrero et al. 1986; Davidov et al. 2006). Based on genomic sequence analysis, mitochondria are most closely related to current *α-proteobacteria* (Andersson et al. 1998; Gray et al. 1999).

Cox et al. (2008) presented evidence for an archaeal origin of eukaryotes based on a phylogenetic analysis of 53 genes. Cotton and McInerney (2010) found that archaea derived genes are significantly more likely than eubacterial genes to be essential to the viability of eukaryotes (the model eukaryote used for comparison was yeast). In addition, archaeal derived genes are more highly expressed and are significantly more highly connected and more central in the yeast protein interaction network than eubacterial genes. These findings indicated that genes of archaeal origin are, in a sense, more important to yeast than genes of eubacterial origin, and that this importance reflects these genes' origin as the ancestral component of the eukaryotic nuclear genome (Cotton and McInerney 2010). Thiergart et al. (2012) concluded, on the basis of a genomic analysis, that the eukaryotic information genes arose from an archael host, whereas the genes related to eukaryotic energy metabolism were of eubacterial (mitochondrial) origin.

5 DNA Damage Likely Promotes Sex in Eukaryotic Microorganisms

In Sect. 2, we cited evidence that mating in the archaea is specifically induced by DNA damaging agents such as UV-irradiation, bleomycin or mitomycin C, suggesting that DNA transfer may be an adaptation for repairing DNA damage.

In this section, we present evidence that meiotic sex in eukaryotic microorganisms is also likely an adaptive response for repairing DNA damage. Extant eukaryotic microbes usually reproduce asexually under favorable conditions, but tend to reproduce sexually under stressful conditions (Bernstein and Bernstein 2013). An important factor in the induction of sex in some eukaryotic microbes appears to be oxidative stress that can cause the production of reactive oxygen species (ROS) leading to DNA damage.

Hydrogen peroxide (H_2O_2) produces oxidative stress that causes a variety of DNA damages including double-strand breaks and modified bases (Slupphang et al. 2003). When cells of the yeast *Schizosaccharomyces pombe* were exposed to (H_2O_2), meiosis was induced (Bernstein and Johns 1989). Mating and formation of sexual spores was increased by 4- to 18-fold upon exposure to this DNA damaging agent.

Sexual reproduction and meiosis are induced by heat shock in the green algae *Volvox carterei* (Kirk and Kirk 1986). This induction is inhibited by anti-oxidants, indicating that heat shock induction of sex and meiosis is mediated by oxidative stress (Nedelcu and Michod 2003). In addition, an inhibitor of the mitochondrial electron transport chain, that induces oxidative stress, was also found to induce sex and meiosis in *V. carteri* (Nedelcu et al. 2004). On the basis of this evidence, Nedelcu and Michod (2003) and Nedelcu et al. (2004) proposed that ROS produced by oxidative stress cause DNA damage leading to the induction of sex and meiosis in *V. carteri*. Thus meiosis in the eukaryotic protists *V. carterei* and *S. pombe* may be part of an adaptive response for coping with DNA damage induced by oxidative stress.

The sexual cycle in eukaryotic microorganisms is often induced in response to stressful environmental conditions. For example, the sexual cycle can be induced by starvation or nutrient limitation in the protozoan *Tetrahymena* (Elliott and Hayes 1953), the green alga *Chlamydomonas reinhardi* (Sager and Granik 1954), the fission yeast *Schizosaccharomyces pombe* (Davey 1998), and in the budding yeast *Saccharomyces cerevisiae* (Herskowitz 1988). The sexual cycle can also be induced in the oomycete *Phytophthora cinnamomi* by mechanical damage to hyphae or by the oxidizing agent H_2O_2 (Reeves and Jackson 1974). Such conditions can increase DNA damage and thus the need for DNA repair.

The findings reviewed in this section, above, illustrate that in eukaryotic microorganisms the sexual cycle is often induced by stressful conditions that may cause DNA damage. Since in archaea the sexual cycle can be induced by DNA damaging agents (see Sect. 2), the induction of the sexual cycle by DNA damaging conditions appears to be a common feature of both the archaea and eukaryotic microorganisms.

6 Similarities Between Archaeal DNA Transfer and Recombination and Eukaryotic Meiosis

Mating in the archaea, like mating in eukaryotes, is a complex adaptation. Nevertheless, both types of mating are similar in their central molecular processes. There are four major steps in archaeal mating: (1) two cells directly contact each other; (2) DNA is transferred between cells, and this may be accompanied by cell fusion; (3) the two homologous chromosomes (or homologous portions of the two chromosomes) derived from the two parental archaea cells align and undergo genetic recombination (exchange of genetic information); (4) the new recombined chromosome is passed on to progeny.

Mating in single-celled haploid eukaryotes such as in the yeast *S. pombe*, can similarly be viewed as occurring by four steps. These steps are: (1) two haploid cells directly contact each other and fuse (syngamy) followed by nuclear fusion (karyogamy); (2) the chromosomes undergo premeiotic replication to form two sister copies; (3) within the nucleus, homologous chromosomes from different parents (i.e. non-sister chromosomes) align in pairs and undergo recombination; (4) two successive cell divisions (without chromosome duplication) lead to haploid progeny cells that can then repeat the cycle.

The key step (step 3) in both mating processes is the intimate alignment of non-sister homologous chromosomes (originating from different parents) that then undergo genetic recombination. In eukaryotes, recombination between non-sister homologous chromosomes is catalyzed by a homolog of the RecA recombinase (i.e. recombinases Dmc1 and/or Rad51). Similarly, in archaea a RecA homolog, RADA1 recombinase, is also employed in recombination (Seitz et al. 1998). Other similarities include proteins Rad50 and Mre11 that catalyze resection of the ends of the DNA at the double strand break, an intermediate step in recombination (White 2011). These observations suggest that archaeal conjugation and eukaryotic meiosis occur by similar processes.

7 Similarities Between Archaeal, Eubacterial and Eukaryotic DNA Transfer Systems

It has been proposed that sex arose very early in the evolution of life, perhaps as early as the RNA world (Bernstein et al. 1984). Thus sex may have already been present in an early DNA ancestor of both eubacteria and archaea. Meiotic sex evolved later with the emergence of eukaryotes. Meiosis, in addition to its similarity to archaeal DNA transfer plus recombination, also has similarities to natural genetic transformation in eubacteria (Bernstein and Bernstein 2010, 2013). Indeed, Gray et al. (2013) have presented evidence that conjugal DNA transfer in mycobacteria has genetic consequences that are strikingly similar to those of meiosis. However, it now appears that genes of archaeal origin play a more important role than genes of eubacterial origin in the central functions of eukaryotes (Cotton and McInerney 2010). As indicated in Fig. 1, the DNA transfer processes of eubacteria and archaea may be derived from a common ancestral transfer process.

8 Emergence of the Eukaryotic Nucleus

During emergence of the eukaryotic cell, the internalized aerobe (proto-mitochondrion) was gradually tamed and exploited. This process likely included protection of the archaeal host genome from the release of reactive oxygen species (ROS) during oxidative phosphorylation and ATP production by the proto-mitochondria. Thus natural selection would have favored the retention of the basic mechanisms for recombinational repair of DNA damage already present in the archaeal ancestor of the eukaryote. Meiosis probably emerged through natural selection as a more effective elaboration of the basic recombinational repair mechanism, able to cope with the increased threat of ROS. Another innovation that likely evolved for coping with DNA damage caused by ROS was the nuclear membrane.

Mans et al. (2004), who studied the comparative genomics, evolution and origins of the nuclear envelope and nuclear pore complex, inferred that the nucleus emerged in the primitive eukaryotic ancestor (the "prekaryote") and was triggered by archaeo-bacterial symbiosis. Martin (2005) reviewed several ideas that have been proposed for the evolutionary origin of the nuclear membrane. These ideas include invagination of the plasma membrane, or the origin of a genuine new membrane system following the establishment of proto-mitochondria in an archaeal host.

In the latter case a possible scenario is that simple cytosolic vesicles, under natural selection, coalesced and acquired increased complexity to form the nuclear membrane and endoplasmic reticulum. Thus, the nuclear envelope, a defining characteristic of the eukaryotic cell, may have arisen as an adaptation for segregating the original archaeal host genome away from the proto-mitochondria, the main source of damaging ROS production.

It is suggested that as archaeal DNA transfer plus recombination evolved into meiosis (meiosis being an improved method for repairing DNA damage) this process was accompanied by the emergence of the nuclear membrane which served the complementary function of protecting the genomic DNA from damaging ROS produced by the proto-mitochondria.

Membrane vesicles may also have been employed in the transport of DNA between cells as observed in some hyperthermophilic archaea (Gaudin et al. 2013).

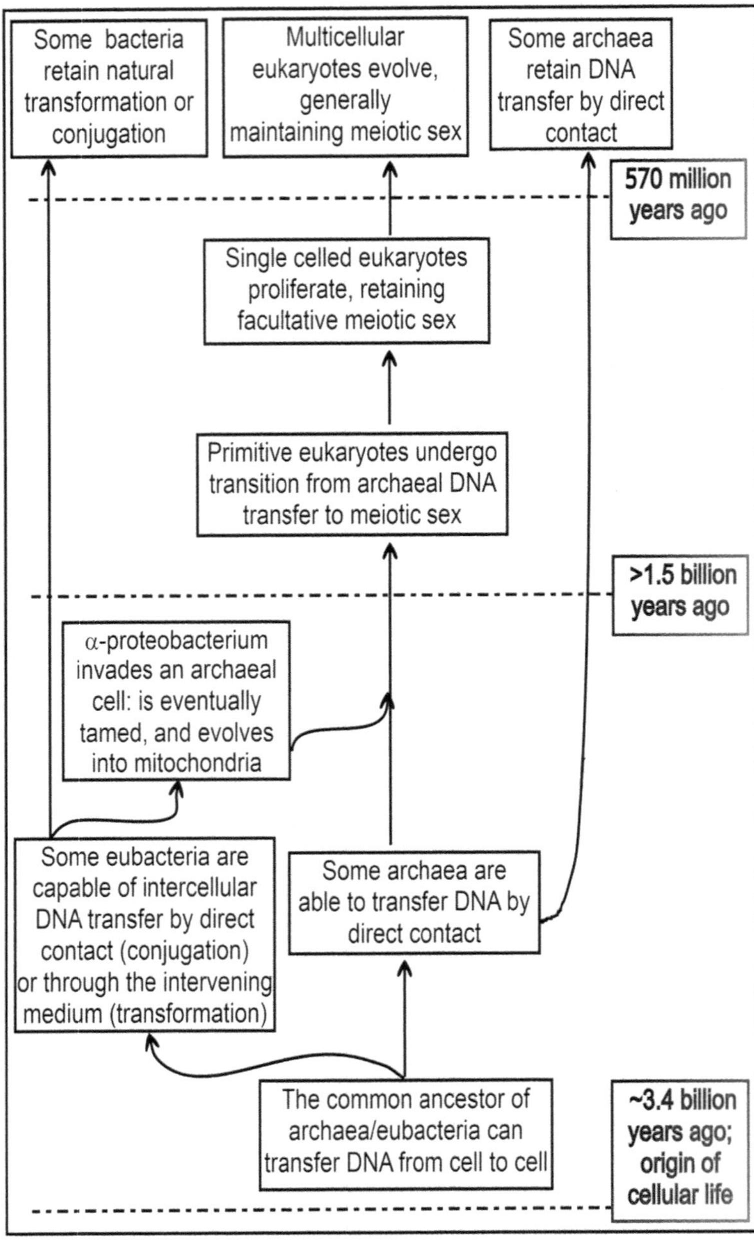

Fig. 1 Evolution of eukaryotic meiotic sex from ancestral archaeal sex. The transition from archaeal sex by conjugal gene transfer to eukaryotic meiotic sex is proposed to have occurred prior to 1.5 billion years ago. DNA transfer processes in a common ancestor of both the eubacteria and archaea likely preceded the emergence of eukaryotic sex. Recent evidence supports a primarily archaeal ancestry for the eukaryotic genome. This suggests that the DNA transfer process of the archaea, rather than that of the eubacteria, is the progenitor of eukaryotic meiotic sex

9 Expansion of the Nuclear Genome

The availability of abundant useful energy (proto-mitochondrial ATP) plus protection of the archaeal genome from damaging ROS by the nuclear membrane may have allowed expansion of genetic information content by the nascent eukaryote leading to further adaptive diversification. Illustrative of this expansion, the *S. solfataricus* genome, representative of the archaea, encodes 2977 proteins (She et al. 2001), whereas the unicellular eukaryotic yeast *Schizosaccharomyces pombe* has 4970 estimated protein-coding genes (Wood et al. 2002). It is presumed that the genomes of eukaryotes are generally larger than the genomes of their prokaryotic ancestors. As genome size increased, this may have driven the evolution of meiosis as a more reliable process for ensuring proper genome processing during such intermediate steps as chromosome alignment, recombination and chromosome segregation than was available in the simpler ancestral DNA transfer processes.

The characteristic nucleosomal organization of the eukaryotic genome may also have arisen as a byproduct of this informational expansion and the need for both compact organization of the genome within the nucleus and protection of the DNA. The fundamental unit of DNA packing inside eukaryotic cells is the nucleosome core particle, in which approximately 150 base pairs of DNA are bound around an octamer of histone proteins. These core particles are assembled into higher order structures, stabilized by other histone- and non-histone proteins, in order to further compact the DNA so that it can fit into the nucleus. In addition to their packaging function, the histone proteins also protect against some DNA damaging agents (Millard 1996).

10 Transfer of Genes from Pre-mitochondrial Genome to the Nuclear Genome

The greater protection of the nuclear genome from ROS by the nuclear membrane may explain the substantial transport of genetic information from the ancestral pre-mitochondrial aerobic bacterium to the nuclear genome of the archaeal host that apparently occurred as the mitochondrion evolved. Phylogenetic studies (Gabaldon and Huynen 2003) suggest that ≥630 genes in the eukaryotic genome had moved from the *α-proteobacterial* to the eukaryotic genome.

Several currently living *α-proteobacteria* are capable of genetic transformation, a form of mating involving DNA transfer and recombination. These include *Methylobacterium organophilum* (O'Connor et al. 1977), *Rhizobium japonicum* (Raina and Modi 1972) and *Agrobacterium tumefaciens* (Demaneche et al. 2001). Thus, among the many genes transported from the *α-proteobacterial* mitochondrial ancestor to the eukaryotic nucleus, there may have been some genes that functioned in genetic transformation. Since meiosis in extant eukaryotes is a complex process involving the integrated action of numerous proteins, some of these proteins may have originally functioned in transformation in the ancestral eubacteria. Although the contribution of archaeal genes was likely more central to the emergence of meiosis, eubacterial genes may have contributed as well.

11 Conclusion

Meiotic sex in eukaryotes likely evolved from archaeal DNA transfer plus recombination, and both processes appear to be adaptations for repair of DNA damage.

References

Ajon M, Fröls S, van Wolferen M, Stoecker K, Teichmann D, Driessen AJ, Grogan DW, Albers SV, Schleper C (2011) UV-inducible DNA exchange in hyperthermophilic archaea mediated by type IV pili. Mol Microbiol 82(4):807–817

Akopyants NS, Kimblin N, Secundino N, Patrick R, Peters N, Lawyer P, Dobson DE, Beverley SM, Sacks DL (2009) Demonstration of genetic exchange during cyclical development of Leishmania in the sand fly vector. Science 324(5924):265–268

Allers T (2011) Swapping genes to survive—a new role for archaeal type IV pili. Mol Microbiol 82(4):789–791

Andersson SG, Zomorodipour A, Andersson JO, Sicheritz-Pontén T, Alsmark UC, Podowski RM, Näslund AK, Eriksson AS, Winkler HH, Kurland CG (1998) The genome sequence of *Rickettsia prowazekii* and the origin of mitochondria. Nature 396(6707):133–140

Bernstein H, Bernstein C (2010) Evolutionary origin of recombination during meiosis. Bioscience 60(7):498–505

Bernstein C, Bernstein H (2013) Evolutionary origin and adaptive function of meiosis. In: Bernstein C, Bernstein H (eds) Meiosis. InTech, Rijeka, Croatia, pp 41–45

Bernstein C, Johns V (1989) Sexual reproduction as a response to H_2O_2 damage in *Schizosaccharomyces pombe*. J Bacteriol 171(4):1893–1897

Bernstein H, Byerly HC, Hopf FA, Michod RE (1984) Origin of sex. J Theor Biol 110(3):323–351

Boubriak I, Ng WL, DasSarma P, DasSarma S, Crowley DJ, McCready SJ (2008) Transcriptional responses to biologically relevant doses of UV-B radiation in the model archaeon, Halobacterium sp. NRC-1. Saline Syst 4:13

Butterfield NJ (2000) *Bangiomorpha pubescens* n. gen., n.sp.: implications for the evolution of sex, multicellularity, and the Mesoproterozoic/Neoproterozoic radiation of eukaryotes. Paleobiology 26(3):386–404

Cotton JA, McInerney JO (2010) Eukaryotic genes of archaebacterial origin are more important than the more numerous eubacterial genes, irrespective of function. Proc Natl Acad Sci USA 107(40):17252–17255

Cox CJ, Foster PG, Hirt RP, Harris SR, Embley TM (2008) The archaebacterial origin of eukaryotes. Proc Natl Acad Sci USA 105(51):20356–20361

Dacks J, Roger AJ (1999) The first sexual lineage and the relevance of facultative sex. J Mol Evol 48(6):779–783

Davey J (1998) Fusion of a fission yeast. Yeast 14(16):1529–1566

Davidov Y, Huchon D, Koval SF, Jurkevitch E (2006) A new alpha-proteobacterial clade of Bdellovibrio-like predators: implications for the mitochondrial endosymbiotic theory. Environ Microbiol 8(12):2179–2188

Demanèche S, Kay E, Gourbière F, Simonet P (2001) Natural transformation of *Pseudomonas fluorescens* and *Agrobacterium tumefaciens* in soil. Appl Environ Microbiol 67(6):2617–2621

Elliott AM, Hayes RE (1953) Mating types in tetrahymena. Biol Bull 105:269–284

Fröls S, Ajon M, Wagner M, Teichmann D, Zolghadr B, Folea M, Boekema EJ, Driessen AJ, Schleper C, Albers SV (2008) UV-inducible cellular aggregation of the hyperthermophilic archaeon *Sulfolobus solfataricus* is meciated by pili formation. Mol Microbiol 70(4):938–952

Fröls S, White MF, Schleper C (2009) Reactions to UV damage in the model archaeon *Sulfolobus solfataricus*. Biochem Soc Trans 37(Pt 1):36–41

Gabaldón T, Huynen MA (2003) Reconstruction of the proto-mitochondrial metabolism. Science 301(5633):609

Gaudin M, Gauliard E, Schouten S, Houel-Renault L, Lenormand P, Marguet E, Forterre P (2013) Hyperthermophilic archaea produce membrane vesicles that can transfer DNA. Environ Microbiol Rep 5(1):109–116

Gray MW, Burger G, Lang BF (1999) Mitochondrial Evolution. Sci 283:1476–1481

Gray TA, Krywy JA, Harold J, Palumbo MJ, Derbyshire KM (2013) Distributive conjugal transfer in mycobacteria generates progeny with meiotic-like genome-wide mosaicism, allowing mapping of a mating identity locus. PLoS Biol 11(7):e1001602. doi:10.1371/journal.pbio.1001602

Grogan DW (1996) Exchange of genetic markers at extremely high temperatures in the archaeon *Sulfolobus acidocaldarius*. J Bacteriol 178(11):3207–3211

Guerrero R, Pedros-Alio C, Esteve I, Mas J, Chase D, Margulis L (1986) Predatory prokaryotes: predation and primary consumption evolved in bacteria. Proc Natl Acad Sci USA 83(7):2138–2142

Halary S, Malik S-B, Lildhar L, Slamovits CH, Hijri M, Corradi N (2011) Conserved meiotic machinery in *Glomus* spp., a putatively ancient asexual fungal lineage. Genome Biol Evol 3:950–958

Hampl V, Hug L, Leigh JW, Dacks JB, Lang BF, Simpson AG, Roger AJ (2009) Phylogenomic analyses support the monophyly of excavata and resolve relationships among eukaryotic "supergroups". Proc Natl Acad Sci USA 106(10):3859–3864

Herskowitz I (1988) Life cycle of the budding yeast *Saccharomyces cerevisiae*. Microbiol Rev 52 (4):536–553

Javaux EJ, Knoll AH, Walter MR (2001) Morphological and ecological complexity in early eukaryote ecosystems. Nature 412:66–69

Johnson A (2003) The biology of mating in *Candida albicans*. Nat Rev Microbiol 1(2):106–116

Kirk DL, Kirk MM (1986) Heat shock elicits production of sexual inducer in Volvox. Science 231 (4733):51–54

Lahr DJ, Parfrey LW, Mitchell EA, Katz LA, Lara E (2011) The chastity of amoebae: re-evaluating evidence for sex in amoeboid organisms. Proc Biol Sci 278(1715):2081–2090

Malik SB, Pightling AW, Stefaniak LM, Schurko AM, Logsdon JM Jr (2008) An expanded inventory of conserved meiotic genes provides evidence for sex in *Trichomonas vaginalis*. PLoS ONE 3(8):e2879. doi:10.1371/journal.pone.0002879

Mans BJ, Anantharaman V, Aravind L, Koonin EV (2004) Comparative genomics, evolution and origins of the nuclear envelope and nuclear pore complex. Cell Cycle 3(12):1612–1637

Martin W (2005) Archaebacteria (Archaea) and the origin of the eukaryotic nucleus. Curr Opin Microbiol 8(6):630–637

McCready S, Müller JA, Boubriak I, Berquist BR, Ng WL, DasSarma S (2005) UV irradiation induces homologous recombination genes in the model archaeon, *Halobacterium* sp. NRC-1. Saline Syst 1:3

Millard JT (1996) DNA modifying agents as tools for studying chromatin structure. Biochimie 78 (10):803–816

Naor A, Gophna U (2013) Cell fusion and hybrids in Archaea: prospects for genome shuffling and accelerated strain development for biotechnology. Bioengineered 4(3):126–129

Nedelcu AM, Michod RE (2003) Sex as a response to oxidative stress: the effect of antioxidants on sexual induction in a facultatively sexual lineage. Proc Biol Sci 7(270 Suppl 2):136–139

Nedelcu AM, Marcu O, Michod RE (2004) Sex as a response to oxidative stress: a twofold increase in cellular reactive oxygen species activates sex genes. Proc Biol Sci 271(1548):1591–1596

O'Connor M, Wopat A, Hanson RS (1977) Genetic transformation in *Methylobacterium organophilum*. J Gen Microbiol 98(1):265–272

Papke RT, Koenig JE, Rodríguez-Valera F, Doolittle WF (2004) Frequent recombination in a saltern population of Halorubrum. Science 306(5703):1928–1929

Peacock L, Bailey M, Carrington M, Gibson W (2014) Meiosis and haploid gametes in the pathogen *Trypanosoma brucei*. Curr Biol 24(2):181–186

Quaiser A, Constantinesco F, White MF, Forterre P, Elie C (2008) The Mre11 protein interacts with both Rad50 and the HerA bipolar helicase and is recruited to DNA following gamma irradiation in the archaeon *Sulfolobus acidocaldarius*. BMC Mol Biol 9:25. doi:10.1186/1471-2199-9-25

Raina JL, Modi VV (1972) Deoxyribonucleate binding and transformation in *Rhizobium jpaonicum*. J Bacteriol 111(2):356–360

Ramesh MA, Malik SB, Logsdon JM Jr (2005) A phylogenomic inventory of meiotic genes; evidence for sex in Giardia and an early eukaryotic origin of meiosis. Curr Biol 15(2):185–191

Reeves RJ, Jackson RM (1974) Stimulation of sexual reproduction in Phytophthora by damage. J Gen Microbiol 84:303–310

Reilly MS, Grogan DW (2002) Biological effects of DNA damage in the hyperthermophilic archaeon *Sulfolobus acidocaldarius*. FEMS Microbiol Lett 208(1):29–34

Rosenshine I, Tchelet R, Mevarech M (1989) The mechanism of DNA transfer in the mating system of an archaebacterium. Science 245(4924):1387–1389

Rzechorzek NJ, Blackwood JK, Bray SM, Maman JD, Pellegrini L, Robinson NP (2014) Structure of the hexameric HerA ATPase reveals a mechanism of translocation-coupled DNA-end processing in archaea. Nat Commun 5:5506. doi:10.1038/ncomms6506

Sager R, Granick S (1954) Nutritional control of sexuality in *Chlamydomonas reinhardi*. J Gen Physiol 37(6):729–742

Seitz EM, Brockman JP, Sandler SJ, Clark AJ, Kowalczykowski SC (1998) RadA protein is an archaeal RecA protein homolog that catalyzes DNA strand exchange. Genes Dev 12(9):1248–1253

She Q, Singh RK, Confalonieri F, Zivanovic Y, Allard G, Awayez MJ, Chan-Weiher CC, Clausen IG, Curtis BA, De Moors A, Erauso G, Fletcher C, Gordon PM, Heikamp-de Jong I, Jeffries AC, Kozera CJ, Medina N, Peng X, Thi-Ngoc HP, Redder P, Schenk ME, Theriault C, Tolstrup N, Charlebois RL, Doolittle WF, Duguet M, Gaasterland T, Garrett RA, Ragan MA, Sensen CW, Van der Oost J (2001) The complete genome of the crenarchaeon *Sulfolobus solfataricus* P2. Proc Natl Acad Sci USA 98(14):7835–7840

Slupphaug G, Kavli B, Krokan HE (2003) The interacting pathways for prevention and repair of oxidative DNA damage. Mutat Res 531(1–2):231–251

Speijer D (2015) Birth of the eukaryotes by a set of reactive innovations: new insights force us to relinquish gradual models. BioEssays 37(12):1268–1276

Speijer D, Lukeš J, Eliáš M (2015) Sex is a ubiquitous, ancient, and inherent attribute of eukaryotic life. Proc Natl Acad Sci USA 112(29):8827–8834

Thiergart T, Landan G, Schenk M, Dagan T, Martin WF (2012) An evolutionary network of genes present in the eukaryote common ancestor polls genomes on eukaryotic and mitochondrial origin. Genome Biol Evol 4(4):466–485

van Wolferen M, Ajon M, Driessen AJ, Albers SV (2013) Molecular analysis of the UV-inducible pili operon from *Sulfolobus acidocaldarius*. Microbiologyopen 2(6):928–937. doi:10.1002/mbo3.128

van Wolferen M, Ma X, Albers SV (2015) DNA Processing proteins involved in the UV induced stress response of *Sulfolobales*. J Bacteriol 197(18):2941–2951

van Wolferen M, Wagner A, van der Does C, Albers SV (2016) The archaeal Ced system imports DNA. Proc Natl Acad Sci USA 113(9):2496–2501

White MF (2011) Homologous recombination in the archaea: the means justify the ends. Biochem Soc Trans. 39(1):15–19. doi:10.1042/BST0390015. Review. PMID: 21265740

Wood ER, Ghané F, Grogan DW (1997) Genetic responses of the thermophilic archaeon *Sulfolobus acidocaldarius* to short-wavelength UV light. J Bacteriol 179(18):5693–5698

Wood V, Gwilliam R, Rajandream MA, Lyne M, Lyne R, Stewart A, Sgouros J, Peat N, Hayles J, Baker S, Basham D, Bowman S, Brooks K, Brown D, Brown S, Chillingworth T, Churcher C, Collins M, Connor R, Cronin A, Davis P, Feltwell T, Fraser A, Gentles S, Goble A, Hamlin N, Harris D, Hidalgo J, Hodgson G, Holroyd S, Hornsby T, Howarth S, Huckle EJ, Hunt S, Jagels K, James K, Jones L, Jones M, Leather S, McDonald S, McLean J, Mooney P, Moule S, Mungall K, Murphy L, Niblett D, Odell C, Oliver K, O'Neil S, Pearson D, Quail MA, Rabbinowitsch E, Rutherford K, Rutter S, Saunders D, Seeger K, Sharp S, Skelton J, Simmonds M, Squares R, Squares S, Stevens K, Taylor K, Taylor RG, Tivey A, Walsh S, Warren T, Whitehead S, Woodward J, Volckaert G, Aert R, Robben J, Grymonprez B, Weltjens I, Vanstreels E, Rieger M, Schäfer M, Müller-Auer S, Gabel C, Fuchs M, Düsterhöft A, Fritzc C, Holzer E, Moestl D, Hilbert H, Borzym K, Langer I, Beck A, Lehrach H, Reinhardt R, Pohl TM, Eger P, Zimmermann W, Wedler H, Wambutt R, Purnelle B, Goffeau A, Cadieu E, Dréano S, Gloux S, Lelaure V, Mottier S, Galibert F, Aves SJ, Xiang Z, Hunt C, Moore K, Hurst SM, Lucas M, Rochet M, Gaillardin C, Tallada VA, Garzon A, Thode G, Daga RR, Cruzado L, Jimenez J, Sánchez M, del Rey F, Benito J, Domínguez A, Revuelta JL, Moreno S, Armstrong J, Forsburg SL, Cerutti L, Lowe T, McCombie WR, Paulsen I, Potashkin J, Shpakovski GV, Ussery D, Barrell BG, Nurse P (2002) The genome sequence of *Schizosaccharomyces pombe*. Nature 415(6874):871–880

Quorum Sensing in Archaea: Recent Advances and Emerging Directions

James C. Charlesworth, Charlotte Beloe, Cara Watters and Brendan P. Burns

1 Introduction

Communication in the microbial environment often occurs over microspatial distances utilizing small messenger molecules to facilitate various changes in the community, often those that confer a competitive advantage. Quorum sensing is a particularly well-studied form of microbial communication in which messenger molecules are released into the environment until a critical threshold of signal is reached triggering phenotypic change(s). These messenger molecules are often subject to diffusion leading quorum sensing to be sometimes referred to as diffusion sensing (Redfield 2002), and typically this means quorum sensing phenotypes are often observed in the latter phases of growth when cell density is high (Llamas et al. 2005).

Quorum sensing systems are present in a range of environments in which bacteria and archaea cohabitate. These include marine environments, waste treatment plants, and even microbial mats where it is suggested quorum sensing could be responsible to maintain and regulate critical microbial interactions within the mat (Decho et al. 2009, 2010). When communities of micro-organisms co-ordinate responses via quorum sensing, it allows for complex phenotypes to result which in some cases can mirror multicellular-like behavior (Waters and Bassler 2005). Phenotypes commonly regulated by quorum sensing that resemble multicellular

J.C. Charlesworth · C. Beloe · C. Watters · B.P. Burns (✉)
School of Biotechnology and Biomolecular Sciences,
University of New South Wales, Sydney 2052, Australia
e-mail: brendan.burns@unsw.edu.au

J.C. Charlesworth · C. Beloe · C. Watters · B.P. Burns
Australian Centre for Astrobiology, University of New South Wales,
Sydney, NSW 2052, Australia

G. Witzany (ed.), *Biocommunication of Archaea*,
DOI 10.1007/978-3-319-65536-9_8

behavior include: luminescence control (Dunlap 1999), antimicrobial production (McGowan et al. 2005), sporulation (Lombardía et al. 2006), regulation of competency (Li et al. 2002), extracellular enzyme production (Chernin et al. 1998), biofilm production (Hammer and Bassler 2003), and morphological changes (Zhang et al. 2012). While only a small number of archaea have been suggested to possess quorum sensing (Paggi et al. 2003; Zhang et al. 2012), several archaea do possess phenotypes which are commonly regulated by quorum sensing in bacteria (Liao et al. 2016). Such phenotypes that could be putatively regulated by quorum sensing in archaea are discussed further in Sect. 3.

1.1 Molecules Involved in Signaling

There are a wide range of molecules employed by quorum sensing systems as signals, including pyrones (Brachmann et al. 2013), quinolones (Diggle et al. 2006), n-acyl homoserine lactones (AHLs) (Dunlap 1999), diketopiperazines (DKPs) (Holden et al. 1999), furansoyl borate diesters (Vilchez et al. 2007), and short auto-inducing peptides (Sturme et al. 2002). In this review, the focus is on the most widely studied class of quorum sensing signals, the AHLs. Initially described in *Aliivibrio fischerii* (Nealson et al. 1970), AHLs were first thought to be constrained to the proteobacterial group until discovered in cyanobacteria (Sharif et al. 2008), flavobacteria (Romero et al. 2010), and of particular relevance to this review, more recently in a single archaeon (Zhang et al. 2012). There is a great degree of variability within the AHL class with a number of potential chemical modifications giving signals unique specificities, and in some cases changing the stability of the molecule (Yates et al. 2002; Decho et al. 2011).

The most common variation within the AHL class is the chain length of carbons, with some bacteria producing shorter chains, approximately 4–8 carbons in length, to some bacteria producing longer chain AHLs, chain lengths >12 (Morohoshi et al. 2008). Other common chemical modifications include addition of oxo or hydroxyl groups (Frey et al. 2010), or finally a carboxyl group, a modification that appears to be unique to archaea (Zhang et al. 2012). The range of variation allows for specificity to LuxR homologues, the receptor protein which binds to AHLs, and also facilitates a degree of cross-talk or one organism producing signal molecules capable of influencing another organism—potentially across domains (Joint et al. 2007). This concept of cross talk is also the principle behind biosensors, specifically engineered strains that do not produce their own AHLs but respond to AHL or AHL mimics producing a phenotypic response (Steindler and Venturi 2007). Several studies have used biosensors with archaeal strains to infer and elucidate AHL or AHL-like activity (Paggi et al. 2003; Tommonaro et al. 2012; Liao et al. 2016), and this is discussed further in Sect. 2. As alluded to earlier, there are also molecules which can mimic the AHL activity for either agonistic (Holden et al. 1999; Pérez-Montaño et al. 2013) or antagonistic effects (Manefield et al. 1999), and such mimics are present across all three domains. One such class of AHL mimics, DKPs have been detected in archaea (Tommonaro et al. 2012) as well as in bacteria (Holden et al. 1999), and have previously been suggested to be a quorum sensing signal (Holden et al. 1999). Other AHL mimics can act as quorum sensing antagonists and can influence microbial communication in other ways including enzymatic degradation, something also present within the archaea (Bzdrenga et al. 2014) and discussed further in Sect. 4.2.

2 Current Evidence for Quorum Sensing in Archaea

Quorum sensing in archaea is an emerging field of research with only a handful of known quorum sensing active archaea present in the literature. The first study linking quorum sensing and archaea occurred in *Natronoccocus occultus,* which linked an extracellular protease to the production of a signal molecule which could activate an AHL biosensor (Paggi et al. 2003). The first confirmation of quorum sensing molecules found in archaea were diketopiperazines (Tommonaro et al. 2012), that are AHL mimics which can in some cases act as signal molecules (Holden et al. 1999). It was suggested this could potentially be responsible for regulating protease activity in *N. occultus* (Tommonaro et al. 2012). However later work in a methanogenic archaeon, *Methanosaeta harudinacea*, described the presence of carboxylated AHLs, a previously unknown class of the molecule (Zhang et al. 2012). This study suggested that carboxylated AHLs when added to earlier phase cultures could trigger morphological change causing the cell to change to filamentous growth. Though these carboxlyated AHLs induced positive responses in bacterial biosensors, bacterial signalling molecules were unable to induce filamentous growth in *M. harudinacea* (Zhang et al. 2012). This potentially hints at these molecules being utilised in one-way inter-domain cross-talk (Zhang et al. 2012; Montgomery et al. 2013). The work on *M. harundinacea* also described a putative AHL synthase, FilI, a histidine kinase different to any known bacterial AHL synthase. This discovery as well as our current understanding of the genetic basis for quorum sensing in archaea is discussed further in Sect. 5. Recent work in the Antarctic haloarchaea *Halorubrum lacusprofundi* also demonstrated AHL activity with AHL biosensors, potentially correlated to biofilm formation (Liao et al. 2016). Beyond these few examples, the direct evidence for quorum sensing in archaea falls frustratingly short. However, given many phenotypes that are commonly regulated in bacteria are also present in archaea, it is likely these behaviors could be easily explained with quorum sensing mechanics.

3 Phenotypes Putatively Regulated by Quorum Sensing in Archaea

3.1 Biofilms

Biofilms are essentially communities of cells which produce exopolysaccarhides and other stabilizing molecules to protect a community from a variety of insults (Stoodley et al. 2002), including anti-bacterials (Stewart and Costerton 2001), grazing (Matz et al. 2008), and environmental pressures such as desiccation (Vanderlinde et al. 2009). Whilst biofilms in bacteria are commonly regulated by quorum sensing (Hammer and Bassler 2003), comparatively less is known about the regulation of biofilm formation within archaea (Fröls 2013). As alluded to earlier in Sect. 2, there is a potential biofilm phenotype which is associated with AHL activity in *H. lacusprofundi* (Liao et al. 2016), however a number of other archaea and archaea/bacteria associations are potentially quorum sensing regulated. *Methanocaldococcus jannaschii* produces a phenotypic response when cultured with the bacterium *Thermys maritima*, with a raised production of exopolysaccharide (EPS) in high cell densities observed (Nichols et al. 2009). However, QS molecules have yet to be detected in association with this phenomenon. Another potential interaction with *T. maritima* occurs with the archaeon *Pyrococcus furiosus*, with both microorganisms producing an AI-2 signal molecule when co-cultured (Nichols et al. 2009). However there is still some debate as to whether AI-2 signals are indeed quorum sensing related (Sun et al. 2004). Another biofilm formed by two archaea, *P. furiosus* and *Methanopyrus kandleri*, has been described although at the time the authors discounted the role of quorum sensing, due to the lack of known AHL homologues in the genome of these archaea (Schopf et al. 2008). However it is possible these signals may be produced by other enzymes (Zhang et al. 2012), especially given the uncertainty around the genetic basis for quorum sensing in archaea (Sect. 5).

The non-methanogenic euryarcheota SM1 has also been observed having unique co-operation with bacteria, forming pearl shaped biofilms (Rudolph et al. 2004). One such association has SM1 forming the majority of the biofilm pearl alongside a bacterial partner *Thiothrix* sp., and it has been suggested the two partner in a syntrophic interaction relating to the sulfur cycle or possible nutrient exchange (Moissl et al. 2002). Another archaeon, *Ferroplasma acidarmanus* Fer1 was proposed to use QS for biofilm formation (Baker-Austin et al. 2010). Though no signalling molecules as such were identified in the latter study, this does not necessarily dismiss the idea that QS may control biofilm formation in this archaeon as the biosensor used, *C. violaceum,* is not able to detect longer chain AHLs (McClean et al. 1997; Baker-Austin et al. 2010). As *F. acidarmanus* is extremely acidophilic it is possible that they employ modified, longer chain molecules for signalling such as the carboxylated variety which were discovered in *M. harundinacea* (Zhang et al. 2012), or may even employ DKPs which were not detected (Tommonaro et al. 2012).

3.2 Extracellular Enzymes

The production of extracellular enzymes such as proteases or chitinases is another phenotype known to be QS controlled in bacteria (Chernin et al. 1998). The production of Nep, a halosin-like protease in the archaeon *Natrialba magadii*, is currently thought to be produced in response to low nutrients (Paggi et al. 2010). However, up regulation of Nep synthesis may be in response to quorum sensing activity. A potential quorum sensing signal was observed in the supernatant from an *N. magadii* culture, however the exact nature and structure of the molecule was not identified (Paggi et al. 2010). Further, AHL-based signalling has been identified in the late phase of *Natronococcus occultus,* where extracts from the late phase were able to both activate AHL biosensors and induce a protease phenotype (Paggi et al. 2003). A number of other archaea which produce extracellular enzymes could be similarly explained by quorum sensing based regulation (Norberg and von Hofsten 1969; Paggi et al. 2003; Charlesworth and Burns 2015).

3.3 Other Potential Quorum Sensing Phenotypes

Membrane vesicle (MV) formation has been shown to be induced by quorum sensing mechanisms in bacteria such as *Pseudomonas aeruginosa.* The quorum sensing molecule PQS interacts with the lipid A portion of the lipopolysaccharide, and consequently alters membrane curvatures, which results in MV formation (Mashburn-Warren et al. 2008). MV formation also occurs in archaea, such as *Sulfolobus* species (Deatherage and Cookson 2012). These MVs are also utilised by these archaea to export bioactive molecules such as the sulfolobicins (Prangishvili et al. 2000). It is proposed that MV formation is quorum sensing regulated in archaea as observed in bacteria, and thus is a viable candidate for future research concerning quorum sensing in archaea (Deatherage and Cookson 2012). A similar scenario is found in gas vesicle production, a process that has been observed in both bacteria and archaea (Ramsay et al. 2011; Pfeifer 2012), yet has only been shown to be quorum sensing-controlled in bacteria due to a lack of similar studies in archaea. For example in *Serratia* sp. ATCC 39006, gas vesicle production is strictly controlled by AHL molecules resulting in gas vesicle production only at high cell density (Ramsay et al. 2011; Pfeifer 2012), and a similar mechanism could potentially occur in archaea.

4 Interference with Quorum Sensing Systems

4.1 *Quorum Quenching*

Quorum quenching is the term given to the process of blocking signalling via quorum sensing (Dong et al. 2002). All the different quorum sensing systems described to date have the same basic mechanism of signal synthesis, accumulation, and detection (LaSarre and Federle 2013). Quorum sensing inhibitors target biosynthesis pathways, degrade or inactivate the signal, or prevent detection by receptors. In the environment these quorum quenching systems could allow for turnover of signal production in communities allowing for effective change between phenotypes, or simply to offer competitive advantages to some organisms by interference with these systems (Manefield et al. 1999). Archaea could potentially influence bacterial AHL systems utilizing these quorum quenching mechanisms (Abed et al. 2013).

Gaining a better understanding of natural quorum quenching mechanisms allows artificial implementation to deter unwanted microbial growth e.g. biofouling. In many situations, the formation of biofilms is detrimental to human activities, including biofouling of ships and pipelines, food and aquaculture contamination, and increased virulence of human pathogens (Natrah et al. 2011; Abed et al. 2013; Montgomery et al. 2013). Disrupting quorum sensing could prevent these biofilms from forming and thus quorum quenching could aid in preventing biofouling (Luo et al. 2015). With increasing resistance to antibiotics among human pathogens, quorum sensing inhibition offers an alternative treatment to reduce pathogen virulence (Bjarnsholt and Givskov 2007). Since quorum sensing is not vital for survival, resistance is less likely to develop to quorum quenching, making it an ideal therapeutic tool (Tommonaro et al. 2015). Whilst the majority of study on quorum quenching has occurred within bacteria, it is likely many systems could be exploited within archaea given archaea often tend to produce more stable enzymes due to environmental stress (Elleuche et al. 2014; Charlesworth and Burns 2016).

4.2 *Enzymatic Degradation of AHLs*

Exogenous signal molecules in the environment can face many challenges from abiotic conditions, however a significant factor in signal turnover can be from enzymatic sources (Decho et al. 2011). Examples can be found in most quorum sensing systems, such as diffusible signal factors (DSF), *Pseudomonas* quinolone system PQS, and AI-2 (LaSarre and Federle 2013). The two main groups of enzymes that function in AHL hydrolysis are lactonases and acylases (Decho et al. 2011). As well as these, oxidoreductases are able to prevent recognition of the AHL signal by inactivating the molecule, achieved through the oxidation or reduction of the acyl side chain (LaSarre and Federle 2013).

Lactonases cleave the ester bond of lactones, resulting in the disintegration of the lactone ring. Since the lactone ring is present in all AHLs, lactonases have a broad substrate specificity (LaSarre and Federle 2013). These lactonases can allow for the use of AHLs as a carbon source instead of a signalling molecule (Decho et al. 2011). Lactonases have previously been identified in a wide range of bacteria occasionally performing important ecological roles such as enhancing plant resistance to root diseases (Dong et al. 2001). There has also been some degree of interest in utilizing lactonases for biotechnological purposes (Chen et al. 2010). The first phosphotriesterase-like lactonases (PLLs) identified in archaea were *Sso*Pox in *Sulfolobus solfataricus* and *Sac*Pox *Sulfolobus acidocaldarius*, and a PLL was later identified in *Sulfolobus islandicus* (Kallnik et al. 2014). *Sso*Pox was shown to inhibit short chain (C_4-HSL) and long chain (3-oxo-C_{12}-HSL) AHLs produced by *Pseudomonas aeruginosa* PAO1 (Ng et al. 2011). This inhibition of *P. aeruginosa* signals prevented the production of virulence factors suggesting potential therapeautic use (Ng et al. 2011). Later research employed the *S. solfataricus* and *S. acidocaldarius* sequences as search queries for similar genes in all known archaeal genomes (Kallnik et al. 2014). The only match outside *Sulfolobus* was for the VmutPLL in the hyper-thermoacidophilic archaeon *Vulcanisaeta mountnovskia* (Kallnik et al. 2014).

The presence of lactonase enzymes in organisms which live in such high temperature environments is intriguing from an ecological perspective, particularly as to whether quorum sensing signalling systems occur in these thermophilic environments. Typically AHLs are heat sensitive though work has found these molecules can occasionally be protected by chemical modifications such as addition of oxo or hydroxyl groups (Yates et al. 2002). It is unknown what effect exotic modifications such as the carboxylated AHLs found in the methanogenic archaea *M. haurdinacea* (Zhang et al. 2012) would have on thermostability of AHLs. In any case if signalling molecules are present in these thermophilic environments, lactonases could quench these systems (Decho et al. 2011; Kallnik et al. 2014). Due to the adaptations of these enzymes to high temperature environments it has been suggested such proteins could be utilized in biotechnology not only for quorum quenching but other uses such as decontamination of organophosphates (Kallnik et al. 2014).

The other common class of AHL degrading enzymes, acylases, hydrolyse the amide bond between the lactone ring and acyl chain, generating a homoserine lactone and fatty acid (Kalia et al. 2011). The substrate specificity of acylases is dependent on chain length and chemical modifications of the AHL (LaSarre and Federle 2013). Acylases are less prevalent than lactonases and only bacterially derived examples have been isolated so far, although some have suggested the potential for archaeal derived acylases (Merone et al. 2008). Examples in bacteria are found in *Variovorax*, *Ralstonia*, *Streptomyces*, and *Pseudomonas* and most belong to the Ntn (*N*-terminal nucleophile) hydrolase superfamily (Tang and Zhang 2014). *Bacillus pumilus* S8-07 contains a heat-stable acylase which retains activity after incubation at 70 °C (Nithya et al. 2010). This finding of acylases in extremophilic environments opens the possibility of these enzyme classes to be present in thermophilic archaea that could putatively have a role in biotic signal degradation (whether archaeal or bacterial).

4.3 Competitive Inhibition of AHLs by AHL Mimics

Blocking the active site on receptor proteins can interrupt detection of signal molecules in a process known as competitive inhibition (O'Neill et al. 2009). Structurally similar molecules, or analogues, can be used to prevent receptor activation and subsequent activity or gene expression (Decho et al. 2011). Some studies have even suggested quorum sensing inhibitors could help address the rise of antibiotic resistances (Bjarnsholt and Givskov 2007).

DKPs are able to activate and/or antagonise AHL-based quorum sensing systems, and it has been suggested this is via competitive inhibition (de Carvalho and Abraham 2012). DKPs have also been shown to activate many quorum sensing systems, including that of the *Escherichia coli* pSB401 biosensor (Holden et al. 1999). However, a higher concentration of DKP was required compared to the natural activator (Holden et al. 1999). The first archaea shown to produce a DKP was the haloarchaeon *Haloterrigena hispanica* (Tommonaro et al. 2012). This DKP also activated AHL biosensors (Tommonaro et al. 2012), showing it is capable of binding to the receptor and therefore may also be able to block activity, however this is yet to be demonstrated. DKPs in bacteria are typically synthesized using one of two pathways, either non-ribosomal peptide synthesis (NRPS) a large gene cluster, or though cyclic dipeptide synthase (Charlesworth and Burns 2015). To date neither system has been uncovered in archaea and therefore the exact genetic basis for DKP signal production remains unknown.

The field of quorum quenching is advancing fast however there are still a number of gaps in our knowledge. There are examples of antagonist mimicry by unknown structures (Decho et al. 2011), quorum quenching organisms using undetermined mechanisms (Abed et al. 2013), and many other uncertainties. A greater focus on archaea and unknown environmental samples could yield promising new forms of quorum quenching including those that could be stable under extreme conditions (Charlesworth and Burns 2015).

5 Genetic Basis for Signaling in Archaea?

AHL synthase is the enzyme responsible for combining the two main constituents of the AHL molecule: the homoserine lactone (HSL) and the acyl side-chain. The HSL moiety is acquired from *S*-adenosylmethionine (SAM) (Jiang et al. 1998). In bacteria, the acyl chain is derived from the fatty acid cycle. The acyl group is often carried by acyl-ACP (acyl carrier protein), which is a cofactor involved in shuttling intermediates between enzymes in fatty acid (FA) biosynthesis. These ACPs are commonly used in a wide range of secondary metabolite synthesis in bacteria such as NRPS clusters (Beld et al. 2014). ACPs are common in bacteria due to the integration of FA in bacterial membrane phospholipids (Lombard et al. 2012).

However unlike FA in bacteria and eukaryotes, archaea use isoprenoids in membrane phospholipids, and possess low levels of FAs (Lombard et al. 2012). Bioinformatic analyses of archaeal genomes to date revealed that archaea do not possess genes for ACP or ACP-processing enzymes. However archaea do possess a near complete FA synthesis pathway, hence it has been proposed they employ an appropriately charged coenzyme A (CoA) in an ACP-independent FA synthesis pathway (Lombard et al. 2012). As a result of this apparent lack of ACP, archaeal AHL synthase would require an alternative source of acyl chains and also uses CoA, an intermediate in fatty acid degradation (Jiang et al. 1998; Lombard et al. 2012).

AHL synthases are members of the acetyl-transferase protein family, however there is very little conservation amongst AHL synthases with only eight residues being found to be conserved, with many of these thought to act on ACP (Watson et al. 2002). Whilst acyl-CoA could function as an acyl donor, it is also possible other analogous proteins such as sterol carrier protein 2 could perform a similar role in archaea (Frolov et al. 1996; Edqvist and Blomqvist 2006). As archaea possess other proteins of the fatty acid pathway which can be involved in AHL synthesis (Hoang et al. 2002), it is possible the fatty acid cycle in archaea feeds the synthesis of archaeal AHLs. Given the lack of conservation among AHL synthases (Watson et al. 2002) and the need for different acyl donors in archaea, it is unlikely a strictly bioinformatic approach will delineate archaeal AHL synthases.

This example of ACP-independent AHL synthesis highlights the biochemical and genetic differences between bacteria and archaea. Work performed on the archaeon *M. harundinacea* suggested a histidine kinase protein termed FilI could be responsible for AHL synthesis, due to similarities between subdomains of the protein and bacterial analogues (Zhang et al. 2012). However as no ACP or ACP analogue have been detected within archaea, further work is needed to confirm this finding.

More recently, research has started to use archaeal genes, such as those encoding the AHL degrading VmutPLL enzyme, when searching other archaeal genomes (Kallnik et al. 2014), however the use of bacterial QS genes as search queries is still commonly employed (Zhang et al. 2012). This approach has been successful in identifying lactonase and acylase genes in Eurarcheota and Crenarcheota (Kalia et al. 2011). However, many archaea that exhibit signaling-dependent behaviours do not contain these gene sequences (Montgomery et al. 2013), suggesting archaea have evolved unique signaling genes (Zhang et al. 2012). Thus, although methodologies based on bacterial genes can be effective, it is not comprehensive and could prove to be a limitation in novel archaeal QS gene detection. Other potential techniques that could be used to detect synthesis and response genes include proteomic work (Sharif et al. 2008), or the use of plasmid based biosensors and clone library approaches (Williamson and Borlee 2005).

6 Conclusions

Whilst well studied in bacteria, quorum sensing research in archaea has only recently begun to be conducted in earnest. The exact nature of signaling molecules within archaea is still to be determined, however these could potentially include exotic AHLs and DKPs, or potentially novel AHL mimics. Future work to determine the exact nature of quorum sensing molecules in archaea will be best undertaken employing analytical chemical techniques such as mass spectrometry and nuclear magnetic resonance. Quorum sensing studies in archaea could reveal a range of control and regulation mechanisms for diverse phenotypes such as biofilm formation which is at present poorly understood within the archaea (Fröls 2013). The exact nature of the genetic and biochemical basis for quorum sensing in archaea is still unclear, however the unique fatty acid pathways found in archaea suggest similarly unique adaptations to enzymes may be present which act on acyl groups (Lombard et al. 2012). It is likely global gene (Guan et al. 2007) and proteomic approaches (Sharif et al. 2008) will be needed to determine both phenotypic action and biochemical basis of quorum sensing in archaea. Although studies on archaeal quorum sensing are limited, it is clear from the few studies that have examined this phenomenon and the emerging directions in the field, that archaea have well and truly 'joined the conversation.'

References

Abed RMM, Dobretsov S, Al-Fori M, Gunasekera SP, Sudesh K, Paul VJ (2013) Quorum-sensing inhibitory compounds from extremophilic microorganisms isolated from a hypersaline cyanobacterial mat. J Ind Microbiol Biotechnol 40:759–772

Baker-Austin C, Potrykus J, Wexler M, Bond PL, Dopson M (2010) Biofilm development in the extremely acidophilic archaeon "*Ferroplasma acidarmanus*" Fer1. Extremophiles 14:485–491

Beld J, Sonnenschein EC, Vickery CR, Noel JP, Burkart MD (2014) The phosphopantetheinyl transferases: catalysis of a post-translational modification crucial for life. Nat Prod Rep 31:61–108

Bjarnsholt T, Givskov M (2007) Quorum-sensing blockade as a strategy for enhancing host defences against bacterial pathogens. Philos Trans R Soc Lond B Biol Sci 362:1213–1222

Brachmann AO, Brameyer S, Kresovic D, Hitkova I, Kopp Y, Manske C et al (2013) Pyrones as bacterial signaling molecules. Nat Chem Biol 1–9

Bzdrenga J, Hiblot J, Gotthard G, Champion C, Elias M, Chabriere E (2014) SacPox from the thermoacidophilic crenarchaeon *Sulfolobus acidocaldarius* is a proficient lactonase. BMC Res Notes 7:333

de Carvalho MP, Abraham W-R (2012) Antimicrobial and biofilm inhibiting diketopiperazines. Curr Med Chem 19:3564–3577

Charlesworth J, Burns BP (2016) An overview of extremophilic adaptations and biotechnological applications in diverse environments. AIMS Microbiol 2:1–11

Charlesworth JC, Burns BP (2015) Untapped resources: biotechnological potential of peptides and secondary metabolites in archaea. Archaea

Chen R, Zhou Z, Cao Y, Bai Y, Yao B (2010) High yield expression of an AHL-lactonase from *Bacillus* sp. B546 in *Pichia pastoris* and its application to reduce *Aeromonas hydrophila* mortality in aquaculture. Microb Cell Fact 9:39

Chernin LS, Winson MK, Thompson JM, Haran S, Bycroft BW, Chet I et al (1998) Chitinolytic activity in *Chromobacterium violaceum*: substrate analysis and regulation by quorum sensing. J Bacteriol 180:4435–4441

Deatherage BL, Cookson BT (2012) Membrane vesicle release in bacteria, eukaryotes, and archaea: a conserved yet underappreciated aspect of microbial life. Infect Immun 80:1948–1957

Decho AW, Frey RL, Ferry JL (2011) Chemical challenges to bacterial AHL signaling in the environment. Chem Rev 111:86–99

Decho AW, Norman RS, Visscher PT (2010) Quorum sensing in natural environments: emerging views from microbial mats. Trends Microbiol 18:73–80

Decho AW, Visscher PT, Ferry J, Kawaguchi T, He L, Przekop KM et al (2009) Autoinducers extracted from microbial mats reveal a surprising diversity of N-acylhomoserine lactones (AHLs) and abundance changes that may relate to diel pH. Environ Microbiol 11:409–420

Diggle SP, Cornelis P, Williams P, Cámara M (2006) 4-quinolone signalling in *Pseudomonas aeruginosa*: old molecules, new perspectives. Int J Med Microbiol 296:83–91

Dong Y, Gusti A, Zhang Q (2002) Identification of quorum-quenching N-acyl homoserine lactonases from Bacillus species. Appl Environ Microbiol 68:1754–1759

Dong YH, Wang LH, Xu JL, Zhang HB, Zhang XF, Zhang LH (2001) Quenching quorum-sensing-dependent bacterial infection by an N-acyl homoserine lactonase. Nature 411:813–817

Dunlap PV (1999) Quorum regulation of luminescence in *Vibrio fischeri*. J Mol Microbiol Biotechnol 1:5–12

Edqvist J, Blomqvist K (2006) Fusion and fission, the evolution of sterol carrier protein-2. J Mol Evol 62:292–306

Elleuche S, Schröder C, Sahm K, Antranikian G (2014) Extremozymes-biocatalysts with unique properties from extremophilic microorganisms. Curr Opin Biotechnol 29:116–123

Frey RL, He L, Cui Y, Decho AW, Kawaguchi T, Ferguson PL, Ferry JL (2010) Reaction of N-acylhomoserine lactones with hydroxyl radicals: rates, products, and effects on signaling activity. Environ Sci Technol 44:7465–7469

Frolov A, Cho TH, Billheimer JT, Schroeder F (1996) Sterol carrier protein-2, a new fatty acyl coenzyme A-binding protein. J Biol Chem 271:31878–31884

Fröls S (2013) Archaeal biofilms: widespread and complex. Biochem Soc Trans 41:393–398

Guan C, Ju J, Borlee BR, Williamson LL, Shen B, Raffa KF, Handelsman J (2007) Signal mimics derived from a metagenomic analysis of the gypsy moth gut microbiota. Appl Environ Microbiol 73:3669–3676

Hammer BK, Bassler BL (2003) Quorum sensing controls biofilm formation in *Vibrio cholerae*. Mol Microbiol 50:101–104

Hoang TT, Sullivan SA, Cusick JK, Schweizer HP (2002) Beta-ketoacyl acyl carrier protein reductase (FabG) activity of the fatty acid biosynthetic pathway is a determining factor of 3-oxo-homoserine lactone acyl chain lengths. Microbiology 148:3849–3855

Holden MT, Ram Chhabra S, de Nys R, Stead P, Bainton NJ, Hill PJ et al (1999) Quorum-sensing cross talk: isolation and chemical characterization of cyclic dipeptides from *Pseudomonas aeruginosa* and other gram-negative bacteria. Mol Microbiol 33:1254–1265

Jiang Y, Camara M, Chhabra SR, Hardie R, Bycroft BW (1998) In vitro biosynthesis of the *Pseudomonas aeruginosa* quorum-sensing signal molecule N-butanoyl-L-homoserine lactone. Mol Microbiol 28:193–203

Joint I, Tait K, Wheeler G (2007) Cross-kingdom signalling: exploitation of bacterial quorum sensing molecules by the green seaweed Ulva. Philos Trans R Soc Lond B Biol Sci 362: 1223–1233

Kalia VC, Raju SC, Purohit HJ (2011) Genomic analysis reveals versatile organisms for quorum quenching enzymes: acyl-homoserine lactone-acylase and-lactonase. Open Microbiol J 5:1–13

Kallnik V, Bunescu A, Sayer C, Bräsen C, Wohlgemuth R, Littlechild J, Siebers B (2014) Characterization of a phosphotriesterase-like lactonase from the hyperthermoacidophilic crenarchaeon *Vulcanisaeta moutnovskia*. J Biotechnol 190:11–17

LaSarre B, Federle MJ (2013) Exploiting quorum sensing to confuse bacterial pathogens. Microbiol Mol Biol Rev 77:73–111

Li Y, Tang N, Aspiras MB, Lau PCY, Lee JH, Ellen RP, Cvitkovitch DG (2002) A quorum-sensing signaling system essential for genetic competence in *Streptococcus mutans* is involved in biofilm formation. J Bacteriol 184:2699–2708

Liao Y, Williams TJ, Ye J, Charlesworth J, Burns BP, Poljak A et al (2016) Morphological and proteomic analysis of biofilms from the Antarctic archaeon, *Halorubrum lacusprofundi*. Sci Rep 6:37454

Llamas I, Quesada E, Martínez-Cánovas MJ, Gronquist M, Eberhard A, González JE (2005) Quorum sensing in halophilic bacteria: detection of N-acyl-homoserine lactones in the exopolysaccharide-producing species of Halomonas. Extremophiles 9:333–341

Lombard J, López-García P, Moreira D (2012) An ACP-independent fatty acid synthesis pathway in archaea: implications for the origin of phospholipids. Mol Biol Evol 29:3261–3265

Lombardía E, Rovetto AJ, Arabolaza AL, Grau RR (2006) A LuxS-dependent cell-to-cell language regulates social behavior and development in *Bacillus subtilis*. J Bacteriol 188: 4442–4452

Luo J, Zhang J, Tan X, McDougald D, Zhuang G, Fane AG et al (2015) Characterization of the archaeal community fouling a membrane bioreactor. J Environ Sci (China) 29:115–123

Manefield M, de Nys R, Kumar N, Read R, Givskov M, Steinberg P, Kjelleberg S (1999) Evidence that halogenated furanones from *Delisea pulchra* inhibit acylated homoserine lactone (AHL)-mediated gene expression by displacing the AHL signal from its receptor protein. Microbiology 145 (Pt 2):283–291

Mashburn-Warren L, Howe J, Garidel P, Richter W, Steiniger F, Roessle M et al (2008) Interaction of quorum signals with outer membrane lipids: insights into prokaryotic membrane vesicle formation. Mol Microbiol 69:491–502

Matz C, Webb JS, Schupp PJ, Phang SY, Penesyan A, Egan S et al (2008) Marine biofilm bacteria evade eukaryotic predation by targeted chemical defense. PLoS ONE 3:1–7

McClean K, Winson M, Fish L, Taylor A (1997) Quorum sensing and *Chromobacterium violaceum*: exploitation of violacein production and inhibition for the detection of N-acylhomoserine lactones. Mol Microbiol 143:3703–3711

McGowan SJ, Barnard AML, Bosgelmez G, Sebaihia M, Simpson NJL, Thomson NR et al (2005) Carbapenem antibiotic biosynthesis in *Erwinia carotovora* is regulated by physiological and genetic factors modulating the quorum sensing-dependent control pathway. Mol Microbiol 55:526–545

Merone L, Mandrich L, Rossi M, Manco G (2008) Enzymes with phosphotriesterase and lactonase activities in archaea. Curr Chem Biol 2:237–248

Moissl C, Rudolph C, Huber R (2002) Natural communities of novel archaea and bacteria with a string-of-pearls-like morphology : molecular analysis of the bacterial partners natural communities of novel archaea and bacteria with a string-of-pearls-like morphology : molecular analysis of the bacterial partners. Appl Environ 68:933–937

Montgomery K, Charlesworth J, LeBard R, Visscher P, Burns B (2013) Quorum sensing in extreme environments. Life 3:131–148

Morohoshi T, Kato M, Fukamachi K, Kato N, Ikeda T (2008) N-acylhomoserine lactone regulates violacein production in *Chromobacterium violaceum* type strain ATCC 12472. FEMS Microbiol Lett 279:124–130

Natrah FMI, Kenmegne MM, Wiyoto W, Sorgeloos P, Bossier P, Defoirdt T (2011) Effects of micro-algae commonly used in aquaculture on acyl-homoserine lactone quorum sensing. Aquaculture 317:53–57

Nealson KH, Platt T, Hastings W, Hastings JW (1970) Cellular control of the synthesis and activity of the bacterial luminescent system. J Bacteriol 104:313–322

Ng FSW, Wright DM, Seah SYK (2011) Characterization of a phosphotriesterase-like lactonase from *Sulfolobus solfataricus* and its immobilization for disruption of quorum sensing. Appl Environ Microbiol 77:1181–1186

Nichols JD, Johnson MR, Chou CJ, Kelly RM (2009) Temperature, not LuxS, mediates AI-2 formation in hydrothermal habitats. FEMS Microbiol Ecol 68:173–181

Nithya C, Aravindraja C, Pandian SK (2010) *Bacillus pumilus* of Palk Bay origin inhibits quorum-sensing-mediated virulence factors in gram-negative bacteria. Res Microbiol 161: 293–304

Norberg P, von Hofsten B (1969) Proteolytic enzymes from extremely halophilic bacteria. J Gen Microbiol 55:251–256

O'Neill J, Lin Q, Geske G, Blackwell H (2009) New and unexpected insights into the modulation of LuxR-type quorum sensing by cyclic dipeptides. ACS Chem Biol 4:1051–1059

Paggi RA, Martone CB, Fuqua C, Castro RE (2003) Detection of quorum sensing signals in the haloalkaliphilic archaeon *Natronococcus occultus*. FEMS Microbiol Lett 221:49–52

Paggi RA, Madrid EA, D'Alessandro CP, Cerletti M, De Castro RE (2010) Growth phase-dependent biosynthesis of Nep, a halolysin-like protease secreted by the alkaliphilic haloarchaeon *Natrialba magadii*. Lett Appl Microbiol 51:36–41

Pérez-Montaño F, Jiménez-Guerrero I, Contreras Sánchez-Matamoros R, López-Baena FJ, Ollero FJ, Rodríguez-Carvajal MA et al (2013) Rice and bean AHL-mimic quorum-sensing signals specifically interfere with the capacity to form biofilms by plant-associated bacteria. Res Microbiol 164:749–760

Pfeifer F (2012) Distribution, formation and regulation of gas vesicles. Nat Rev Microbiol 10: 705–715

Prangishvili D, Holz I, Stieger E, Nickell S, Kristjansson JK, Zillig W (2000) Sulfolobicins, specific proteinaceous toxins produced by strains of the extremely thermophilic archaeal genus Sulfolobus. J Bacteriol 182:2985–2988

Ramsay JP, Williamson NR, Spring DR, Salmond GPC (2011) A quorum-sensing molecule acts as a morphogen controlling gas vesicle organelle biogenesis and adaptive flotation in an enterobacterium. Proc Natl Acad Sci 108:14932–14937

Redfield R (2002) Is quorum sensing a side effect of diffusion sensing? Trends Microbiol 10: 365–370

Romero M, Avendaño-Herrera R, Magariños B, Cámara M, Otero A (2010) Acylhomoserine lactone production and degradation by the fish pathogen *Tenacibaculum maritimum*, a member of the Cytophaga-Flavobacterium-Bacteroides (CFB) group. FEMS Microbiol Lett 304:131–139

Rudolph C, Moissl C, Henneberger R, Huber R (2004) Ecology and microbial structures of archaeal/bacterial strings-of-pearls communities and archaeal relatives thriving in cold sulfidic springs. FEMS Microbiol Ecol 50:1–11

Schopf S, Wanner G, Rachel R, Wirth R (2008) An archaeal bi-species biofilm formed by *Pyrococcus furiosus* and *Methanopyrus kandleri*. Arch Microbiol 190:371–377

Sharif DI, Gallon J, Smith CJ, Dudley E (2008) Quorum sensing in cyanobacteria: N-octanoyl-homoserine lactone release and response, by the epilithic colonial cyanobacterium Gloeothece PCC6909. ISME J 2:1171–1182

Steindler L, Venturi V (2007) Detection of quorum-sensing N-acyl homoserine lactone signal molecules by bacterial biosensors. FEMS Microbiol Lett 266:1–9

Stewart PS, Costerton JW (2001) Antibiotic resistance of bacteria in biofilms. Lancet 358:135–138

Stoodley P, Sauer K, Davies DG, Costerton JW (2002) Biofilms as complex differentiated communities. Ann Rev Microbiol 56:187–209

Sturme MHJ, Kleerebezem M, Nakayama J, Akkermans ADL, Vaugha EE, de Vos WM (2002) Cell to cell communication by autoinducing peptides in gram-positive bacteria. Antonie Van Leeuwenhoek 81:233–243

Sun J, Daniel R, Wagner-Döbler I, Zeng A-P (2004) Is autoinducer-2 a universal signal for interspecies communication: a comparative genomic and phylogenetic analysis of the synthesis and signal transduction pathways. BMC Evol Biol 4:36–46

Tang K, Zhang XH (2014) Quorum quenching agents: resources for antivirulence therapy. Mar. Drugs 12:3245–3282

Tommonaro G, Abbamondi GR, Iodice C, De Rosa S et al (2012) Diketopiperazines produced by the halophilic archaeon, *Haloterrigena hispanica*, activate AHL bioreporters. Environ Microbiol 63:490–495

Tommonaro G, Roberto AG, Ebru TO, Barbara N (2015) Investigating the quorum sensing system in halophilic bacteria. In: Halophiles: biodiversity and sustainable exploitation. Springer, pp 189–207

Vanderlinde EM, Muszynski A, Harrison JJ, Koval SF, Foreman DL, Ceri H et al (2009) *Rhizobium leguminosarum* biovar viciae 3841, deficient in 27-hydroxyoctacosanoate-modified lipopolysaccharide, is impaired in desiccation tolerance, biofilm formation and motility. Microbiology 155:3055–3069

Vilchez R, Lemme A, Thiel V, Schulz S, Sztajer H, Wagner-Döbler I (2007) Analysing traces of autoinducer-2 requires standardization of the *Vibrio harveyi* bioassay. Anal Bioanal Chem 387:489–496

Waters CM, Bassler BL (2005) Quorum sensing: cell-to-cell communication in bacteria. Ann Rev Cell Dev Biol 21:319–346

Watson WT, Minogue TD, Bodman SB, Von Val DL, Churchill MEA, Von Bodman SB (2002) Structural basis and specificity of acyl-homoserine lactone signal production in bacterial quorum sensing. Mol Cell 9:685–694

Williamson L, Borlee B (2005) Intracellular screen to identify metagenomic clones that induce or inhibit a quorum-sensing biosensor. Appl Environ Microbiol 71:6335–6344

Yates EA, Philipp B, Buckley C, Atkinson S, Chhabra SR, Sockett RE et al (2002) N-acylhomoserine lactones undergo lactonolysis in a pH-, temperature-, and acyl chain length-dependent manner during growth of *Yersinia pseudotuberculosis* and *Pseudomonas aeruginosa*. Infect Immun 70:5635–5646

Zhang G, Zhang F, Ding G, Li J, Guo X, Zhu J et al (2012) Acyl homoserine lactone-based quorum sensing in a methanogenic archaeon. ISME J 1–9

Biofilm Lifestyle of Thermophile and Acidophile Archaea

Alvaro Orell, Simone Schopf, Lennart Randau and Mario Vera

1 Introduction

Biofilms represent a common and widespread microbial lifestyle on earth. This sessile mode of growth is enabled by communities of microorganisms that are embedded in a self-produced matrix of extracellular polymeric substances (EPS). Biofilms can be found self-associated either as "floating mats" in air-water interfaces or as surface-associated communities of microorganisms. Microbial diversity within a biofilm can vary from few to hundreds of different species.

The intrinsic advantages of the biofilm lifestyle are largely attributed to the presence of EPS. EPS consist of macromolecules such as polysaccharides, proteins, nucleic acids, and lipophilic compounds. Secretion of EPS provides several goods to the community: (i) a protective barrier against environmental changing conditions, (ii) a highly hydrated environment (as EPS can retain up to 90% water), (iii) a nutrient source and an extracellular redox reactive space, (iv) co-metabolism and

A. Orell (✉) · L. Randau
Max Planck Institute for Terrestrial, MicrobiologyKarl-von-Frischs Strasse
10, 35043 Marburg, Germany
e-mail: alvaro.orell@mpi-marburg.mpg.de

L. Randau
e-mail: lennart.randau@mpi-marburg.mpg.de

S. Schopf
Institute of Biosciences, Environmental Microbiology, Technische,
Universität Bergakademie Freiberg, Leipziger Str 29, 09599 Freiberg, Germany
e-mail: simone.schopf@ioez.tu-freiberg.de

M. Vera (✉)
Institute for Biological and Medical Engineering, Schools of Engineering, Medicine and Biological Sciences, Department of Hydraulic and Environmental Engineering, Pontificia Universidad Católica de Chile, Av. Vicuña Mackenna, 4860 Santiago, Chile
e-mail: mariovera@uc.cl

G. Witzany (ed.), *Biocommunication of Archaea*,
DOI 10.1007/978-3-319-65536-9_9

synergistic processes by providing a physical matrix where cell-cell interactions can easily occur, (v) a prerequisite for the exchange of genetic material (Flemming and Wingender 2010). In this chapter we review recent findings on biofilm formation in extremophilic archaea with a focus on acidophilic mineral dissolving species.

2 Biofilms Formed by Hyperthermophilic Anaerobic Archaea at Neutral PH

Pyrococcus furiosus DSM 3638^T was isolated from geothermally heated marine sediments at the beach of Porto di Levante, Vulcano Island (Italy), and exhibits an optimal growth temperature of 100 °C at neutral pH (Fiala and Stetter 1986). From one pole of the coccoid cells, tufts of up to 50 monoplolar polytrichous archaella emanate with a single diameter of 10 nm (Näther-Schindler et al. 2014). In this hyperthermophilic anaerobic archaeal model organism it has been shown that archaella mediate several functions, including swimming motility, formation of cell-cell connections, and surface attachment (Näther et al. 2006). According to the published genome database of *P. furiosus* (Robb et al. 2001), the archaellum (formerly archaeal flagellum) only consists of the two flagellins *fla*B1 and *fla*B2. However, re-sequencing of the *fla*-operon revealed that a 771 bp segment had been missed (Näther-Schindler et al. 2014). This sequence contains a novel third gene encoding the major flagellin, which for consecutive numbering, was named *fla*B0.

It has been shown that during stationary phase growth up to 5% of *P. furiosus* cells are interconnected by a bundle of archaella forming cable like structures with a total diameter of 100 to 200 nm. The physiological function of this cell-cell connection is still unclear, but it has been discussed as a prerequisite for horizontal gene transfer (Näther et al. 2006). *P. furiosus* attaches to surfaces via its archaella, which are essential for the establishment of a permanent adhesion. A number of potential substrate surfaces and materials have been tested including sand, metals, glass, and plastics due to biotechnological application of the adhesive archaellum as molecular glue are envisioned. It was demonstrated that adhesion of *P. furiosus* is surface-dependent displaying two patterns: microbial growth occurred either as very dense microcolonies (e.g. on gold, nickel, copper, and plexiglass) or dispersed as single cells (e.g. on aluminium, enamel, and wood). Notably, only very little attachment could be attested on glass. Interestingly, it was determined that cells detached from the surface when treated with antibodies against whole archaella, demonstrating that surface attachment is mediated by these cell structures (Näther-Schindler et al. 2014). Surface attachment and formation of cell-cell connections via flagella has later been demonstrated to occur also for the hyperthermophilic methanogen *Methanocaldococcus villosus* (Bellack et al. 2011).

In order to determine if *P. furiosus* was able to interact with other archaeal species, *Methanopyrus kandleri* was co-cultured. Both archaea thrive in similar marine biotopes characterised by hot (around 100 °C) and anaerobic conditions

(Kurr et al. 1991). Furthermore, visual discrimination between both species was easily possible (cocci vs. rods up to 10 μm long). A stable co-culture could be established, based on an interspecies hydrogen transfer (Schopf et al. 2008). *P. furiosus* produced H_2 by fermentation and *M. kandleri* consumed it as sole energy donor for the reduction of CO_2. It was questioned if *M. kandleri* was able to pre-colonize glass cover slides to trigger subsequent attachment of *P. furiosus*. This was followed by epifluorescence microscopy with DAPI staining. After 4 h of incubation only few rods (*M. kandleri*) were attached to a glass cover slip. After 8 h of incubation, the cocci (*P. furiosus*) became reliably visible, assumingly "sitting" on the rod shaped *M. kandleri* cells. Thereafter the development of the biofilm was rapid. After 3 h the colonies had already reached a diameter of 10–20 μm, whereas after another 3 h the diameter was up to 50 μm with a "fried egg" appearance. The bottom layer of the biofilm grown on glass consisted of *M. kandleri* as pre-colonizer and dense top layer of *P. furiosus* cells, resulting in a structured archaeal bi-species biofilm. Within the biofilm, three types of interactions were observed in electron micrographs: (i) *P. furiosus* as well as *M. kandleri* cells attached themselves to the surface with the help of cell appendages, (ii) intra- and interspecies connections via archaella occurred, and (iii) direct cell-cell contacts between the two species emerged. Additionally, it was determined that *P. furiosus* interacted with other methanogens (*Methanocaldococcus villosus*, *Methanocaldococcus infernus*), as the growth of both archaea in co-culture was stimulated with respect to cell counts in pure culture (Weiner et al. 2012). On the contrary, in a co-culture between *P. furiosus* and *Methanotorris igneus*, an inhibitory effect on the growth of the latter archaeon was reported. The biofilm forming lifestyle of the *P. furiosus*/*M. villosus* co-culture was analyzed on a variety of surfaces. Both strains in co-culture colonized all tested surfaces (e.g. aluminium, cellophane, PVC, and mica). On glass, *M. villosus* seemed to first colonize the surface, following adhesion by *P. furiosus*, which as aforementioned did not colonize this surface in pure culture. Biofilm cells were predominantly evenly distributed over the surface rather than forming large aggregates, which was strikingly different from the structured *P. furiosus*/*M. kandleri* biofilms. As commonality, cells interacted heavily via their archaella, interconnecting and attaching themselves to the surface.

3 Biofilm Studies in Acidophilic Archaea

3.1 Natural and Man-Made Acidophilic Archaeal Biofilms

Numerous species of acidophilic archaea thrive in natural or man-made environments where sulfur oxidation and/or weathering of metal sulfides such as pyrite (FeS_2) or chalcopyrite ($CuFeS_2$) take place, generating acid mine/rock drainage sites (AMD/ARD) (i.e. biomining operations). This is an indirect consequence of the ability of microorganisms to oxidize iron (II)-ions and/or reduced inorganic sulfur compounds (RISC), producing iron(III)-ions and protons, which are the attacking agents for the oxidation of metal sulfides. Archaea employed in bioleaching/biooxidation processes mostly belong to the *Sulfolobales*, a group of thermophilic, sulfur- and/or iron(II)-oxidizers including genera such as *Sulfolobus*, *Acidianus*, *Metallosphaera* and *Sulfurisphaera* (Norris et al. 2000). Also, within the *Thermoplasmales* three mesophilic iron(II)-oxidizing species, *Ferroplasma acidiphilum* (Golyshina et al. 2000), *Ferroplasma acidarmanus* (Dopson et al. 2004; Edwards et al. 2000) and *Ferroplasma thermophilum* (Zhou et al. 2008) are known.

AMD sites are probably the most extensively studied acidic ecosystems to date. Even though these environments usually display extreme acidity and high concentrations of toxic metals, a diverse range of autotrophic and heterotrophic archaea and bacteria can thrive within AMDs and often form mixed-community biofilms. Occasionally, these biofilm communities exhibit relatively low microbial diversity as frequently no more than five genera can be distinguished. AMDs with milder physicochemical conditions generate a much greater biodiversity of biofilms/streamers. For instance, AMDs have been characterized in the Iron Mountain range in California. The investigated biofilms revealed that *Leptospirillum* group II corresponds to the predominant bacterial taxon, encompassing around 75% of the microbial fraction within the biofilms, and archaea (mostly *Ferroplasma*) account for ~10%, (Baker and Banfield 2003; Comolli et al. 2009). Archaeal abundance increases with biofilm age and thickness, dominating the top layer in stratified, mature biofilms (Wilmes et al. 2009). Archaea are also found within submerged suboxic biofilms in these systems, acting as relevant players in carbon and nutrient cycling within these sessile communities (Justice et al. 2014). Besides *Ferroplasma*, a deep branching clade of euryarchaea termed ARMAN (Archaeal Richmond Mine Acidophilic Nanoorganisms) has also been detected in the Iron Mountain biofilms (Baker and Banfield 2003; Comolli et al. 2009). Furthermore, other uncultivated archaea, including methanogens, have been detected in other subterranean AMD-impacted environments (Johnson 2012).

Ferroplasma contributes to the generation of AMD and is involved in the global sulfur and iron cycling (Golyshina et al. 2000; Golyshina and Timmis 2005). Interestingly, two distinct biofilm morphologies were described for the moderate thermophilic archaeon *F. acidarmanus* Fer1. A multilayer biofilm was developed on glass and pyrite surfaces, and up to 5 mm-long filaments were found on sintered glass spargers in gas lift bioreactors (Baker-Austin et al. 2010). Proteomic profiling of these biofilms revealed the up-regulation of proteins involved in the adaptation to anoxic conditions indicating the existence of anaerobic zones within the *F. acidarmanus* biofilm structure (Baker-Austin et al. 2010). Cells of *Ferroplasma* usually promote growth and metabolic activity of other leaching bacteria by detoxifying leached waters and, thereby maintaining robust bioleaching microbial communities (Golyshina and Timmis 2005; Zhang et al. 2015a; Zhou et al. 2008)

It has been proposed that microbial attachment to ores and subsequent biofilm formation might increase initial leaching activities. This is mainly due to a distinctive microenvironment formed between the microorganisms and the surface, where metal dissolution takes place. In general, the majority of leaching microorganisms can grow attached to metal sulfide or sulfur surfaces. The sites of attachment of archaea on metal sulfides and how these specific location(s) are sensed by archaea remain as open questions. Several lines of evidence from leaching bacteria suggest that this does not occur randomly (Rohwerder et al. 2003; Sanhueza et al. 1999). Interactions of three axenic cultures of the thermophiles *Acidianus brierleyi*, *M. sedula* and *S. metallicus* with pyrite were first documented using SEM and TEM. Progression of deposited structures, ranging from sub-micron precipitates and disc-shaped structures on the pyrite surface, was observed (Mikkelsen et al. 2007). Interestingly, cell attachment of *M. sedula* and *S. metallicus* did not exhibit any preferential orientation. However, pyrite oxidation and pit etching were influenced by surface symmetries (Etzel et al. 2008). Cells of *M. sedula* were found to be moving along the metal ore by Epifluorescence Microscopy (EFM), suggesting that cell appendages are involved in cell attachment to the ore (Huber et al. 1989).

Some geothermal springs are moderately or extremely acidic, hosting thermoacidophilic microorganisms that act as the principal primary producers in these ecosystems. The Yellowstone National Park (USA) geothermal ecosystem has received particular attention. Unlike the biofilm communities of AMD environments, acidic ferric iron mat-biofilms in Yellowstone acidic hot springs are predominantly inhabited by archaea, and their bacterial diversity increases only at temperatures below 60 °C (Kozubal et al. 2012). Biofilms formed on the surface of iron oxide mats exhibit a high diversity of crenarchaeal representatives of *Sulfolobales, Thermoproteales,* and *Desulfurococcales*, with *Metallosphaera yellowstonensis* being the dominant species.

3.2 Microscopy and Molecular Studies of Archaeal Acidophile Biofilms

Advanced microscopy techniques have been applied to visualize archaeal cells and their components during biofilm formation. Confocal Laser Scanning Microscopy (CLSM) in combination with fluorescent probes provides detailed 3-dimensional structure and compositional information (Lawrence et al. 2003; Neu and Lawrence 2009). EPS represent a crucial part of microbial biofilms and a key element in terms of biofilm functionality (Neu and Lawrence 2009). Fluorescently-labeled lectin binding assays (FLBA) (Zippel and Neu 2011) have been applied to archaeal leaching and sulfur oxidizing species (see further). FLBA allows simultaneous visualization and characterization of EPS glycoconjugates combined with the use of other cell stains specific for proteins, nucleic acids or lipids, which allow visualizing cells and some additional EPS components (Neu and Lawrence 2014). Apart from labelled lectins, other fluorochromes can be simultaneously applied to visualize and characterize additional EPS compounds. For instance, the Syto and Sypro fluorochromes series are used to detect cells via their nucleic acids and cellular proteins, respectively. FM dyes (FM1-43 and FM4-64) and Nile red are specific to stain membranes and lipophilic compounds. DDAO (7-hydroxy-9H-1, 3-dichloro-9,9-dimethylacridin-2-one) stains nucleic acids and normally does not penetrate cell membranes, thereby allowing for extracellular DNA (eDNA) visualization (Koerdt et al. 2010; Zhang et al. 2015b) (Fig. 1).

The process of biofilm formation of acidophile archaea on their natural surfaces is not thoroughly understood. Under laboratory conditions, several leaching archaeal species are able to form monolayer biofilms on metal sulfide surfaces (i.e. pyrite/chalcopyrite), especially when they are grown under chemolitotrophic conditions. Since most species of *Acidianus*, *Sulfolobus* and *Metallosphaera* are also able to grow under mixotrophic and heterotrophic conditions, in these cases, their biofilms can exhibit more complex structures. This may be due to several factors, including an increased amount of carbon sources for building EPS polymers, the use of plastic, glass or other type of surfaces, and the cultivation conditions (static or hydrodynamics conditions). Biofilm formation has been studied in great detail for thermoacidophilic *Sulfolobus* species. *Sulfolobus acidocaldarius*, *Sulfolobus solfataricus* and *Sulfolobus tokodaii* exhibit diverse biofilm morphologies: simple carpet-like structures in *S. solfataricus* and tall, dense tower-like arrangements in

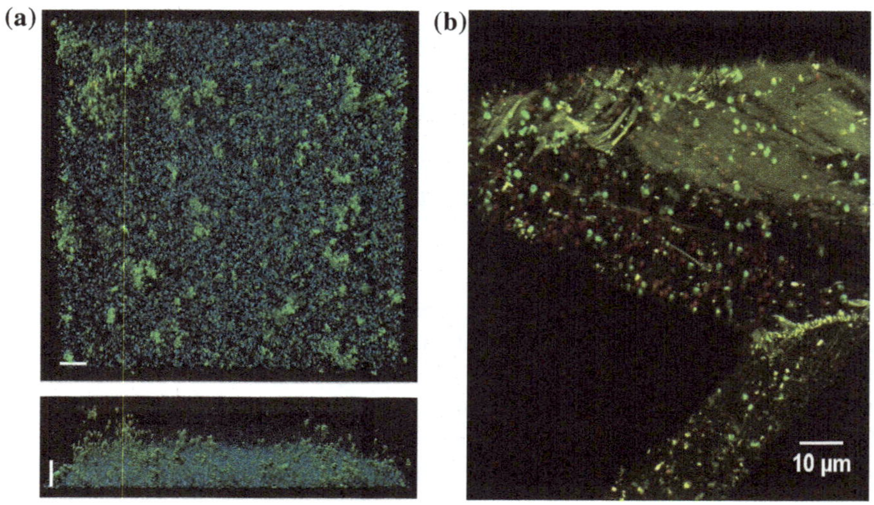

Fig. 1 Biofilms formed by thermoacidophilic archaea. **a** Confocal laser scanning microscopy (CLSM) analysis of 3-day-old biofilms formed by *Sulfolobus acidocaldarius* MW001 on polystyrene surfaces. Cells were stained with DAPI (*blue signals*) and fluorescently labelled lectins ConA (*green signals*) and IB4 (*yellow signals*) to visualize glucose/mannose residues and *N*-galactosyl-residues, respectively. An overlay image of all three channels is shown. *White bars* indicate 20 µm length (Orell et al., unpublished). **b** CLSM image of biofilms formed by *Acidianus copahuensis* DSM 29099 on the surface of a pyrite grain. Cells were stained with Syto 64 (*red*) and the lectin GS-I (*green*). Note the presence of microbial footprints on the pyrite surface (Zhang et al., unpublished)

S. acidocaldarius (Koerdt et al. 2010). *S. acidocaldarius* was shown to engage most readily in biofilm growth. Additionally, pH and temperature changes promote the formation of biofilms in these organisms, suggesting that this sessile lifestyle represents a protection mechanism against environmental fluctuations. During biofilm development, *S. acidocaldarius, S. solfataricus and S. tokodaii* produce an extracellular matrix containing glucose, galactose, mannose and *N*-acetylglucosamine (Koerdt et al. 2010). In order to identify genetic factors that contribute to biofilm formation, genome-wide proteome and transcriptome changes were studied in *S. acidocaldarius*, *S. solfataricus* or *S. tokodaii* biofilms (Koerdt et al. 2011). Intriguingly, expression profiling of biofilms versus planktonic cells for these three closely related species revealed evidence of species-specific responses, where approximately 15% of the *S. acidocaldarius* genes were differentially expressed, whereas *S. solfataricus* and *S. tokodaii* displayed only 3.4 and 1% of altered gene expression, respectively. The regulated genes encompassed a variety of basic cellular functions, comprising energy production/conversion, amino acid metabolism, lipid and carbohydrate metabolism, transport related functions and cell surface modifications. Even though the transcriptomic and proteomic responses displayed notably differences between species, one relevant common response gene was uncovered—the Lrs14 candidate transcriptional regulator, which is hypothesized to play a key role in the regulation of biofilm development in *Sulfolobus* (see Sect. 4.2).

Detailed knowledge on biofilms formed by thermo-acidophiles, especially on some of their natural substrates, is still rather limited. Nevertheless, some progress in the visualization and characterization of biofilms produced by acidophilic archaea has been recently reported. A screening by fluorescence lectin-binding analysis (FLBA) with 75 commercially available lectins of *F. acidiphilum* DSM 28986, *Sulfolobus metallicus* DSM 6482^T and a novel isolate *Acidianus* sp. DSM 29099 showed several glycoconjugates to be present upon biofilm formation. Several glycoconjugates containing monosaccharides such as fucose, glucose, galactose, mannose, *N*-acetyl glucosamine (GlcNAc) and *N*-acetyl galactosamine (GalNAc) were detected in these biofilms. Lectins with positive signals upon biofilm formation on pyrite and sulfur substrates of *Acidianus* sp. DSM 29099 were mostly different, suggesting changes in the EPS glycoconjugates composition. Major binding patterns as tightly bound EPS and loosely bound EPS, were detected on both substrates (Zhang et al. 2014).

The EPS composition of *S. metallicus*^{T} when grown as biofilms on elemental sulfur (S^0) was investigated by CLSM, Fourier Transform Infrared Spectroscopy (FT-IR), and biochemical analyses. In order to visualize cell and EPS distributions, biofilm cells were stained with various dyes specific for glycoconjugates, proteins, nucleic acids and lipids. *S. metallicus*^{T} biofilm cells were shown to be heterogeneously distributed and characterized as individual cells, microcolonies, and large clusters up to hundred micrometers in diameter. EPS were also extracted and analyzed from both planktonic and biofilm cell subpopulations. Planktonic cells showed capsular EPS mainly composed of carbohydrates and proteins, whereas carbohydrates dominated colloidal EPS. Proteins were found to be major components in EPS from biofilm cells. Using specific probes combined with CLSM, it was also shown that extracellular proteins and nucleic acids were present in the EPS matrix. Time series of biofilm pictures revealed that *S. metallicus*^{T} biofilm cells were embedded in a flexible EPS matrix (Zhang et al. 2015b).

Little is known about cell interactions among bioleaching archaea and their bacterial counterparts in mixed cultures. Cell attachment and biofilm formation patterns of *S. metallicus*^{T} and *Acidianus* isolates DSM 29038 and DSM 29099 were studied in pure and mixed cultures on sulfur or pyrite surfaces. CLSM combined with cell fluorescent dyes as well as fluorescently labeled lectins were used to visualize different components (e.g. DNA, proteins or glycoconjugates) of these species. The data indicate that cell attachment and biofilm formation were species- and substrate dependent. Pyrite leaching experiments coupled to pre-colonization and further inoculation with a second species suggest that both species may adversely influence each other during pyrite leaching with respect to initial attachment and pyrite dissolution rates. In addition, the investigation of binary biofilms on pyrite showed that both species were heterogeneously distributed on pyrite surfaces in the form of individual cells or microcolonies. Using lectins that are able to specifically bind single species within mixed cultures revealed physical contact between both species (Castro et al. 2016).

4 Regulatory Mechanisms of Archaeal Biofilm Formation

4.1 Extra- and Intra-cellular Signaling Pathways

The regulatory network that governs the transition from planktonic (free-living) lifestyle to a biofilm mode of growth is just beginning to be elucidated for biofilm-forming archaea. It is apparent that there are fundamental differences to the bacterial biofilm-forming signalling pathways that are particularly intriguing and challenging. The community-wide behaviour in bacteria (e.g. biofilm formation), is regulated through a well-characterized, tightly coordinated interplay between the extracellular cell-cell communication system Quorum Sensing (QS) and c-di-GMP mediated signalling.

Very few reports on archaeal cell-to-cell communication are available. It has been shown that neutrophilic methanogenic and halophilic archaea can activate AHL bioreporters (Tommonaro et al. 2012; Zhang et al. 2012), suggesting that they are able to produce AHL-like QS inducers. Similar to Gram-negative bacteria, *Methanosaeta harundinacea* 6Ac, *Methanosarcina mazei* and *Methanothermobacter thermautotrophicus* produce a type of carboxylated AHLs via the synthetase FilI, a bacterial-like LuxI orthologue (Zhang et al. 2012). However, a QS-based biofilm developmental process for these methanogenic archaeal strains has yet to be evaluated. On the other hand, the possibility of AHL-based QS in hydrothermal systems appears rather unlikely. This is mainly due to: (i) the intrinsic thermolability of AHL molecules, and (ii) the absence of orthologues to mesophilic AHL synthetases in the genomic sequences of thermo-acidophiles. Nevertheless, the thermo-acidophilic archaea *S. solfataricus* and *Sulfolobus islandicus* were shown to produce lactonase enzymes, which are involved in the hydrolysis of AHLs, most likely preventing cell-cell QS communication. Therefore, one could hypothesize that these archaeal lactonases might play a role in niche competition against bacterial species.

A boronated furanone-based QS autoinducer (AI-2) was found to be produced by the neutrophilic hyperthermophile *P. furiosus*. It was proposed that the AI-2 molecule is synthesized through a combination of biotic and abiotic reaction steps (Nichols et al. 2009). Regulatory implications of AI-2 mediated signalling have not yet been investigated in any biofilm-forming thermoacidophilic archaea. A peptide-based quorum sensing in a co-culture between *Thermotoga maritima* and *Methanococcus jannaschii* has been reported, inducing, among others, a putative EPS production pathway (Johnson et al. 2005). Considering the greater heat stability of peptides, they appear as more suitable candidates to mediate cell-to-cell communication in hyperthermophilic environments (Nichols et al. 2009). In an unpublished study (not including *M. harundinacea*) which tested >800 archaeal isolates for the production of quorum sensing signals (using two detection systems for homoserine lactones and peptide signalling each) no indication for such molecules was obtained (Wirth, personal communication).

In bacteria, c-di-GMP is a well-studied second messenger that promotes biofilm formation via mechanisms that can differ even among related microorganisms. Thus far, the occurrence of c-di-GMP has not been described for any archaeon (Baker-Austin et al. 2010), indicating different intracellular signal transduction pathways for biofilm development in the archaeal domain. The recently discovered c-di-AMP has been involved in processes that allow *Bacillus subtilis* cells to manage extreme membrane and cell wall stress (Witte et al. 2008). Emerging data also indicates that increased cellular levels of c-di-AMP stimulate biofilm development in the bacteria *Streptococcus mutans* and *B. subtilis* (Gundlach et al. 2016; Peng et al. 2016). Bioinformatic analyses have revealed the occurrence of DAC (diadenylyl cyclase) domain-containing genes in a subset of archaeal genomes, suggesting that c-di-AMP synthesis and signalling occurs in both the bacterial and archaeal domains (Romling 2008). It remains to be assessed whether c-di-AMP might also play a role in the regulation of archaeal biofilm formation.

4.2 Transcriptional Regulation of Archaeal Biofilms

Two novel biofilm-related transcriptional regulators have been identified in the thermoacidophilic crenarchaeon *S. acidocaldarius* (Koerdt et al. 2010; Orell et al. 2013b). They are characterized as small winged helix-turn-helix (wHTH) DNA-binding proteins that significantly influence biofilm formation. Functional studies have suggested that one of these Lrs14-like regulators (for Leucine-responsive regulator of *Sulfolobus*) acts as a global biofilm activator in *S. acidocaldarius*, since its absence led to notable impairment of biofilm development. In contrast, the AbfR1 regulator (for archaeal biofilm regulator 1) is proposed to function as a biofilm repressor, since its deletion favoured EPS production and leads to the formation of thick biofilms. Remarkably, in the *S. acidocaldarius abfR1* deletion mutant strain, the expression of the adhesive pili (*aap*) encoding genes was found to be up-regulated, whereas transcript levels of the motility structure genes, encoding the archaellum (*fla*), were down-regulated, rendering the mutant strain non-motile. The AbfR1 protein was demonstrated to bind to the promoter regions of *fla* and *aap* genes, presumably controlling the expression of both cell surface structures (Orell et al. 2013b). The dual role of AbfR1 in controlling cell motility and formation of biofilms is reminiscent of the *E. coli* biofilm master regulator CsgD. CsgD was shown to directly repress genes coding for flagella components and to activate the synthesis of EPS, thereby regulating the switch from planktonic growth to the biofilm-lifestyle (Pesavento et al. 2008). Unlike the *E. coli* CsgD regulator, *S. acidocaldarius* AbfR1 appears to work as an activator of archaellum synthesis and as a repressor of the genes required for EPS biosynthesis and biofilm formation (Orell et al. 2013b).

5 Future Perspectives

Archaeal metabolic capabilities imply a major impact on the biogeochemical cycling of essential elements on earth (Frols et al. 2012; Justice et al. 2014). This wide repertoire of metabolic activities are in some cases unique to archaea and has allowed employing them within a number of biotechnologies that range from the production of industrial enzymes to whole-cell applications (Cowan and Fernandez-Lafuente 2011). Congruently, archaeal biofilms have proven to be operational catalytic platforms for industrial processes. They are applied as biofilters in wastewater treatment processes or aquaculture systems and to mediate bio-degradation of methanol, bio-oxidation of H_2S contaminated waste gas as well as metal-bioleaching processes (Barcon et al. 2012; Brown et al. 2013; Gomez-Silvan et al. 2010; Morales et al. 2011; Orell et al. 2013a).

Under thermophilic conditions, the passivation of chalcopyrite ($CuFeS_2$) surfaces by the presence of RISCs can be strongly reduced in comparison with chalcopyrite leaching at lower temperatures. This has importance for the biomining industry, in which thermoacidophilic archaea such as *Acidianus*, *Sulfolobus* and *Metallosphera* may be considered for applications in which high temperatures are generated. The understanding of biofilm formation may contribute to improve reactor designs and conditions for influencing oxidation processes of metal sulfides in stirred-tank reactors.

Our current knowledge on chemical composition and microbial dynamics of acidophilic archaeal biofilms is still rather limited. The presence and functional roles of macromolecules and metabolites within EPS from these biofilms remain to be analyzed in greater depth. Many proteins currently classified as 'hypothetical' are expected to fulfill important roles in controlling the dynamics of acidophilic biofilm phenotypes and facilitate interactions with other microbial populations. The advent of improved archaeal genetics will allow for the study of these proteins and regulatory networks that are fundamentally different from bacterial pathways are awaiting their discovery. The elucidation of possible archaeal QS and cell signaling processes could be useful for several biotechnological applications, e.g. during the bioleaching of metal sulfides. A detailed understanding of the archaeal biofilm lifestyle and cell-cell interactions will be helpful to develop strategies for bioleaching manipulation both in metal recovery, where enhancement of leaching rates is necessary, and the mitigation of AMD, where the reduction or inactivation of biofilms might reduce environmental problems.

References

Baker-Austin C, Potrykus J, Wexler M, Bond PL, Dopson M (2010) Biofilm development in the extremely acidophilic archaeon '*Ferroplasma acidarmanus*' Fer1. Extremophiles 14(6):485–491

Baker BJ, Banfield JF (2003) Microbial communities in acid mine drainage. FEMS Microbiol Ecol 44(2):139–152

Barcon T, Alonso-Gutierrez J, Omil F (2012) Molecular and physiological approaches to understand the ecology of methanol degradation during the biofiltration of air streams. Chemosphere 87(10):1179–1185

Bellack A, Huber H, Rachel R, Wanner G, Wirth R (2011) *Methanocaldococcus villosus* sp. nov., a heavily flagellated archaeon that adheres to surfaces and forms cell-cell contacts. Int J Syst Evol Microbiol 61(Pt 6):1239–1245

Brown MN, Briones A, Diana J, Raskin L (2013) Ammonia-oxidizing archaea and nitrite-oxidizing nitrospiras in the biofilter of a shrimp recirculating aquaculture system. FEMS Microbiol Ecol 83(1):17–25

Castro C, Zhang R, Liu J, Bellenberg S, Neu TR, Donati E, Sand W, Vera M (2016) Biofilm formation and interspecies interactions in mixed cultures of thermo-acidophilic archaea *Acidianus* spp. and *Sulfolobus metallicus*. Res Microbiol 167(7):604–612

Comolli LR, Baker BJ, Downing KH, Siegerist CE, Banfield JF (2009) Three-dimensional analysis of the structure and ecology of a novel, ultra-small archaeon. ISME J 3(2):159–167

Cowan DA, Fernandez-Lafuente R (2011) Enhancing the functional properties of thermophilic enzymes by chemical modification and immobilization. Enzyme Microb Technol 49(4):326–346

Dopson M, Baker-Austin C, Hind A, Bowman JP, Bond PL (2004) Characterization of *Ferroplasma* isolates and *Ferroplasma acidarmanus* sp. nov., extreme acidophiles from acid mine drainage and industrial bioleaching environments. Appl Environ Microbiol 70(4):2079–2088

Edwards KJ, Bond PL, Gihring TM, Banfield JF (2000) An archaeal iron-oxidizing extreme acidophile important in acid mine drainage. Science 287(5459):1796–1799

Etzel K, Klingl A, Huber H, Rachel R, Schmalz G, Thomm M, Depmeier W (2008) Etching of 111 and 210 synthetic pyrite surfaces by two archaeal strains *Metallosphaera sedula* and *Sulfolobus metallicus*. Hydrometallurgy 94(1):116–120

Fiala G, Stetter KO (1986) *Pyrococcus furiosus* sp. nov. represents a novel genus of marine heterotrophic archaebacteria growing optimally at 100 °C. Arch Microbiol 145(1):56–61

Flemming HC, Wingender J (2010) The biofilm matrix. Nat Rev Microbiol 8(9):623–633

Frols S, Dyall-Smith M, Pfeifer F (2012) Biofilm formation by haloarchaea. Environ Microbiol 14 (12):3159–3174

Golyshina OV, Pivovarova TA, Karavaiko GI, Kondrateva TF, Moore ER, Abraham WR, Lunsdorf H, Timmis KN, Yakimov MM, Golyshin PN (2000) *Ferroplasma acidiphilum* gen. nov., sp. nov., an acidophilic, autotrophic, ferrous-iron-oxidizing, cell-wall-lacking, mesophilic member of the *Ferroplasmaceae* fam. nov., comprising a distinct lineage of the Archaea. Int J Syst Evol Microbiol 50 Pt 3:997–1006

Golyshina OV, Timmis KN (2005) *Ferroplasma* and relatives, recently discovered cell wall-lacking archaea making a living in extremely acid, heavy metal-rich environments. Environ Microbiol 7(9):1277–1288

Gomez-Silvan C, Molina-Munoz M, Poyatos JM, Ramos A, Hontoria E, Rodelas B, Gonzalez-Lopez J (2010) Structure of archaeal communities in membrane-bioreactor and submerged-biofilter wastewater treatment plants. Bioresour Technol 101(7):2096–2105

Gundlach J, Rath H, Herzberg C, Mader U, Stulke J (2016) Second messenger signaling in *Bacillus subtilis*: accumulation of cyclic di-AMP inhibits biofilm formation. Front Microbiol 7:804. doi:10.3389/fmicb.2016.00804

Huber G, Spinnler C, Gambacorta A, Stetter KO (1989) *Metallosphaera sedula* gen, and sp. nov. represents a new genus of aerobic, metal-mobilizing, thermoacidophilic archaebacteria. Syst Appl Microbiol 12(1):38–47

Johnson DB (2012) Geomicrobiology of extremely acidic subsurface environments. FEMS Microbiol Ecol 81(1):2–12

Johnson MR, Montero CI, Conners SB, Shockley KR, Bridger SL, Kelly RM (2005) Population density-dependent regulation of exopolysaccharide formation in the hyperthermophilic bacterium *Thermotoga maritima*. Mol Microbiol 55(3):664–674

Justice NB, Pan C, Mueller R, Spaulding SE, Shah V, Sun CL, Yelton AP, Miller CS, Thomas BC, Shah M, VerBerkmoes N, Hettich R, Banfield JF (2014) Heterotrophic archaea contribute to carbon cycling in low-pH, suboxic biofilm communities. Appl Environ Microbiol 78 (23):8321–8330

Koerdt A, Godeke J, Berger J, Thormann KM, Albers SV (2010) Crenarchaeal biofilm formation under extreme conditions. PLoS ONE 5(11):e14104. doi:10.1371/journal.pone.0014104

Koerdt A, Orell A, Pham TK, Mukherjee J, Wlodkowski A, Karunakaran E, Biggs CA, Wright PC, Albers SV (2011) Macromolecular fingerprinting of *Sulfolobus* species in biofilm: a transcriptomic and proteomic approach combined with spectroscopic analysis. J Proteome Res 10(9):4105–4119

Kozubal MA, Macur RE, Jay ZJ, Beam JP, Malfatti SA, Tringe SG, Kocar BD, Borch T, Inskeep WP (2012) Microbial iron cycling in acidic geothermal springs of yellowstone national park: integrating molecular surveys, geochemical processes, and isolation of novel fe-active microorganisms. Front Microbiol 3:109. doi:10.3389/fmicb.2012.00109

Kurr M, Huber R, König H, Jannasch HW, Fricke H, Trincone A, Kristjansson JK, Stetter KO (1991) *Methanopyrus kandleri*, gen. and sp. nov. represents a novel group of hyperthermophilic methanogens, growing at 110 °C. Arch Microbiol 156(4):239–247

Lawrence J, Swerhone G, Leppard G, Araki T, Zhang X, West M, Hitchcock A (2003) Scanning transmission X-ray, laser scanning, and transmission electron microscopy mapping of the exopolymeric matrix of microbial biofilms. Appl Environ Microbiol 69(9):5543–5554

Mikkelsen D, Kappler U, Webb RI, Rasch R, McEwan AG, Sly LI (2007) Visualisation of pyrite leaching by selected thermophilic archaea: Nature of microorganism-ore interactions during bioleaching. Hydrometallurgy 88:143–153

Morales M, Arancibia J, Lemus M, Silva J, Gentina JC, Aroca G (2011) Bio-oxidation of H_2S by *Sulfolobus metallicus*. Biotechnol Lett 33(11):2141–2145

Näther-Schindler DJ, Schopf S, Bellack A, Rachel R, Wirth R (2014) *Pyrococcus furiosus* flagella: biochemical and transcriptional analyses identify the newly detected flaB0 gene to encode the major flagellin. Front Microbiol 5:695. doi:10.3389/fmicb.2014.00695

Näther DJ, Rachel R, Wanner G, Wirth R (2006) Flagella of *Pyrococcus furiosus*: multifunctional organelles, made for swimming, adhesion to various surfaces, and cell-cell contacts. J Bacteriol 188(19):6915–6923

Neu T, Lawrence J (2009) Extracellular polymeric substances in microbial biofilms. In: Moran ABP, Holst O, von Itzstein M (eds) Microbial glycobiology: Structures, relevance and applications. Elsevier, San Diego, pp 735–758

Neu TR, Lawrence JR (2014) Investigation of microbial biofilm structure by laser scanning microscopy. Adv Biochem Eng Biotechnol 146:1–51

Nichols JD, Johnson MR, Chou CJ, Kelly RM (2009) Temperature, not LuxS, mediates AI-2 formation in hydrothermal habitats. FEMS Microbiol Ecol 68(2):173–181

Norris PR, Burton NP, Foulis NA (2000) Acidophiles in bioreactor mineral processing. Extremophiles 4(2):71–76

Orell A, Frols S, Albers SV (2013a) Archaeal biofilms: the great unexplored. Annu Rev Microbiol 67:337–354

Orell A, Peeters E, Vassen V, Jachlewski S, Schalles S, Siebers B, Albers SV (2013b) Lrs14 transcriptional regulators influence biofilm formation and cell motility of Crenarchaea. ISME J 7(10):1886–1898

Peng X, Zhang Y, Bai G, Zhou X, Wu H (2016) Cyclic di-AMP mediates biofilm formation. Mol Microbiol 99(5):945–959

Pesavento C, Becker G, Sommerfeldt N, Possling A, Tschowri N, Mehlis A, Hengge R (2008) Inverse regulatory coordination of motility and curli-mediated adhesion in *Escherichia coli*. Genes Dev 22(17):2434–2446

Robb FT, Maeder DL, Brown JR, DiRuggiero J, Stump MD, Yeh RK, Weiss RB, Dunn DM (2001) Genomic sequence of hyperthermophile, *Pyrococcus furiosus*: implications for physiology and enzymology. Methods Enzymol 330:134–157

Rohwerder T, Gehrke T, Kinzler K, Sand W (2003) Bioleaching review part A: progress in bioleaching: fundamentals and mechanisms of bacterial metal sulfide oxidation. Appl Microbiol Biotechnol 63(3):239–248

Romling U (2008) Great times for small molecules: c-di-AMP, a second messenger candidate in Bacteria and Archaea. Sci Signal 1(33):pe39

Sanhueza A, Ferrer IJ, Vargas T, Amils R, Sánchez C (1999) Attachment of *Thiobacillus ferrooxidans* on synthetic pyrite of varying structural and electronic properties. Hydrometallurgy 51(1):115–129

Schopf S, Wanner G, Rachel R, Wirth R (2008) An archaeal bi-species biofilm formed by *Pyrococcus furiosus* and *Methanopyrus kandleri*. Arch Microbiol 190(3):371–377

Tommonaro G, Abbamondi GR, Iodice C, Tait K, De Rosa S (2012) Diketopiperazines produced by the halophilic archaeon, *Haloterrigena hispanica*, activate AHL bioreporters. Microb Ecol 63(3):490–495

Weiner A, Schopf S, Wanner G, Probst A, Wirth R (2012) Positive, neutral and negative interactions in cocultures between *Pyrococcus furiosus* and different Methanogenic Archaea. Microbiology Insights 5:1–10

Wilmes P, Remis JP, Hwang M, Auer M, Thelen MP, Banfield JF (2009) Natural acidophilic biofilm communities reflect distinct organismal and functional organization. ISME J 3(2):266–267

Witte G, Hartung S, Buttner K, Hopfner KP (2008) Structural biochemistry of a bacterial checkpoint protein reveals diadenylate cyclase activity regulated by DNA recombination intermediates. Mol Cell 30(2):167–178

Zhang G, Zhang F, Ding G, Li J, Guo X, Zhu J, Zhou L, Cai S, Liu X, Luo Y, Shi W, Dong X (2012) Acyl homoserine lactone-based quorum sensing in a methanogenic archaeon. ISME J 6 (7):1336–1344

Zhang L, Zhou W, Li K, Mao F, Wan L, Chen X, Zhou H, Qiu G (2015a) Synergetic effects of *Ferroplasma thermophilum* in enhancement of copper concentrate bioleaching by *Acidithiobacillus caldus* and *Leptospirillum ferriphilum*. Biochem Eng Journal 93:142–150

Zhang R, Neu T, Bellenberg S, Kuhlicke U, Sand W, Vera M (2014) Use of lectins to in situ visualize glycoconjugates of extracellular polymeric substances in acidophilic archaeal biofilms. Microb Biotechnol. doi:10.1111/1751-7915.12188

Zhang R, Neu TR, Zhang Y, Bellenberg S, Kuhlicke U, Li Q, Sand W, Vera M (2015b) Visualization and analysis of EPS glycoconjugates of the thermoacidophilic archaeon *Sulfolobus metallicus*. Appl Microbiol Biotechnol 99(17):7343–7356

Zhou H, Zhang R, Hu P, Zeng W, Xie Y, Wu C, Qiu G (2008) Isolation and characterization of *Ferroplasma thermophilum* sp. nov., a novel extremely acidophilic, moderately thermophilic archaeon and its role in bioleaching of chalcopyrite. J Appl Microbiol 105(2):591–601

Zippel B, Neu T (2011) Characterization of glycoconjugates of extracellular polymeric substances in tufa-associated biofilms by using fluorescence lectin-binding analysis. Appl Environ Microbiol 77(2):505–516

The Compressed Vocabulary of the Proteins of Archaea

Gustavo Caetano-Anollés, Bushra F. Minhas, Fayez Aziz, Fizza Mughal, Khuram Shahzad, Guy Tal, Jay E. Mittenthal, Derek Caetano-Anollés, Ibrahim Koç, Arshan Nasir, Kelsey Caetano-Anollés and Kyung Mo Kim

1 Introduction

Where there is communication, there is language. Where there is language, there are sets of signs and rules drawn from some alphabet that define codes or ciphers in the communication system (Table 1). Language involves conveying a meaning, the creation of messages, and their encoding in entities or objects (lexemes, i.e. elementary units of lexical meaning), the collective of which form vocabularies

G. Caetano-Anollés (✉) · B.F. Minhas · F. Aziz · F. Mughal · G. Tal · I. Koç · A. Nasir
Department of Crop Sciences, University of Illinois, Urbana, IL 61801, USA
e-mail: gca@illinois.edu

G. Caetano-Anollés · B.F. Minhas · F. Mughal · K. Shahzad
Illinois Informatics Institute, University of Illinois, Urbana, IL 61801, USA

J.E. Mittenthal
Department of Cell and Developmental Biology,
University of Illinois, Urbana, IL 61801, USA

D. Caetano-Anollés
Department of Evolutionary Genetics, Max-Planck-Institut Für Evolutionsbiologie,
24306 Plön, Germany

I. Koç
Molecular Biology and Genetics, Gebze Technical University, Kocaeli, Turkey

A. Nasir
Department of Biosciences, COMSATS Institute of Information Technology,
Islamabad 45550, Pakistan

K. Caetano-Anollés
Department of Agricultural Biotechnology and Research Institute for Agriculture and Life
Sciences, Seoul National University, Seoul 151-921, Republic of Korea

K.M. Kim
Division of Polar Life Sciences, Korea Polar Research Institute,
Incheon 21-990, Republic of Korea

G. Witzany (ed.), *Biocommunication of Archaea*,
DOI 10.1007/978-3-319-65536-9_10

Table 1 Definitions

Term	Definition
Communication	The intentional act of conveying meaning from one entity to another through the use of messages
Message	A discrete unit of communication of information
Language	A formal symbolic system of complex communication that uses signs and rules to communicate meaning
Sign	A semiotic entity or unit of lexical meaning (lexeme) whose existence bears causal relation to its object and acts as likeness, token, indication or symbol of that relationship (e.g. word with semantic or pragmatic content drawn from some alphabet)
Rule	A logical form, abstraction or function to produce a syntax governing the structure of signs
Grammar	An approach to specifying language that is based on re-writing rules
Cipher	An algorithm to encrypt and decrypt a message at elemental level
Code	A codebook of ciphers that convey meaning
Vocabulary	The set of signs used by the communication system (e.g. the lexicon of a language)
Syntaxis	The systematic combination of the inventory of signs according to rules of grammar
Semantics	Meaning of signs and their relationships at different hierarchical systems of the message
Pragmatics	Context dependent relation of signs and sign user according to pragmatic rules, i.e. rules that govern and discriminate successful from non-successful interactional motifs

(lexicons). Holistic views of language are not new and have been discussed for example within the context of semiotics (meaningful communication; Kravchenko 2006) and realism (truth as correspondence between cognition and reality; Elder-Vass 2013). For example, language can be considered a biological phenomenon that emerges from interpreting signs in complex structure and organization (Witzany 2014, 2016). From a biological point of view, communication appears to be a central and undisputed activity of living systems. What it not clear, however, is whether 'meaningful' communication has been already expressed at the molecular level when different macromolecules and molecular parts interact to produce coordinated outcomes within the context of the cell. There are good indications that the orchestration of these outcomes is limited by energy and information dissipation, and by the constraints that evolving and hierarchical systems impose on their structure (Caetano-Anollés et al. 2010). Here we explore the existence of a vocabulary in the molecular repertoires of proteins (proteomes) and their associated molecular functions (functionomes). We are after sets of signs and rules defining a general biological language. We focus on the molecules of organisms belonging to superkingdom Archaea. Necessarily, the exercise involves developing both comparative and retrodictive genomic frameworks capable of dissecting the origin and evolution of proteomes and functionomes through meaningful data-driven hypotheses of ontology and change. First, we showcase evidence to discuss that there is active communication between molecules that follow language laws. Second, we discuss how these laws manifest within constraints imposed by the engineering of the emerging biological systems and trade-offs between economy, flexibility and robustness. Third, we focus on evolution of entire archaeal molecular repertoires and show that there are clear patterns of origin and molecular diversification that are consistent with these trade-offs. These patterns favor economy, and to some extent robustness, an outcome that probably stems from the very early evolutionary origin of Archaea in the harsh environments of primordial Earth.

2 Protein Domains: The Structural, Functional and Evolutionary Units of Proteins

Proteins are the major functional workhorses of the cell (Caetano-Anollés et al. 2009). They are large biological macromolecules consisting of one or more long chains of amino acid residues, the sequences of which define their *primary structures*. However, the functions of proteins are largely determined by their 3-dimensional atomic structures, which materialize when amino acid sequences collapse into a folded and functionally active state that abandons the benefits of interactions with water. These folded structures represent spatial arrangements of helical segments that are more or less wound (3_{10}-helices, α-helices, π-helices and polyproline helices, respectively) and extended chain segments called β-strands that form flat, curved or barrel sheet structures (Caetano-Anollés et al. 2010). The preponderant helical and sheet components are stabilized through short-range or long-range hydrogen-bond interactions, respectively. These *secondary structures* are spatial motifs that often contain or are separated by segments that are non-structured. Collectively, they fold into *tertiary structures* by minimizing the energy conformations of individual amino acid residues in a chain, fostering hydrogen-bonding of polar groups, and maximizing packing and structural compactness. Folding units in separate chains then associate to form *quaternary structures* in multi-subunit complexes. The Dictionary of Protein Secondary Structure (DSSP) uses single letter codes to describe the succession of secondary structures in a polypeptide chain: 3_{10}-helices (G), α-helices (H), π-helices (I), extended strand (E), hydrogen bonded turns (T), β-bridges (B), bends (S) and coils (C) (Kabsch and Sander 1983). These letter abstractions, which resemble those of amino acid residues, can be used to align secondary structure and help overlap closely related 3-dimensional structures (Holm and Rosenström 2010). Note that the folded atomic structures provide a fundamental scaffold for favorable stable conformations of small subsets of amino acid residues responsible for protein functions. These residues are generally lodged in pockets on the surface of the protein, making active, regulatory or binding motifs that enable the molecular functions of gene products. However, networks of residues throughout the molecule can also play important roles in allosteric regulation and protein stability.

Molecular stability can be enhanced through some form of modularization. Modules are sets of integrated parts of a system that cooperate to perform a task. Parts in modules interact more extensively with each other than with other parts and modules of the system. Within proteins, there are functionally important higher-level structural modules, the *structural domains*. These modules are arrangements of elements of secondary structure that fold into well-packed and compact structural units of the polypeptide chain (Wetlaufer 1973; Richardson 1981). They can fold and function largely independently and contribute to overall protein stability by establishing a multiplicity of intramolecular interactions (Bhaskara and Srinivasan 2011). More importantly, domains have been shown to be evolutionarily conserved and present in different molecular and functional contexts throughout the protein world (Janin and Wodak 1983; Murzin et al. 1995; Riley and Labedan 1997). They are thus considered the structural, functional and evolutionary units of proteins. Since domains behave as evolutionary modules, they can combine to form multidomain proteins (Chothia et al. 2003; Moore et al. 2008). Similar or different domains can fuse in evolution or excise from multidomain proteins through fission, driven mostly by the forces of genome rearrangement (Wang and Caetano-Anollés 2009). Fusions often occur directionally in the protein chains (Ekman et al. 2005; Toll-Riera and Alba 2013). The rise of *domain organization* appears massive, resulting in an arrangement of protein 'architectures' that enhances the functional toolkit of the cell and enables a molecular vocabulary (Bashton and Chothia 2007). This additional level of structural organization has been shown to impact evolutionary constraints imposed by folding speed, protein flexibility (Debès et al. 2013) and the length of domains and proteins (Wang et al. 2011). For example, the length of domains in multidomain proteins, which are highly represented in eukaryotes, have on average substantially smaller domains than single domain proteins. We note that the evolutionary mechanics of domain combination in multidomain proteins contribute significantly to global patterns of domain gain and loss (Nasir et al. 2014a). These patterns materialize differently in the proteomes of the three superkingdoms of life, Archaea, Bacteria and Eukarya.

3 Levels of Organization in Protein Structure

Linderstrøm-Lang and Schellman (1959) realized more than half a century ago that protein structure was hierarchical when they proposed their four-tiered hierarchy of structural organization: primary, secondary, tertiary, and quaternary. These levels can be illustrated with ferritin (Fig. 1), and in particular with the ferritin complex of the hyperthermophilic archaeon *Pyrococcus furiosum* (Tatur et al. 2007). Ferritin is a large multisubunit protein with molecular functions of storage and time-release delivery of iron that is widely distributed in life (Bai et al. 2015). The protein typically forms globular structures that are large (450 kDa) and hollow. These *quaternary structures* are composed of 24 small subunit monomers (19–21 kDa), each representing a single polypeptide chain. In Bacteria and Archaea, subunits are of one kind. In complex organisms, however, there could be up to 4 different kinds of subunits. Each subunit holds a single structural domain. The structure of each subunit is typical of the ferritin-like fold, which embodies a 4-helix closed bundle with a left-handed twist. The bundle contains 4 rungs of α-helices in groups of two, i.e., two αα-hairpin motifs linked by a crossover connection that sandwiches the dimetal-binding site and a ferroxigenase center that is located in the middle of the domain. An extra carboxy-terminal α-helix provides support to the 432 point-group symmetry of the quaternary tetraicosamer complex.

A careful examination of the ferritin structure however shows the limitations of the Linderstrøm-Lang and Schellman proposal. To begin, the αα-hairpins of the ferritin bundles represent an example of *supersecondary structure* defining the overall fold topologies of structural domains. These αα-hairpins have short connection conformations and very similar patterns of hydrophobic, hydrophilic and glycine residues in their sequences (Efimov 1991). While domains are considered the evolutionary units of proteins, this other level of structural granularity could also hold deep evolutionary history. In fact, supersecondary motifs such as αα-hairpins, as well as ββ-hairpins and βαβ-elements are recurrent elements in numerous folds, and can combine in evolution to form more complex topologies (Söding and Lupas 2003; Trifonov and Frenkel 2009). Many of these motifs can form shorter loop structures ~25–30 amino acid residues long that are stabilized by the formation of van der Waals locks (Berezovsky and Trifonov 2001). Some of these structures hold molecular functions and are evolutionarily highly conserved (Goncearenco and Berezovsky 2015). These so-called 'elementary functional loops' combine, typically in groups of two or three, to bind cofactors and create active, binding or regulatory sites. When analyzed with phylogenomic approaches, the loops associate with very old structural domains and are recruited for central metabolic functions (Aziz et al. 2016). Other loop structures are not combined but are repeated in different fold contexts (Alva et al. 2015). They are proposed to represent a primordial vocabulary responsible for the creation of the very first folded domains. Even lower hierarchical levels between primary and secondary structure could also hold repositories of historical information. Phylogenomic analysis of protein structure show patterns of dipeptide composition and enrichment characteristic of individual domains that is suggestive of their ancient evolutionary combination to form higher level protein structure (Caetano-Anollés et al. 2013). Thus, even the amino acid compositions of protein sequences could carry a hidden vocabulary for the structure of proteins.

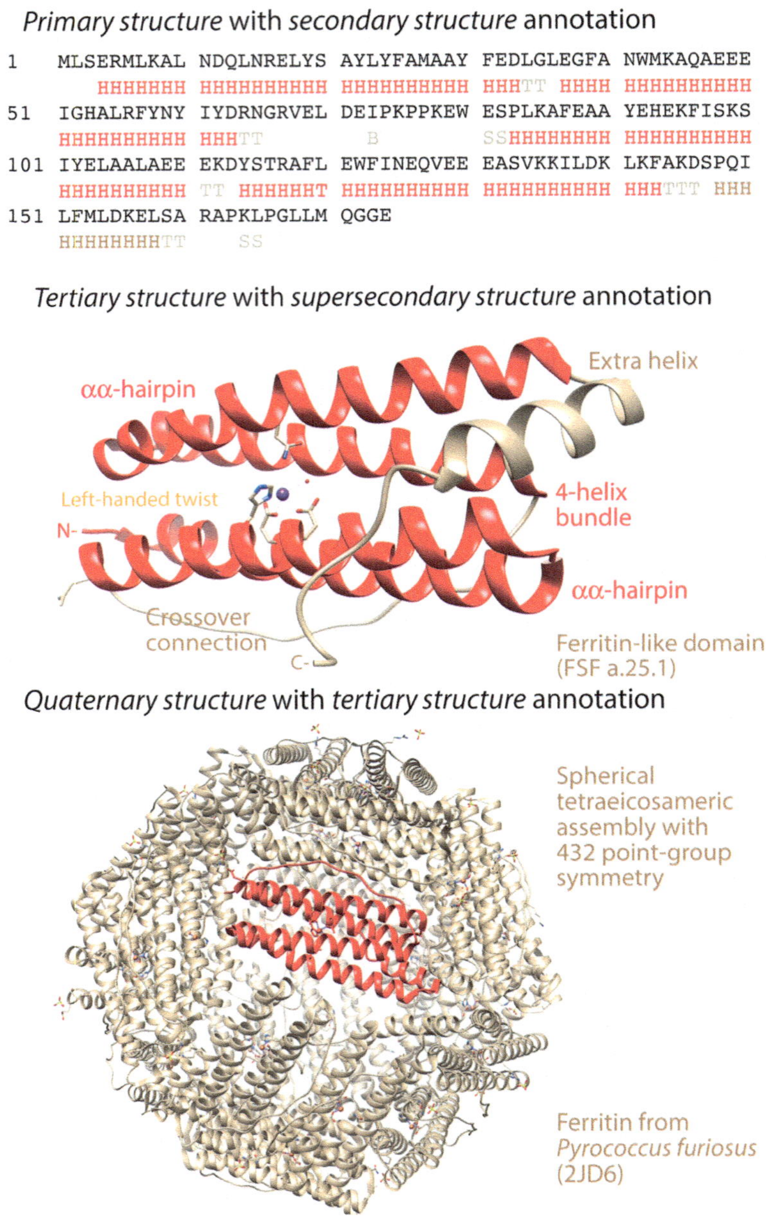
Primary structure with secondary structure annotation
1 MLSERMLKAL NDQLNRELYS AYLYFAMAAY FEDLGLEGFA NWMKAQAEEE
HHHHHHH HHHHHHHHHH HHHHHHHHHH HHHTT HHHH HHHHHHHHHH
51 IGHALRFYNY IYDRNGRVEL DEIPKPPKEW ESPLKAFEAA YEHEKFISKS
HHHHHHHHHH HHHTT B SSHHHHHHHH HHHHHHHHHH
101 IYELAALAEE EKDYSTRAFL EWFINEQVEE EASVKKILDK LKFAKDSPQI
HHHHHHHHHH TT HHHHHHT HHHHHHHHHH HHHHHHHHHH HHHTTT HHH
151 LFMLDKELSA RAPKLPGLLM QGGE
HHHHHHHHTT SS
Tertiary structure with supersecondary structure annotation
αα-hairpin
Extra helix
Left-handed twist
N-
4-helix bundle
αα-hairpin
Crossover connection
C-
Ferritin-like domain (FSF a.25.1)
Quaternary structure with tertiary structure annotation
Spherical tetraeicosameric assembly with 432 point-group symmetry
Ferritin from Pyrococcus furiosus (2JD6)

◂**Fig. 1 Levels of protein organization in the structure of ferritin from the archaeaon *Pyrococcus furiosus* (PDB entry 2JD6)**. The primary structure of the ferritin subunit monomer and its associated 'ferritin-like' structural domain is shown in 'string' format at the top with secondary structures indexed using DSSP notation along its sequence, from amino (N)- to carboxy (C)-terminus. The tertiary structure of the domain defined at fold superfamily (FSF) level (a.25.1) of the Structural Classification of Proteins (SCOP) is shown in 'cartoon' format in the middle with indexed supersecondary structures. The ferritin-like fold harbors a dimetal binding site (the ferroxidase center) in the middle of two αα-hairpin supersecondary structures (with *red* colored *backbones*) linked by a 'crossover connection' (in tan). This makes a closed 4-helix bundle with left-handed twist. The structure contains only one iron atom (*purple sphere*) in site A (the other sites are empty). The complete quaternary structure of the archaeal ferritin is shown at the bottom with its 24 subunits (each representing a separate polypeptide chain holding a ferritin-like domain; subunit 0 is colored *red*)

4 A Survey of the World of Structural Domains and the Construction of Hierarchies

The linguistic metaphor has dominated molecular biology from its beginnings, with the use of central terms such as 'translation' and 'transcription' and the numerous lexical and syntactic similarities that connect linguistics and life (Searls 2001, 2002). For example, there are clear correspondences between the progressions of language and molecular biology (modified from Searls 2001; see Table 1 for definitions):

$$\textit{Lexicon} \Rightarrow \textit{Syntaxis} \Rightarrow \textit{Semantics} \Rightarrow \textit{Pragmatics}$$
$$\textit{Structure} \Rightarrow \textit{Function} \Rightarrow \textit{Fitness}$$

The lexicon of a language (the inventory of words) and its syntaxis (the systematic combination of the inventory according to grammar rules) correspond to the hierarchy of molecular structure, from molecular sequences to folds and beyond, and to the combinatorics of corresponding structural modules. The semantics of a language (the meaning of linguistic expressions) corresponds to the biological function of molecular structures that are sometimes annotated to genes [e.g. the Gene Ontology (GO) project (Ashburner et al. 2000)]. The pragmatics of language (the context in which language contributes to meaning) corresponds to the biological fitness (the persistence potential of functions within different or changing environments). Note how progressions become increasingly transformative in their abstraction. In biology, they transform from static elements of biological language to dynamic evolving entities. Using a Saussurean definition of language, Ji (1997) proposed a language analog of biology that he defined as a "*self-organizing system of molecules, some of which encode, act as signs for, or trigger, gene-directed cellular processes*". The definition incorporates the complexity of self-organization in biology but lacks the pragmatics of biology, the effect which the environment has on the survival of the fittest molecular vocabularies for the sake of both the persistence of the molecular language and the organismal system.

It is highly likely that one central language of molecular biology is the modular assembly of proteins, especially the combination of domains in protein evolution (Wang and Caetano-Anollés 2009). It resembles the combination of atomic linguistic units (morphemes) to form higher level linguistic units such as words or phrases (lexemes) (Searls 2002; Gimona 2006). In recent years, the atomic structures of protein domains have been massively surveyed and summarized in hierarchical classifications schemes (Chothia et al. 2003). Domains have been unified into homologous groups and hierarchically organized in several classifications, including the Structural Classification of Proteins (SCOP) (Murzin et al. 1995) and CATH (Orengo et al. 1997) databases, which are now considered gold standards of protein taxonomy. The basic assumption is that domains are evolutionary building blocks and that they collectively represent sets of signs for a basic vocabulary of structure and function. Indeed, phylogenomic and numerical analyses together with the use of machine learning approaches strongly support the existence of significant phylogenetic signal in the structure of these modules (Caetano-Anollés et al. 2009), which validates their use. For example, the SCOP classification groups domains into fold families (FFs), fold superfamilies (FSFs), folds (Fs) and protein classes (Cs) in a hierarchy of decreasing granularity (Fig. 2). Domains with pairwise amino acid sequence identities of more than 30% (exceptionally with lower sequence identities but conserved functions and structures) are unified into FFs, which by definition are phylogenetically closely related at the sequence level. Those FFs that share similar structural and functional features suggesting a common origin are

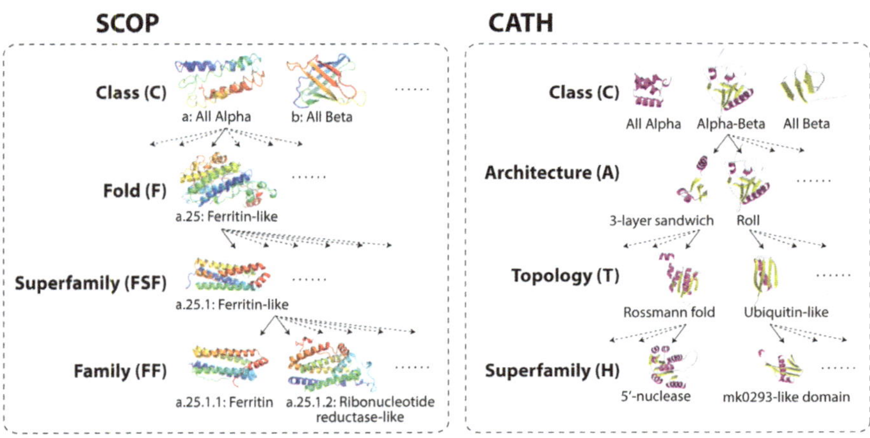

Fig. 2 Top hierarchical levels of the SCOP and CATH classifications of structural domains. The SCOP hierarchy is illustrated with the ferritin-like fold a.25 (core: 4 helices; bundle, closed, left-handled twist; one crossover connection), ferritin-like superfamily a.25.1 (contains bimetal-ion center in the middle of the bundle), and ferritin family c.25.1.1 (10 protein domains; e.g. apoferritin, bacterioferritin). The CATH classification is illustrated with the Rossmann and ubiquitin-like fold = topology lineages. Note that the CATH classification is more structurally grained than SCOP. Cartoons of domain structure follow thumbnail visualizations of SCOPe and CATH

further unified into FSFs. FSFs sharing secondary structures that are similarly arranged and connected in 3-dimensional space are further unified into folds. Finally, those that share similar overall designs are further grouped into protein classes (*a* through *l*; only 7 are true classes). Similarly, the hierarchy of structural abstractions in CATH, from top to bottom, involves classes (C), architectures (A), topologies (T), homologous superfamilies (H) and sequence families. Since structure is far more evolutionarily conserved than sequence, these taxonomic strategies make good use of the levels of organization we previously described. They are capable of abstracting history from structural data. Deep phylogenetic signal increases with higher levels of structural abstraction. In turn, modern definitions of molecular functions increase their significance.

The original SCOP classification that originated in 1995 ended in its version 1.75 (June 2009) but has been extended by SCOPe, a database that preserves the original hierarchical scheme of SCOP (Fox et al. 2014). SCOPe version 2.05 (February 2016) classifies 244,326 domains (77,439 PDB entries) into 1,221 Fs, 2,008 FSFs and 4,851 FFs. In turn, its predecessor SCOP version 1.75 classified its 110,080 domains (38,221 PDB entries) into 1,195 Fs, 1,962 FSFs, and 3,902 FFs. Thus, high SCOP abstraction levels are approaching a plateau, i.e. a finite set of universal structures for protein domains. Similar tendencies can be found in CATH (Sillitoe et al. 2015). The latest version, CATH v.4.1 (January 2015), classifies 308,999 (108,378 PDB entries) into 40 A, 1,373 Ts, and 2,737 Hs.

5 Laws of Language in the Structure and Organization of Proteins

If there is meaningful language in the structure of proteins, there must be quantitative linguistic patterns in molecules, i.e. an exact science of language in biology. To begin with, protein molecules and their levels of organization must comply with statistical laws of language (Altmann et al. 2006), many of which are meaningfully interrelated and universal. There are three general groups of laws (recently reviewed by Altmann and Gerlach 2016):

(i) *Probability distributions*. These statistical laws make predictions of possible outcomes, such as the Zipf's law—the frequency of a word in a given text being inversely proportional to its rank ordering in a frequency table (Zipf 1948). The Zipf-Mandelbrot generalization assumes that the lexicons of languages organize in such a way that the most frequent words become the shortest. Mathematically, the frequency of a word $f(k)$ at rank k follows Equation (1)

$$f(k) = K/(k+q)^{\gamma} \tag{1}$$

where K is a normalizing constant and q and γ are parameters. The Zipf's law is a special case of the scale-free distribution that it explains. The scale-free property pervades the make up of many networks, including biological networks such as those describing metabolism and protein-protein interactions (Strogatz 2001). Scale-free networks follow power-law distributions $P(k) \sim k^{-\gamma}$, with k representing the number of links of the network and $P(k)$ the probability that the network has exactly k links. The power-law behavior of such scale-free distributions can be explained by the 'rich get richer' motto that follows the 'preferential attachment' or 'Yule' processes that are typical of network growth, in which nodes with many connections are prone to get more connected (Barabási 2009). For example, a Zipf law explains patterns of occurrence of Pfam domains in proteins as well as words in Shakespeare's Romeo and Juliet (Searls 2002). The Zipf law is followed by structural domains at fold and FSF levels (Qian et al. 2001; Caetano-Anollés and Caetano-Anollés 2003), with γ power law decay values of ~ 2 for Bacteria and Archaea and ~ 1.4 for Eukarya (Caetano-Anollés and Caetano-Anollés 2003) matching values for the English and Chinese languages, respectively (Li et al. 2016). Thus, protein domains and literary languages share the same probability distribution but with different scaling regimes distinguishing akaryotic microbes from higher organisms (Nasir et al. 2017). We note that power-law behavior is not intrinsic to biological networks. For example, a phylogenomic tracing of age in bipartite networks of protein domains and elementary functional loops (and their projections) revealed that the scale-free property of these networks was transferred from loops to domains $\sim$ 3.4 billion years ago (Aziz et al. 2016). Thus, scale-free probability distributions are emergent (not intrinsic) properties of biological networks. A number of other probability distributions are expected to be incorporated in molecular structure as emergent properties. For example, quantitative linguistic distribution models of decreasing returns accurately describe the distribution of the length of proteins in proteomes (Eroglu 2014). Thus, language-like behavior exists in proteins and its significance can be explained. This has been demonstrated by using a probabilistic approach borrowed from linguistic patterns in speech recognition and incorporating it for detecting protein domains via Hidden Markov Models (HMMs) (Coin et al. 2003).

(ii) *Functional type laws*. These laws link two kinds of variables, such as word frequency and word length or the frequency of syntactic constructions and their complexity. One of the most notorious is the Menzerath-Altmann (MA) law. Menzerath (1928) found that longer syllables contained shorter articulated sounds, and later, that words with more syllables were phonetically shorter. He summarized this generality for language constructs with the motto "the greater the whole, the smaller its constituents" (Menzerath 1954). These qualitative statements were elaborated mathematically by Altmann (1980). Mathematical elaborations were later supported by statistical analyses of many languages, and linguistic and phonetic relationships of many types, including vocal communications outside humans (gelada primates; Gustison et al. 2016). Recently, the elaborations have been extended to genomes (e.g., Ferrer-i-Cancho and Forns 2010; Baixeries et al. 2012). One general formulation of the accepted MA law, which adds the effect of hierarchy in the makeup of the parts of the system being studied (Altmann 1980), follows Equation (2)

$$y(x) = Ax^b e^{-cx} \tag{2}$$

with $y(x)$ being the length of the parts, x representing the length of the system (or constructs of parts), and A, b and c fitting parameters. x can also represent a discrete variable describing the number of parts that make up the system. A more general formulation adds dependencies on additional variables (Strauss and Altmann 2006). $y(x)$ is generally measured by counting parts defined at a deeper level of the system's organization (e.g. amino acids of domains). This general formulation of the law accommodates the effects of multi-level structure that is typical of a language.

In a recent study, we showed that a MA law governs the organization of domains in proteomes (Shahzad et al. 2015). We used a highly curated dataset of ~3 million sequences from 745 proteomes covering superkingdoms Archaea, Bacteria and the four main kingdoms of Eukarya (Protista, Plants, Fungi and Metazoa) (Wang et al. 2011). Structural domains were carefully annotated with domain and linker lengths of proteins using hidden Markov models (HMMs) of structural recognition, which guarantee ~98% success in structural assignments when atomic structures are known. Protein entries were retrieved trusting the low probability of cryptic domains matching non-domain linker sequences ($P < 0.0001$) and the absence of biases imposed on length estimates by superkingdom-specific Markovian models (Wang et al. 2011). Plotting the lengths of structural domains against their number in the proteins of proteomes belonging to 60 organisms showed significant negative correlations that are typical of a general two-parameter power law form of the MA law. Figure 3a shows an example log-log plot for the proteome of the extremely halophilic marine archaeon *Halobacterium salinarum*. The linear regression lines in the log-log plots show high coefficients of determination (R^2) with values ranging 0.85–1.00, and significant F test-derived correlations ($p < 0.0001$–0.133). We also observed an MA law governing the relationship between the length of secondary structures defined by the DSSP classification and their number in structural domains (Minhas, Mittenthal and Caetano-Anollés, ms. in preparation). Figure 3b shows an example log-log plot describing a proteomic MA law for protein secondary structures of *H. salinarum*. Thus, universal MA laws of language govern the size of domains and secondary structures in proteins at these two levels of structural abstraction.

(iii) *Developmental type laws.* These laws relate language properties to time, with time sometimes expressed as accumulating innovation. Generally, they describe the growth and evolution of vocabularies. They can also describe language variation at all levels, from lexical and phonetic to semantic and syntactic. The Heaps law (also known as the Herdan law) is a typical example. It describes how the number of different words V in a document or corpora (the vocabulary) scales with the database of words N measured as the number of words (the length of text) (Herdan 1964, Heaps 1978). Equation 3 describes this empirical scaling

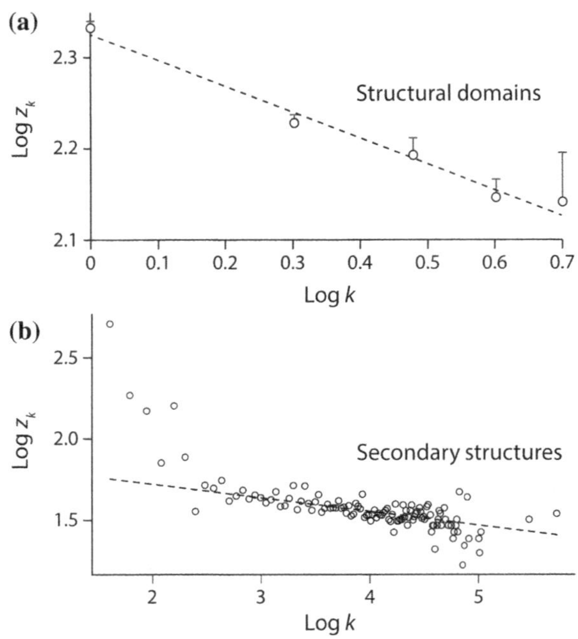

Fig. 3 The Menzerath-Altmann (MA) law of language in the structural domains and secondary structures of the *Halobacterium salinarum* proteome. **a** A log-log plot describes how the average length (z_k) of structural domains in a protein decreases with the number (k) of structural domains it contains for the entire proteome. The horizontal bars on the circles depict the standard error of the means. The hatched regression line indicates the weighted linear fitting ($y = -0.281x + 2.324$; $R^2 = 0.980$; $p = 0.0012$). Data from Shazhad et al. (2015). **b** The log-log plot describes how the average length (z_k) of secondary structure elements 1 of a protein domain decreases with the number of secondary structure elements it contains for all proteins in the proteome. The hatched lines indicate the weighted regression line ($y = -0.084x + 1.892$; $R^2 = 0.414$; $p = 6 \times 10^{-14}$). The $y(x) = Ax^b$ two-parameter power-law formulation of the MA law facilitates fitting parameters in the log-log plots to a straight line

$$V(n) = KN^{\beta} \tag{3}$$

with K and β being empirically determined parameters. Note that the Heaps law can be interpreted in the context of a Zipf distribution when $\beta = 1/\gamma$, and that constituents of vocabularies are constant over centuries if they represent 'core' words that appear with high frequency (Gerlach and Altmann 2013). Thus, vocabulary growth (a dynamic process) can be described in terms of database size (a static view). There are a number of scaling relationships of these kinds in the vocabulary of genomes, some reflecting the scaling of the number of genes in different functional categories as a function of genome size (Molina and van Nimwegen 2009). For the vast majority of these functional categories, the scaling appears quite universal. An analysis of 30 archaeal organisms shows for example that the vocabulary of FSF domains scales with genome size and total FSFs with β = 0.30–0.36 and K = 33–416 (Fig. 4). These values are not far away from those of the English text corpora ($\beta = 0.4$–0.8 and $K = 10$–200) (e.g. the Gutenberg Project e-book collection, $\beta = 0.45$; Tria et al. 2014). Since sublinear growth ($\beta < 1$) is a signature of the Heaps law, the structural domain vocabulary of Archaea follows closely the scaling laws of English language.

One typical law of historical linguistics is the Pietrowski law. The law describes a process of accumulation of change with rates first increasing and then decreasing, i.e. accumulation starts with an exponential form and is followed by an S-shaped (sigmoid curve) form. When developmental processes continue for very long time, the accumulation can sometimes end with a reverse S-shape. The elaboration of Beöthy and Altmann (1982) of the Pietrowski law, with assumptions borrowed from Weinreich et al. (1968), arrives at a logistic expression that describes the growth of the proportion p of new forms developed in time. This is described in Equation (4),

$$p = 1/1 + ae^{-bt} \tag{4}$$

where a in an integration constant and b a constant parameter. A phylogenomic analysis of the accumulation of structural domains in evolution shows complex logistic behavior (Caetano-Anollès et al. 2011). This is illustrated in Figure 5a. The accumulation of FSF domains in time was annotated with Vogel's classification of molecular functions, which assigns 7 major functional categories and 50 subcategories to SCOP domains based on SCOP, Interpro, Swiss-Prot and literature

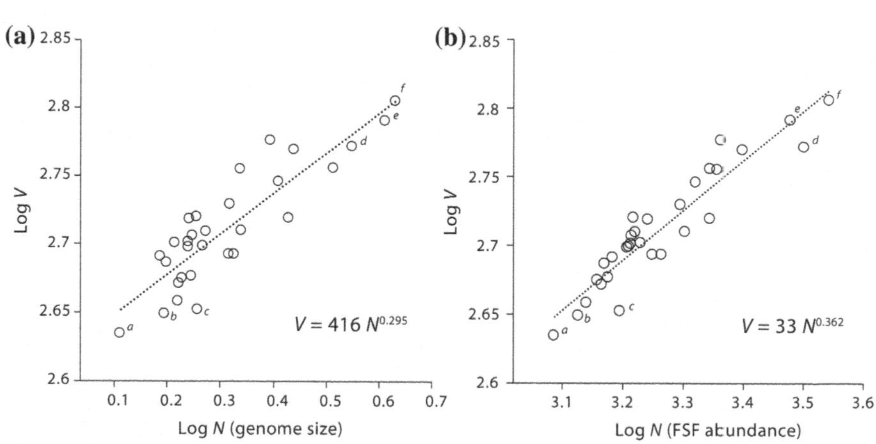

Fig. 4 A Heaps law in the structural domain vocabulary of the proteomes of Archaea. The vocabulary (V) of FSF domains (their occurrence) in a proteome scales linearly with database size (N), measured as genome length in millions of base pairs **a** (y = 0.295x + 2.619; $R^2 = 0.77$; $p = 5 \times 10^{-14}$) or FSF number **b** (y = 0.362x + 1.531; $R^2 = 0.88$; $p = 3 \times 10^{-16}$), in a log-log plot of the variables. The scaling relationship follows Heaps formulations. Example organisms with smallest and largest genomes: *a*, *Ignicoccus hospitalis*; *b*, *Staphylotermus marinus*; *c*, *Thermophilus pendens*; *d*, *Methanospirillum hungatei*; *e*, *Methanosarcina mazei*; *f*, *Haloarcula marismortui*

sources (Vogel 2005; Vogel and Chothia 2006). Accumulation was expressed as a cumulative frequency distribution. S-like curves were obtained for the seven general categories suggesting a match of structural domain and language organization. The growth of FSFs in the 'General' and 'Metabolism' categories occurred before all others; 'General' includes the 'Small molecule binding' subcategory, a set of functions associated with FSFs typical of early metabolism (Fig. 5a). These initial FSFs were followed closely by FSFs in 'Information', 'Regulation', and 'Intracellular processes'. As expected, molecular functions that are more evolutionarily derived and typical of multicellular organisms ('Extracellular processes', which include cell adhesion, immune response, toxins and defense, and blood clotting) or viruses ('Viral proteins' in category 'Other') grew later in evolution. Note that the more ancient categories showed clear double S-curves that are typical of paths of high performance in diffusion of innovation models (Rogers 1962). These double S-curves and the rationale of their generation is illustrated in Figure 5b. The innovation diffusion model is inspired by Gabriel Tarde's proposals developed in the 1890s of the spread of cultural innovations by diffusion from one society to another (Kinnunen 1996). Innovation is communicated among entities as time progresses, some of which are prone to adopt the innovation (Rogers 1962). The innovation lifecycle involves a 5-step decision-making process that involves 'innovators', 'early adopters' (trendsetters), an 'early majority', a 'late majority', and 'laggards', in that order. If language change is considered a 'communication' process that adopts innovations of this kind, then S-curves can be better modelled using the socioeconomic-focused Bass method (Bass 1969). The method considers a homogeneous population with a fraction of innovation adopters (ρ) increasing

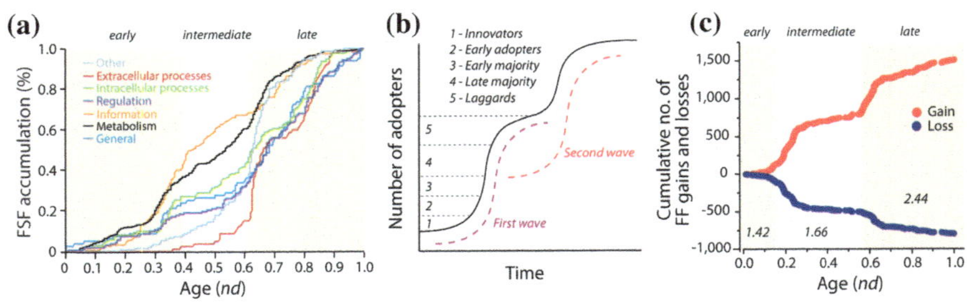

Fig. 5 S-shaped growth of structural domain vocabularies in proteomes. a Accumulation of SCOP FSFs linked to Vogel's seven broad functional categories along an evolutionary timeline (chronology), which was derived from a phylogenomic tree of structural domains that was reconstructed from 184 proteomes representing all three superkingdoms of life. The age of domains is given as node distance (*nd*) values, in a scale from 0 (origin) to 1 (the present). Data from Caetano-Anollés et al. (2011). **b** Double S-curve typical of paths of high performance in diffusion of innovation models. **c** Cumulative numbers of gains and losses in the occurrence of SCOP FFs in Archaea. Scatter plots reveal growth trends in the accumulation of FF gains and losses in the proteomes of 48 archaeal organisms. Gains are identified in red while losses are identified in blue. The three evolutionary epochs of the protein world (early, intermediate, late) are marked with corresponding average gain-to-loss ratios in italics. A similar pattern is obtained when studying abundance of FFs. Data from Nasir et al. (2014)

when they meet non-adopters (1- ρ) under intrinsic forces with rate q (coefficient of imitation) and under external forces with rate p (coefficient of innovation). The adoption process follows Equation (5)

$$d\rho(t)/dt = (p + q\rho(t))(1 - \rho(t)) \tag{5}$$

and its derivation leads to the differential Equation (6)

$$d\rho(t) = p + (q - p)\rho(t) - q[\rho(t)]^2 \tag{6}$$

which corresponds to a Ricatti form with constant coefficient, a first order ordinary differential equation that is quadratic in the unknown function. Note that the Bass diffusion model becomes a logistic distribution (complying with the Pietrowski law) when $p = 0$, i.e. when only imitation by non-adopters governs the diffusion process. In other words, the logistic and Bass models become isomorphic (but not identical; see Guseo 2016). Conversely, the model reduces to the exponential distribution when $q = 0$, i.e. when the emergence of innovation drives change. Under the Bass model, S-curves of vocabulary replacement can be modeled to show differential effects of intrinsic imitation and exogenous innovation factors on the replacement process (Ghanbarnejad et al. 2014). The Bass model can also be used to explain successive generations (waves) of innovation (Norton and Bass 1987). These waves of innovations can compete with each other, sometimes leading to substitutions of innovations. However, when innovation occurs in parts of more complex wholes, their substitution enhances the entire system, which successfully continues adoption.

This second alternative explains the double S-curves that describe domain innovation in proteins (Fig. 5a). In fact, we recently reconstructed ancestral states in the branches of phylogenomic trees of structural domains at FF level and phylogenomic trees of proteomes derived from the same datasets (Nasir et al. 2014a, 2014b). This allowed to calculate the number of gains and losses of FFs for every time step of the chronology of domains, which should be regarded as chains of continued adoptions and substitutions of innovations. Figure 5c shows an analysis of this kind for all proteomes of Archaea. Scatter plots reveal clear double S-curves of accumulation of both FF gains and losses. Gains however override losses at increasing ratios of 1.42, 1.66 and 2.44 for the early (architectural diversification), intermediate (superkingdom specification) and late (organismal diversification) epochs of domain evolution, which are defined for example in Caetano-Anollés et al. (2011). The first S-curve covers exactly the early and intermediate epochs, while the beginning of the second wave starts at the boundary of the intermediate and late epochs. Note that the boundary is defined by a burst of structural domains that establish domain combinations, and are specific to lineages in a diversifying world of proteomes (Wang and Caetano-Anollés 2009). The boundary is quickly followed in evolutionary time by a massive combinatorial 'big bang' in domain organization. It is clear that the first wave of innovation and its associated S-curve (Fig. 5b) involves the discovery of new domains while the second wave involves

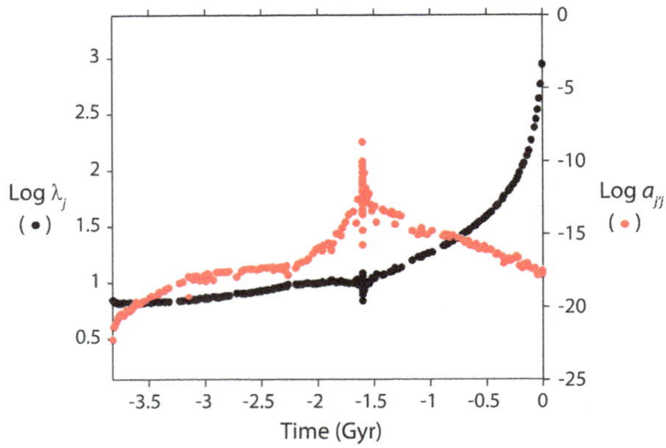

Fig. 6 Change of parameters specifying rates of generation of new variants of an FSF domain (λ_j) and rates of generation of new FSFs (a_{ij}) along the chronology of structural domains, with time expressed in billions of years (Gyr). λ_j starts at 7 Gyr^{-1}, ranges 7–12 Gyr^{-1} during the transition occurring at 1.5 Gyr, and grows super-exponentially to 867 Gyr^{-1} due to the combinatorial explosion of domains in multidomain proteins. In contrast, a_{ij} increases towards the transition and then decreases, showing typical values of 10^{-17} Gyr^{-1}, which matches the 10^{-11}–10^{-77} fraction of sequence space that the structure occupies. Data from Tal et al. (2016)

domains that have a capacity to combine with others to enhance the functional repertoire of the cell. Thus, two innovation modes appear in time as protein structural domains adopt new folds and molecular functions. The first mode involves domain makeup while the second involves domain organization in proteins.

Remarkably, these same two innovation modes became emergent properties of a dynamic model of protein evolution. The evolution of structural domains at fold and FSF levels was modeled with 'global' birth-death differential equations and domain abundances as state variables (Tal et al. 2016). The 'irreversible tree-hugging' model of abundance growth of domain structures that was used, allowed only 'forward' transitions from ancestral domain structures to neighboring descendants, along branches of a tree of structural domains. The model is governed by Equation 7.

$$dN_j/dt = (\lambda_j - \sum_{i \neq j} a_{ij})N_j + \sum\nolimits_{i \neq j} a_{ji} N_i \tag{7}$$

N_j are global genomic abundances of domain structures of type j, λj is the rate (birth minus death) of generating new variants of structure j, and a_{ij} is the rate of 'forward' transition from structure j to structure i. In the model, each fold or FSF diversifies by producing domain variants in approximately exponential manner with rate λ. The accumulating variants produce a growing 'funnel'. At some point, given rate a, the growing funnel generates a new fold or FSF structure. This new structure is a unique and rare innovation (since $a << \lambda$) that creates a new growing funnel of variants. The growth of protein domains in evolution is thus governed by two parameter classes, λ and *a*, which are analogous to imitation and innovation parameters *q* and *p* of the Bass model, respectively. Figure 6 plots rate constants *a* and λ against time of origin of corresponding domains, i.e. their age. Remarkably, the plots reveal trends that are generative of a biphasic pattern, with λ always increasing and *a* increasing and then decreasing in evolutionary time. The inflection point occurs 1.5 billion years ago (nd $\sim$ 0.55), a time that coincides with the rise of the second wave of innovations of Figure 5c and the 'big bang' of domain combinations. The biphasic pattern that is evident by linking phylogenomic data and the birth-death differential equations of protein evolution can be explained by a general model of evolution of modules (Mittenthal et al. 2012). In this model, modules emerge through a biphasic process of diversification. In the first phase, parts are weakly linked and associate variously. As they diversify, parts compete with each other and are selected for performance. The emerging interactions constrain their structure and associations. This causes parts to self organize into modules with tight linkages. In the second phase, variants of the modules evolve and become new parts for a new generative cycle of higher-level organization. This model is compatible with double S-curves in diffusion of innovation models. It also explains the existence of a transition in rate constants of structural diversification of domains (Fig. 6).

6 Language and the persistence triangle

Altmann (1980) suspected that the size-dependent minimization principle of the MA law was "somehow connected with the principle of least effort or with some not yet known principle of balance recompensating lengthening on one hand with shortening on the other". The law was later interpreted as the result of either 'compression of information' or self-organization (Köhler 1987; Cramer 2005; Gustison et al. 2016). We recently interpreted the law differently, claiming the minimization principle resulted from trade-offs between economy, flexibility and robustness (Shahzad et al. 2015). For that purpose, we used a general framework of persistence strategies for biological systems that is supported by ecological and molecular data (Yafremava et al. 2013). Figure 7 illustrates how economy, flexibility and robustness (defined in Table 2) impose constraints on the engineering of biological systems when strategies impinge on the matter-energy budget and unfold information (suitably defined in Yafremava et al. 2013). Different trade-offs define a 'persistence triangle' for those systems, with vertices representing the three main strategies. We illustrate the framework by describing processes of diminishing returns in the molecular makeup of proteins (Fig. 7a). We discovered that the MA law represents a power law relationship when molecular persistence (*P*) is made a function of domain accretion, measured as number of domains *k* in proteins (Shahzad et al. 2015). *P* holds two terms, one reflecting the matter-energy cost of adding domains and extending their length in proteins (PME), and the other describing the influence of information present in domain length and domain count on the flexibility and robustness of the molecular system (PFR). Since P depends only on the length and number of domains, the derivative of P with respect to k, when set equal to zero, gives the power law version of the MA law of Equation 2, $z_k = Ak^b$, where z_k is domain length, and b and A are the slope and intercept of the log-log plot linear fittings, respectively. Thus, P describes a frustrated landscape of trade-off relationships that unfolds when an axis of matter-energy budget is complemented by an axis of information. Information is here stored in intramolecular

Table 2 Three principles driving the persistence of a system

Term	Definition
Economy	The budget of the system in terms of matter-energy costs to maintain its functioning mechanisms.
Flexibility	The structural and functional mechanisms responding to *internal* and *external* changes imposed on the molecular system. These mechanisms require processing of information and delimit the potential of the system to adapt to environmental change. More flexible systems are generally larger, harbor more complex functionalities, and are more diverse in finding trade-off solutions.
Robustness	The mechanisms that use information to maintain structure and function despite external influence and protect molecules from malfunction. These mechanisms provide the ability to resist damage and change. Robustness includes stability but refers to the broader processes that are passive from an information point of view.

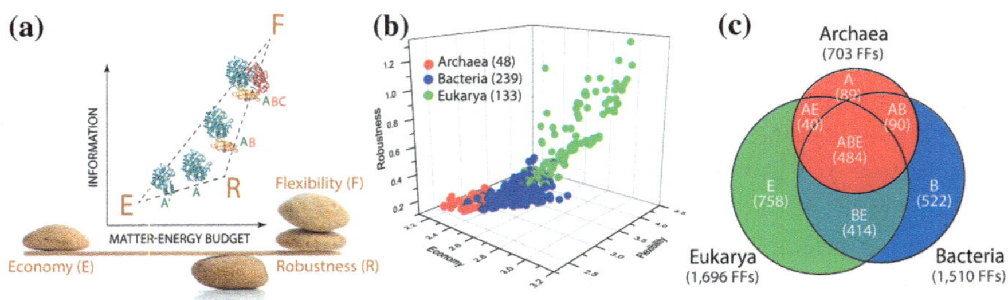

Fig. 7 The persistence triangle and its impact on the proteomic vocabulary of superkingdoms. a Diagram showing how domain size and organization in proteins affects molecular persistence. Domain A' extends its length in A, and A combines with B and C in evolution to form multi-domain proteins with different propensities towards economy (E), flexibility (F) and robustness (R), finding optimal trade-off solutions given a matter-energy budget and information. **b** Scatter plot describing the persistence strategies of the organisms of superkingdoms, measured by the occurrence and abundance of FF domains. Axes are in logarithmic scales and the number in parentheses indicate number of proteomes that were sampled in the study. Data from Nasir et al. (2014a). **c** Venn diagram of the diversity and distribution of 2,397 FFs in 420 proteomes of superkingdoms Archaea (A), Bacteria (B) and Eukarya (E) describing seven possible taxonomic groups of FF sharing (Kim and Caetano-Anollés 2012). The Venn taxonomic groups of Archaea (ABE, AE, AB and A) are highlighted, showing the high levels of compression in the generation of fold novelties

and intermolecular interactions necessary for molecular function and stability. For example, information in the combinations of the domains of multi-domain proteins generates an additional vocabulary that can be equated with language information (Bashton and Chothia 2007). Plots of *b* versus *A*, and *b* and *A* versus effective protein length (sum of length of domains in a protein without considering linkers and terminal sequences) for the 60 proteomes of Archaea, Bacteria and Eukarya that were examined showed distinct trade-off solutions (Shahzad et al. 2015). Archaea, Fungi and to a lesser degree Plants fostered molecular economy, each at their own economic stratum, as evidenced by the number of domains they harbor. Fungi increased domain size in single domain proteins while reinforcing the MA pattern of diminishing returns. In contrast, Metazoa, and to lesser degrees Protista and Bacteria, relaxed economy, each at their own economic stratum. Metazoa achieved maximum flexibility and robustness by harboring compact molecules and complex domain organization.

Similar trade-off dynamics materializes in the domain makeup of Proteomes (Nasir et al. 2014a). The diversity (use) and abundance (reuse) of structural domains defined at the FF level of the SCOP hierarchy were used as proxies for the economy and flexibility of proteomes. These proxies were then plotted against a reuse-to-use ratio of FFs (proportional to the scaling coefficient β of the Heaps law), which was previously shown to represent an index of robustness (Yafremava et al. 2013). The proteomes of Eukarya traded economy for maximal flexibility and robustness. In contrast, the proteomes of archaeal and bacterial microbes relaxed flexibility and robustness in favor of economy (Fig. 7b). The proteomes of Archaea fostered the greatest levels of economy. Figure 7c compares the structural diversity of FF domains in the three superkingdoms and confirms that archaeal proteomes are the least diverse. The Venn diagram shows that Archaea holds the smallest vocabulary (703 FFs versus 1,510 FFs for Bacteria and 1696 FFs for Eukarya) with the least number of superkingdom-specific FFs (89 FFs versus 522 FFs for Bacteria and 758 FFs). While a substantial number of domains defined at this very low abstraction level of SCOP (20%) is common to all three superkingdoms (the ABE Venn group), over two thirds of the archaeal vocabulary (69%) belongs to this group. Thus, the domain vocabularies of the proteomes of Archaea are the most 'compressed' of the cellular world.

7 A Semantic Compression of the Domain Vocabulary of Archaea

Semantic compression is a process of compacting a vocabulary by reducing language heterogeneity, while at the same time maintaining its semantics. In other words, a process of conveying a same message with a smaller set of words (Chomsky 1995; Sayood 1996). The substantial push towards economy manifested in the domain makeup of the proteins and proteomes of Archaea is also apparent in

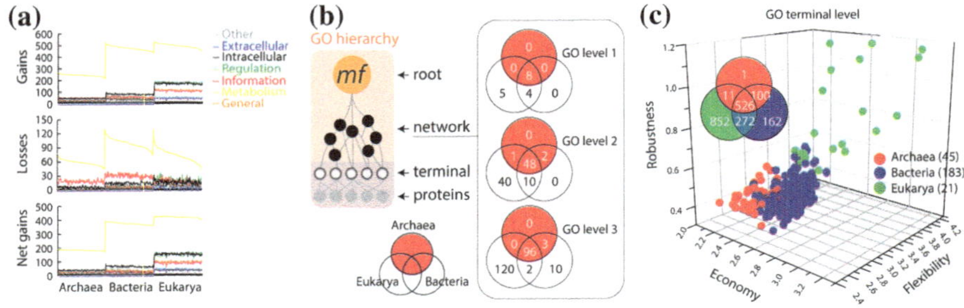

Fig. 8 Semantic and pragmatic compression in the proteomes of Archaea. a Functional characterization of FF domain gains and losses in a set of 150 proteomes representing the three superkingdoms of life. The proteomes of organisms drawn from those examined in panel B were analyzed using the same phylogenomic methods that are described in Fig. 5c. Their FFs were annotated using Vogel's seven broad categories of molecular functions and proteomes ordered in the plots according to the FF abundance levels of the most abundant category for net gains, gains and losses. Data from Nasir et al. (2014a). **b** The GO hierarchy associates proteins to molecular, biological and cellular roles. For example, 'terminal' GO terms of molecular functions (*mf*) that are associated with protein-encoding genes are further grouped into a hierarchy through an ontological 'network'. The Venn diagrams describe the sharing patterns of GO levels 1, 2 and 3 terms in the 249 genomes of the three superkingdoms. Data from Koç and Caetano-Anollés (2017). **c** Scatter plot describing the persistence strategies of organisms measured by the occurrence and abundance of their associated terminal GO terms. Axes are in logarithmic scales and the number in parentheses indicate number of proteomes that were examined in the study. The Venn diagram describes the diversity and distribution of 1,924 terminal GO terms in the superkingdoms. Data from Nasir et al. (2014b)

the growing semantics of its molecular vocabulary (Kim and Caetano-Anollés 2012; Yafremava et al. 2013; Nasir et al. 2014a, 2014b; Shahzad et al. 2015). A functional characterization of FF domains that are gained and lost along the evolutionary timeline of domain discovery revealed significant variations in FF abundance levels annotated to Vogel's seven functional categories at single proteome level (Fig. 8a). Consistent trends of domain recruitment were present in all three superkingdoms for all functional categories matching extant patterns described in Nasir et al. (2011), with gain and loss of metabolic domains dominating those of the other functional categories. Final recruitment of domains (their net gains) increased in the order Archaea, Bacteria and Eukarya for all categories. An analysis of the 'metabolism' category (yellow lines in Fig. 8a) showed that the individual proteomes of Archaea exhibited minimal variation in both gains and losses of FF abundance when compared to the proteomes of Bacteria and Eukarya. In fact, variation in the gain and loss of FFs between proteomes increased in the order Archaea, Bacteria and Eukarya, with variation of FF loss being particularly significant. Thus, the reductive evolutionary compression of the molecules of Archaea had the additional effect of maximally reducing the heterogeneity of the metabolic vocabulary in all proteomes of this microbial superkingdom. Interestingly, loss of informational domains in Archaea and Bacteria (associated with replication, transcriptions, etc.) were compensated by gains in Archaea but not in Bacteria. Thus, Archaea salvages molecular functions linked to information, increasing its robustness relative to Bacteria, and supporting additional evidence for this trend (Yafremava et al. 2013). In contrast, no clear patterns of loss were observed in Eukarya; the high variability of losses relative to those of metabolism indicates it represents a random process. Notable patterns of gain-and-loss dynamics in Eukarya showed increases in net gains of FFs involved in 'Metabolism', 'Regulation'–'Intracellular processes', 'General'–'Information', and 'Extracellular processes'–'Other', in that order (Fig. 8a). Their abundance mismatches the preponderance of metabolism and information in eukaryotic annotations.

The archaeal semantic compression also resulted in highly reduced vocabularies of molecular functions. Here we propose that this occurs through a process of 'pragmatic compression' that reduces the semantics of molecular function without sacrificing its pragmatics. This new compression conveys all possible messages with a smaller set of meanings. In other words, the compressed message uses fewer meanings while preserving the context of the vocabulary, which in biology has the goal of increasing the fitness of an organism in interaction with its environment. Pragmatic compression becomes evident when surveying GO annotations of genes at different levels of the GO hierarchy (Fig. 8b). The GO project generates a directed acyclic graph (DAG) for each of three ontological groups, molecular functions (*mf*), biological processes (*bp*) and cellular compartments (*cc*)(Ashburner et al. 2000). Protein-encoding genes are for example annotated with terminal GO terms of molecular functions and these are further grouped into different levels of ontological abstraction, the highest being GO level 1. The distribution and sharing of GO levels 1, 2 and 3 terms between superkingdoms were made explicit with Venn diagrams (Fig. 8b). Out of 17 level 1 terms, 47% were universal (present in at least one organism of Archaea, Bacteria and Eukarya) and 29% were Eukarya-specific. Similarly, out of 101 level 2 terms, 48 were universal (48%) and 40 were Eukarya-specific (40%). Finally, out of 231 level 3 terms, 96 were universal (42%), 10 were Bacteria-specific (4%) and 120 were Eukarya-specific (52%). Remarkably, 100, 94 and 97% of level 1, 2 and 3 terms, respectively, were universal in Archaea. In contrast, 67, 80 and 86% were universal in Bacteria and 47, 48 and 44% were universal in Eukarya, respectively. One tantalizing conclusion is that the repertoire of high-level GO terms of Archaea is highly enriched in terms that were already present in a hypothetical common ancestor of life that existed billions of years ago. It is particularly noteworthy that there were no Archaea-specific GO terms at these top three levels of GO classification. Thus, the pragmatic compression of archaeal functional repertoires (functionomes) is maximal.

To showcase trade-off dynamics of the functionomes of superkingdoms, we compared the diversity (use), abundance (reuse) and reuse-to-use ratios of GO terminal terms for a set of 249 functionomes (Nasir et al. 2014a, 2014b). These statistics were used as proxies for functionome economy, flexibility and robustness and were displayed in a 3-dimensional plot describing a persistence triangle for molecular semantics and pragmatics (Fig. 8c). The patterns resembled those already described with FF domains (Fig. 7b). Archaeal and bacterial functionomes favored economy (Fig. 8c). In turn, eukaryotic functionomes favored flexibility and robustness. The functionomes of Archaea showed the highest levels of economy-driven compression but their levels of robustness were not far away from those of Bacteria. A Venn diagram (inset of Fig. 8c) that compares the diversity of GO terminal terms in the three superkingdoms confirms that archaeal functionomes are the least diverse at this lowest level of ontological classification. The 638 terminal GO terms of Archaea are substantially lower than those of Bacteria (1,060) and Eukarya (1,661). Most terms were universal in Archaea, 82% versus 49% and 32%, for Bacteria and Eukarya, respectively. Only one GO terminal term was Archaea-specific ("methylenetetrahydromethanopterin dehydrogenase activity [GO: 0030268]"). Thus, the vocabularies of molecular functions of Archaea should be considered the most evolutionarily compressed and homogeneous of the cellular world.

8 The Natural History of Semantic and Pragmatic Compression

All language is a set of symbols whose use among its speakers assumes a shared past

—The Aleph and Other Stories, Jorge Luis Borges.

Natural history is the study of the origin, evolution and structure of natural objects. Its goal is to uncover patterns and processes unfolding in nature that are responsible for the complexities of life. We have been studying the natural history of proteomes and functionomes for almost two decades and done so at global levels. The history of diversification of structural domains at different levels of SCOP and CATH abstraction provided considerable insight into the origin and diversification of the organismal world (e.g., Wang and Caetano-Anollés 2009; Kim and Caetano-Anollés 2012; Bukhari and Caetano-Anollés 2013). Tables of ages for the origin of each structural domain group are available. Similarly, phylogenies of molecular functions and their associated chronologies provide evolutionary ages of GO terms at different levels of the GO hierarchy (e.g. Nasir et al. 2014b; Koç and Caetano-Anollés 2017). Phylogenomic analysis of these terms also produced a picture of organismal diversification that is congruent with that obtained from structural domains (Kim et al. 2014). In all of these deep phylogenomic studies, Archaea represents the first domain of diversified life that appeared in evolution (reviewed in Caetano-Anollés et al. 2014). Since domain gains override losses in proteomes (Nasir et al. 2014a), structural domains and their associated molecular functions grow in time, complying with the expectations of the Heaps law of language (Nasir et al. 2017). Therefore, the accumulation of domain and their associated GO terms can inform about the natural history of semantic and pragmatic compression in molecular evolution. As an example, Fig. 9 presents a heat diagram

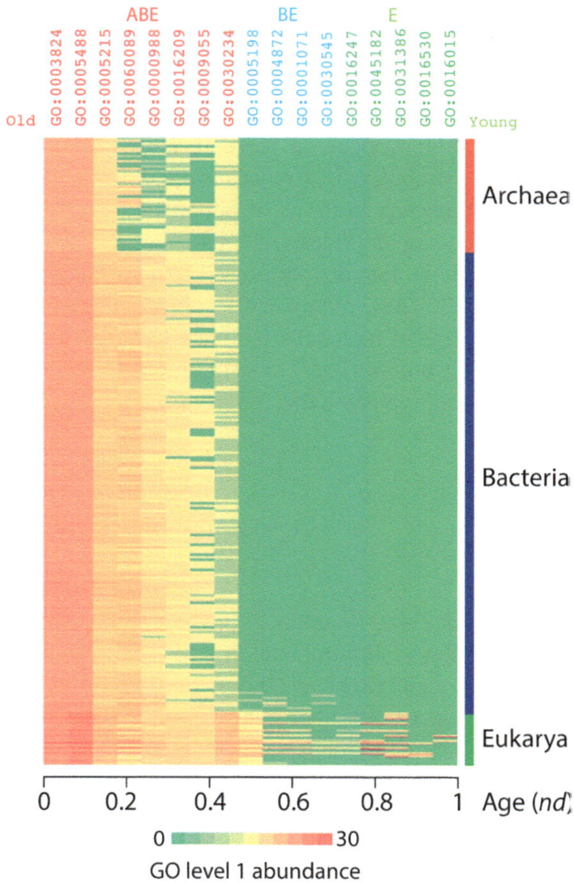

Fig. 9 A natural history of genomic abundance of level 1 GO terms reveals the early origin of semantic and pragmatic compression. The heat map shows the distribution of genomic abundances of 17 level 1 GO terms in 249 organisms and maps them to an evolutionary timeline of molecular function innovation derived from a phylogenomic reconstruction. The age of GO terms is expressed in a relative scale of node distance (*nd*) values, from 0 (origin of molecular functions) to 1 (the present). GO terms: GO:0003824, catalytic activity; GO:0005488, binding; GO:0005215, transporter activity; GO:0060089, molecular transducer activity; GO:0000988, protein binding transcription factor activity; GO:0016209, antioxidant activity; GO:0009055, electron carrier activity; GO:0030234, enzyme regulator activity; GO:0005198, structural molecule activity; GO:0004872, receptor activity; GO:0001071, nucleic acid binding transcription factor activity; GO:0016247, channel regulator activity; GO:0045182, translation regulator activity; GO:0030545, receptor regulator activity; GO:0031386, protein tag; GO:0016530, metallochaperone activity; GO:0016015, morphogen activity. Data from Koç and Caetano-Anollés (2017)

that shows genomic abundances of level 1 GO terms for a set of 249 organisms with columns organized according to the age of the molecular function at this top level of the ontological hierarchy. As expected, the first 3 GO terms were the most abundant. They were linked to binding, metabolism and transport, functions that are truly universal, i.e. present in all 249 genomes examined. The next 5 GO terms were also universal (they belonged to the ABE Venn category) but were not present in all functionomes. They were specially absent in some Archaea. The rest were either present in Eukarya and very few lineages of Bacteria or were Eukarya-specific, all of which appeared more recently. The heat diagram also shows that the functional vocabulary of Archaea is the oldest and most compressed. It is also the most homogeneous in terms of abundance levels of the most basal functions (binding, metabolism and transport), matching our previous observations (Fig. 8a). These same patterns manifest at other levels of the GO hierarchy (Koç and Caetano-Anollés 2017). Thus, the diversification of the metabolic vocabulary of Archaea along the timeline of molecular innovation appears to arise from strong evolutionary reductive constraints. These constraints were likely imposed by the microbial lifestyle and the typical harsh environments in which the archaeal organisms live.

9 Conclusions

We have shown that the structural domains of proteins can be considered the signs of a molecular language and their combination a set of hidden rules defining the molecular functions of the cell. However, biomolecules are highly structured and hold multiple layers of organization that comply with the laws of language and diversify in ways that are consistent with language change. Of all cellular superkingdoms, Archaea holds the smallest, oldest and most homogeneous vocabulary. This molecular vocabulary has been compressed in evolution with semantic and pragmatic rationales. This "generalized encoding" endows biological systems with the relative ability to evolve faster (Cariani 1998).

We end by noting that the understanding of the makeup and meaning of this molecular language does not mean we understand the drivers of molecular change or the hidden codes that are responsible for molecular innovations. Our classification schemes of structure and ontology are scientific community efforts that are important but can be defective. At present, the community continues to strive towards a code that attempts to describe how the structure and molecular functions of proteins are linked to the genetic code.

References

Altmann EG, Gerlach M (2016) Statistical laws of linguistics. In: Degli Esposti M, Altmann EG, Pachet F (eds) Creativity and Universality in Language. Lecture Notes in Morphogenesis. Springer, Berlin, pp 7–26

Altmann G (1980) Prolegomena to Menzerath's law. Glottometrika 1980(2):1–10

Altmann G, Köhler R, Vulanović R (2006) Encyclopedia of linguistic laws. http://lql.uni-trier.de/index.php/Main_Page. Accessed 2 Jan 2017

Ashburner M, Ball CA, Blake JA, Botstein D, Butler H, Cherry JM, Davis AP, Dolinski K, Dwight SS, Eppig JT (2000) Gene Ontology: tool for the unification of biology. Nat Genet 25:25–29

Alva V, Söding J, Lupas AN (2015) A vocabulary of ancient peptides at the origin of folded proteins. eLife 4:e09410

Aziz MF, Caetano-Anollés K, Caetano-Anollés G (2016) The early history and emergence of molecular functions and modular scale-free network behavior. Sci Rep 6:25058

Bai L, Xie T, Hu Q, Deng C, Zheng R, Chen W (2015) Genome-wide comparison of ferritin family from Archaea, Bacteria, Eukarya, and Viruses: its distribution, characteristic motif, and phylogenetic relationship. Sci Nat 102:64

Baixeries J, Hernández-Fernández A, Ferrer-i-Cancho R (2012) Random models of Menzerath-Altmann law in genomes. Biosystems 107:167–173

Barabási A-L (2009) Scale-free networks: a decade and beyond. Science 325:412–413

Bashton M, Chothia C (2007) The generation of new protein functions by the combination of domains. Structure 15:85–99

Bass F (1969) A new product growth for model consumer durables. Manag Sci 15(5):215–227

Beöthy E, Altmann G (1982) Das Piotrowski-Gesetz und der Lehnwortschatz. Zs Sprachwissenschaft 1:171–178

Berezovsky IN, Trifonov EN (2001) Van der Waals locks: Loop-n-lock structure of globular proteins. J Mol Biol 307:1419–1426

Bhaskara RM, Srinivasan N (2011) Stability of domain structures in multi-domain proteins. Sci Rep 1:40

Bukhari SA, Caetano-Anollés G (2013) Origin and evolution of protein fold designs inferred from phylogenomic analysis of CATH domain structures in proteomes. PLoS Comp Biol 3: e1003009

Caetano-Anollés G, Caetano-Anollés D (2003) An evolutionary structured universe of protein architecture. Genome Res 13:1563–1571

Caetano-Anollés G, Wang M, Caetano-Anollés D, Mittenthal J (2009) The origin, evolution and structure of the protein world. Biochem J 417:621–637

Caetano-Anollés G, Yafremava L, Mittenthal JE (2010) Modularity and dissipation in evolution of macromolecular structures, functions, and networks. In: Caetano-Anollés G (ed) Evolutionary bioinformatics and systems biology. Wiley-Blackwell, Hoboken, NJ, USA, pp 443–449

Caetano-Anollés D, Kim KM, Mittenthal JE, Caetano-Anollés G (2011) Proteome evolution and the metabolic origins of translation and cellular life. J Mol Evol 72:14–33

Caetano-Anollés G, Wang M, Caetano-Anollés D (2013) Structural phylogenomics retrodicts the origin of the genetic code and uncovers the evolutionary impact of protein flexibility. PLoS ONE 8:e72225

Caetano-Anollés G, Nasir A, Zhou K, Caetano-Anollés D, Mittenthal JE, Sun F-J, Kim KM (2014) Archaea: the first domain of diversified life. Archaea 2014:590214

Cariani P (1998) Towards an evolutionary semiotics: the emergence of new sign-functions in organisms and devices. In: Van der Vijver G, Salthe S, Delpos M (eds) Evolutionary systems. Kluwer, Dordretch, pp 359–376

Chomsky N (1995) The minimalist program. MIT Press, Cambridge, MA

Chothia C, Gough J, Vogel C, Teichmann SA (2003) Evolution of the protein repertoire. Science 300:1701–1703

Coin L, Bateman A, Durbin R (2003) Enhanced protein domain discovery by using language modeling techniques from speech recognition. Proc Natl Acad Sci USA 100(8):4516–4520

Cramer I (2005) The parameters of the Altmann-Menzerath law. J Quant Linguist 12(1):41–52

Debès C, Wang M, Caetano-Anollés G, Gräter F (2013) Evolutionary optimization of protein folding. PLoS Comput Biol 9:e1002861

Efimov AV (1991) Structure of αα-hairpins with short connections. Protein Eng 4(3):245–250

Ekman D, Bjorklund AK, Frey-Skott J, Elofsson A (2005) Multi-domain proteins in the three kingdoms of life: orphan domains and other unassigned regions. J Mol Biol 348:231–243

Elder-Vass D (2013) Debate: seven ways to be a realist about language. J Theor Soc Behav 44:3

Eroglu S (2014) Language-like behavior of protein length distribution in proteomes. Complexity 20:12–21

Ferrer-i-Cancho R, Forns N (2010) The self-organization of genomes. Complexity 15:34–36

Fox NK, Brenner SE, Chandonia J-M (2014) SCOPe: structural classification of proteins—extended, integrating SCOP and ASTRAL data and classification of new structures. Nucleic Acids Res 42:D304–D309

Gerlach M, Altmann EG (2013) Stochastic model for the vocabulary growth in natural languages. Phys Rev 3:021006

Ghanbarnejad F, Gerlack M, Miotto JM, Altmann EG (2014) Extacting information from S-curves of language change. J R Soc Interface 11:20141044

Gimona M (2006) Protein linguistics—a grammar for modular protein assembly? Nature Rev Mol Cell Biol 7:68–73

Goncearenco A, Berezovsky IN (2015) Protein function from its emergence to diversity in contemporary proteins. Phys Biol 12:045002

Guseo R (2016) Diffusion of innovations dynamics, biological growth and catenary function. Phys A 464:1–10

Gustison ML, Semple S, Ferrer-i-Cancho R, Bergman TJ (2016) Gelada vocal sequences follow Menzerath's linguistic law. Proc Natl Acad Sci USA 113(19):E2750–E2758

Heaps HS (1978) Information retrieval: computational and theoretical aspects. Academic Press, New York

Herdan G (1964) Quantitative linguistics. Butterworth Press, Oxford

Holm L, Rosenström P (2010) Dali server: conservation mapping in 3D. Nucleic Acids Res 38: W545–W549

Janin J, Wodak SJ (1983) Structural domains in proteins and their role in the dynamics of protein function. Prog Biophys Mol Biol 42:21–78

Ji S (1997) Isomorphism between cell and human languages: molecular, biological, bioinformatics and linguistics implications. BioSystems 44:17–39

Kabsch W, Sander C (1983) Dictionary of protein secondary structure: pattern recognition of hydrogen-bonded and geometrical features. Biopolymers 22(12):2577–2637

Kim KM, Caetano-Anollés G (2012) The evolutionary history of protein fold families and proteomes confirms that the archaeal ancestor is more ancient than the ancestors of other superkingdoms. BMC Evol Biol 12:13

Kim KM, Nasir A, Caetano-Anollés G (2014) A tree of cellular life inferred froma genomic census of molecular functions. J Mol Evol 79:240–262

Kinnunen J (1996) Gabriel Tarde as a founding father of innovation diffusion research. Acta Sociol 39(4):431

Koç I, Caetano-Anollés G (2017) The natural history of molecular functions inferred from an extensive phylogenomic analysis of Gene Ontology data. Manuscript submitted under revision

Köehler R (1987) System theoretical linguistics. Theor Linguist 14(2–3):241–258

Kravchenko A (2006) Cognitive linguistics, biology of cognition and biosemiotics: bridging the gaps. Lang Sci 28:51–75

Li S, Lin R, Bian C, Ma QDY, Ivanov PC (2016) Model of the dynamic construction process of texts and scaling laws of words organization in language systems. PLoS ONE 11(12): e01688971

Linderstrøm-Lang K, Schellman JA (1959) Protein structure and enzymatic activity. In: Lardy H, Myrback K (eds) The Enzymes, 2nd edn. Academic Press, New York, pp 443–510

Menzerath P (1928) Uber einige phonetische probleme. Actes du Premier Congrès International de Linguists. Sijthhof, Leiden, pp 104–105

Menzerath P (1954) Die Architektonik des Deutschen Wortschatzes. Dümmler, Bonn, Germany

Mittenthal JE, Caetano-Anollés D, Caetano-Anollés G (2012) Biphasic patterns of diversification and the emergence of modules. Front Genet 3:147

Molina N, van Nimwegen E (2009) Scaling laws in functional genome content across prokaryotic clades and lifestyle. Trends Genet 25(6):243–247

Moore AD, Bjorklund AK, Ekman D, Bornberg-Bauer E, Elofsson A (2008) Arrangements in the modular evolution of proteins. Trends Biochem Sci 33:444–451
Murzin A, Brenner SE, Hubbard T, Clothia C (1995) SCOP: a structural classification of proteins for the investigation of sequences and structures. J Mol Biol 247:536–540
Nasir A, Naeem A, Khan MJ, Lopez-Nicora HD, Caetano-Anollés G (2011) Annotation of protein domains reveals remarkable conservation in the functional make up of proteomes across superkingdoms. Genes 2:869–911
Nasir A, Kim KM (2014) Caetano-Anollés G (2014a) Global patterns of domain gain and loss in superkingdoms. PLoS Comput Biol 10:e1003452
Nasir A, Kim KM, Caetano-Anollés G (2014) A phylogenomic census of molecular functions identifies modern thermophilic Archaea as the most ancient form of cellular life. Archaea 2014:706468
Nasir A, Kim KM, Caetano-Anollés G (2017) Phylogenetic tracings of proteome size support the gradual accretion of protein structural domains and the early origin of viruses from primordial cells. Front Microbiol 8:1178
Norton JA, Bass FM (1987) A diffusion theory model of adoption and substitution for successive generations of high-technology products. Manage Sci 33(9):1069–1086
Orengo C, Michie A, Jones S, Jones D, Swindells M, Thornton JM (1997) CATH—a hierarchic classification of protein domain structures. Structure 5:1093–1109
Qian J, Luscombe NM, Gersten M (2001) Protein family and fold occurrence in genomes: power-law behavior and evolutionary model. J Mol Biol 313:673–681
Richardson JS (1981) The anatomy and taxonomy of protein structure. Adv Protein Chem 34:167–339
Riley M, Labedan B (1997) Protein evolution viewed through Escherichia coli protein sequences: introducing the notion of a structural segment of homology, the module. J Mol Biol 268:857–868
Rogers EM (1962) Diffusion of innovations. Free Press, New York
Sayood K (1996) Introduction to data compression. Morgan Kauffman, San Francisco
Searls DB (2001) Reading the book of life. Bioinformatics 17(7):579–580
Searls DB (2002) The language of genes. Nature 420:211–217
Shahzad K, Mittenthal KE, Caetano-Anollés G (2015) The organization of domains in proteins obeys Menzerath-Altmann's law of language. BMC Syst Biol 9:44
Sillitoe I, Lewis TE, Cuff A, Das S, Ashford P, Dawson NL, Furnham N, Laskowski RA, Lee D, Lees JG, Lehtinen S, Studer RA, Thornton J, Orengo CA (2015) CATH: comprehensive structural and functional annotations for genome sequences. Nucleic Acid Res 43:D376–D381
Söding J, Lupas AN (2003) More than the sum of their parts: on the evolution of proteins from peptides. BioEssays 25:837–846
Strauss S, Altmann G (2006) Hierarchic relations. In: Altmann G, Köhler R, Vulanović R (eds) Encyclopedia of linguistic laws. http://lql.uni-trier.de/index.php/Main_Page
Strogatz SH (2001) Exploring complex networks. Nature 410:268–276
Tal G, Boca SM, Mittenthal JE, Caetano-Anollés G (2016) A dynamic model for evolution of protein structure. J Mol Evol 82:230–243
Tatur J, Hagen WR, Matias PM (2007) Crystal structure of the ferritin from the hyperthermophilic archaeal anaerobe *Pyrococcus furiosum*. J Biol Inorg Chem 12(5):615–630
Tria F, Loreto V, Servedio VDP, Strogatz SH (2014) The dynamics of correlated novelties. Sci Rep 4:5890
Trifonov EN, Frenkel ZM (2009) Evolution of protein modularity. Curr Opin Struct Biol 18:335–340
Toll-Riera M, Alba MM (2013) Emergence of novel domains in proteins. BMC Evol Biol 13:47
Vogel C (2005) Function annotation of SCOP domain superfamilies 1.69. Superfamily—HMM library and genome assignments server. http://supfam.org/SUPERFAMILY/function.html. Accessed 2 Jan 2017
Vogel C, Chothia C (2006) Protein family expansions and biological complexity. PLoS Comp Biol 2:e48
Wang M, Caetano-Anollés G (2009) The evolutionary mechanics of domain organization in proteomes and the rise of modularity in the protein world. Structure 17:66–78

Wang M, Kurland CG, Caetano-Anollés G (2011) Reductive evolution of proteomes and protein structures. Proc Natl Acad Sci USA 108:11954–11958

Weinreich U, Labov W, Herzog M (1968) Empirical foundations for a theory of language change. University of Texas Press, Austin, Texas

Wetlaufer DB (1973) Nucleation, rapid folding, and globular intrachain regions in proteins. Proc Natl Acad Sci USA 70:697–701

Witzany G (2014) Language and communication as Universal requirements for life. In: Kolb V (ed) Astrobiology: an evolutionary approach. CRC Press, pp. 349–370

Witzany G (2016) The biocommunication method: on the road to an integrative biology. Comm Integr Biol 9(2):e1164374

Yafremava LS, Wielgos M, Thomas S, Nasir A, Wang M, Mittenthal JE, Caetano-Anollés G (2013) A general framework of persistence strategies for biological systems helps explain domains of life. Front Genet 4:16

Zipf GK (1948) Human behavior and the principle of least effort. Addison-Wesley, Boston, MA

KaiC-like ATPases as Signal Transduction Hubs in Archaea

Kira S. Makarova and Eugene V. Koonin

1 Introduction

Signal transduction systems are essential for all life forms as information channels between the organism and the environments as well as developmental regulators. The general organization of a signal transduction system includes at least three components, namely, sensor (input), transmitter and effector (output). Bacterial signal transduction is dominated by the two-component systems, in which the information is transmitted by histidine kinases, whereas in eukaryotes, the central role in signal transduction belongs to Ser/Thr/Tyr protein kinases. Both bacterial and eukaryotic signal transduction networks have been characterized in detail as described in numerous reviews (Whitworth and Cock 2009; Wuichet et al. 2010; Galperin 2010; Taylor and Kornev 2011; Jin and Pawson 2012; Zschiedrich et al. 2016). Archaeal signal transduction systems have not been studied in comparable detail (Esser et al. 2016). Two-component systems homologous to bacterial counterparts have been experimentally characterized in *Halobacteria* in the context of chemotaxis and phototaxis (Schlesner et al. 2012). It has been shown that many of these components are involved in complex protein-protein interaction network and contribute to the regulation of the archaellum (archaeal rotary motor) activity (Schlesner et al. 2012). In the methanogenic mesophilic archaeon *Methanosaeta harundinacea*, histidine kinase FilI controls cell morphological transitions and affects the methane production (Li et al. 2014). Phylogenetic analysis suggests that archaea acquired two-component signal transduction system components from bacteria on multiple occasions (Koretke et al. 2000). This conclusion is compatible with the results of the last archaeal common ancestor (LACA) based on arCOG (archaeal clusters of orthologous groups) database that estimated the probability for

K.S. Makarova · E.V. Koonin (✉)
National Center for Biotechnology Information, National Library of Medicine, Bethesda, MD 20894, USA
e-mail: koonin@ncbi.nlm.nih.gov

G. Witzany (ed.), *Biocommunication of Archaea*,
DOI 10.1007/978-3-319-65536-9_11

histidine kinases to be present in LACA at less than 19% (Wolf et al. 2012). Notably, the two-component systems are mostly found in mesophilic archaea that appear to have captured numerous bacterial genes via horizontal gene transfer (HGT) (Koonin et al. 2001; Nelson-Sathi et al. 2012; Deschamps et al. 2014; Garushyants et al. 2015; Papke et al. 2015; Lopez-Garcia et al. 2015) (Table 1).

Protein phosphorylation, mostly attributed to Ser/Thr/Tyr protein kinases, apparently plays important role in archaea, but details on the specific roles of protein phosphorylation in signal transduction are sparse (Esser et al. 2016). Unlike histidine kinases, three Ser/Thr/Tyr protein kinases families, RIO1 (COG1718), RIO2 (COG0478) and SPS1 (COG0515), have been traced back to LACA (Wolf et al. 2012), and at least some of these kinases appear to be key regulators of archaeal cell cycle and membrane remodeling (Makarova et al. 2010). However, Ser/Thr/Tyr protein kinases are not particularly prone to expansion in archaea and therefore do not appear to be involved in versatile signal transduction pathways (Table 1).

Given the relative paucity of identifiable signal transduction systems in Archaea, search for new, perhaps, archaea-specific ones, is an attractive goal. One way to identify potential new signal transduction machineries involves in depth analysis of paralogous gene families with known or predicted activities compatible with roles in signal transduction. Due to the high similarity to the eukaryotic recombinase component Rad55 (homologs of bacterial RecA and archaeal RadA), until recently, the COG0467 family (Galperin et al. 2015) in archaea has been implicated in DNA recombination pathways. One of the proteins from this family has been experimentally studied in this context. Specifically, it has been shown that the SSO2452 protein from *Sulfolobus solfataricus* is not an active recombinase but binds single-stranded DNA and inhibits D-loop formation by RadA (McRobbie et al. 2009). However, outside the archaea, the best-studied protein in this family is cyanobacterial circadian clock ATPases KaiC, which does not appear to be involved in DNA metabolism (Axmann et al. 2014; Johnson et al. 2017). Phylogenetic analysis suggested that COG0467 family forms a separate clade within the RecA ATPase superfamily (Leipe et al. 2000) (Fig. 1a). It has been hypothesized that, given the wide spread and major expansion of this family in archaea, which contrasts the patchy distribution in bacteria, the KaiC component of the cyanobacterial circadian clock was acquired by HGT from archaea (Leipe et al. 2000). Recently, the structure of the archaellum subunit FlaH, which belongs to an archaea-specific family (COG2874), has been solved and shown to be similar to the N-terminal domain of KaiC (Meshcheryakov and Wolf 2016). Another structure for a distinct family of an archaea-specific uncharacterized proteins (DUF835, pfam05763) shows that these proteins are inactivated ATPases that are most closely related to KaiC (Kang et al. 2009). Thus, currently, at least two additional families can be included into KaiC group (Fig. 1a). The list of arCOGs that belong to the KaiC superfamily is given in Table 2. The evolutionary reconstruction indicates that KaiC-like ATPases from arCOG01171, arCOG001174 and arCOG04148 (FlaH) were likely present in LACA (Wolf et al. 2012). Members of the KaiC superfamily are present even in archaea with the smallest genomes, such as

Table 1 Three major protein superfamilies involved in signal transduction in selected archaea and bacteria

Genome	Serine/threonine protein kinase	KaiC-like	Signal transduction histidine kinase[a]
	COG0515, COG0478, COG1718	COG0467	COG0642, COG0643, COG2202
Aeropyrum pernix K1	2	3	0
Desulfurococcus kamchatkensis 1221n	3	3	0
Ignicoccus hospitalis KIN4 I	3	3	0
Hyperthermus butylicus DSM 5456	2	5	0
Pyrolobus fumarii 1A	5	2	0
Sulfolobus acidocaldarius DSM 639	7	2	0
Pyrobaculum aerophilum IM2	4	12	0
Archaeoglobus fulgidus DSM 4304	2	14	21
Halobacterium salinarum R1	4	10	14
Haloferax volcanii DS2	3	11	37
Methanothermobacter thermautotrophicus Delta H	2	6	11
Methanocaldococcus jannaschii DSM 2661	2	2	0
Methanococcus maripaludis C6	2	1	3
Methanocella conradii HZ254	3	10	19
Methanosarcina acetivorans C2A	2	7	49
Methanosarcina mazei Go1	2	7	32
Pyrococcus furiosus COM1	3	21	0
Thermococcus kodakarensis KOD1	2	32	2
Thermoplasma acidophilum DSM 1728	2	3	0
Candidatus Korarchaeum cryptofilum OPF8	2	3	0
Nanoarchaeum equitans Kin4 M	2	1	0
Nitrosoarchaeum koreensis MY1	1	2	8
Candidatus Caldiarchaeum subterraneum	1	5	0
Escherichia coli K12 MG1655	2	0	25
Bacillus subtilis subtilis 168	3	0	24
Nostoc sp. PCC 7120	50	2	128
Thermus thermophilus HB8	4	0	1

Note Archaea are highlighted by bold. The numbers of respective COGs were retrieved from recent updates of COG (Galperin et al. 2015) and arCOG (Makarova et al. 2015) databases

[a]the data for histidine kinases domains were obtained using PSI_BLAST search with histidine kinase domain as a query against complete genome database (as of March 2016)

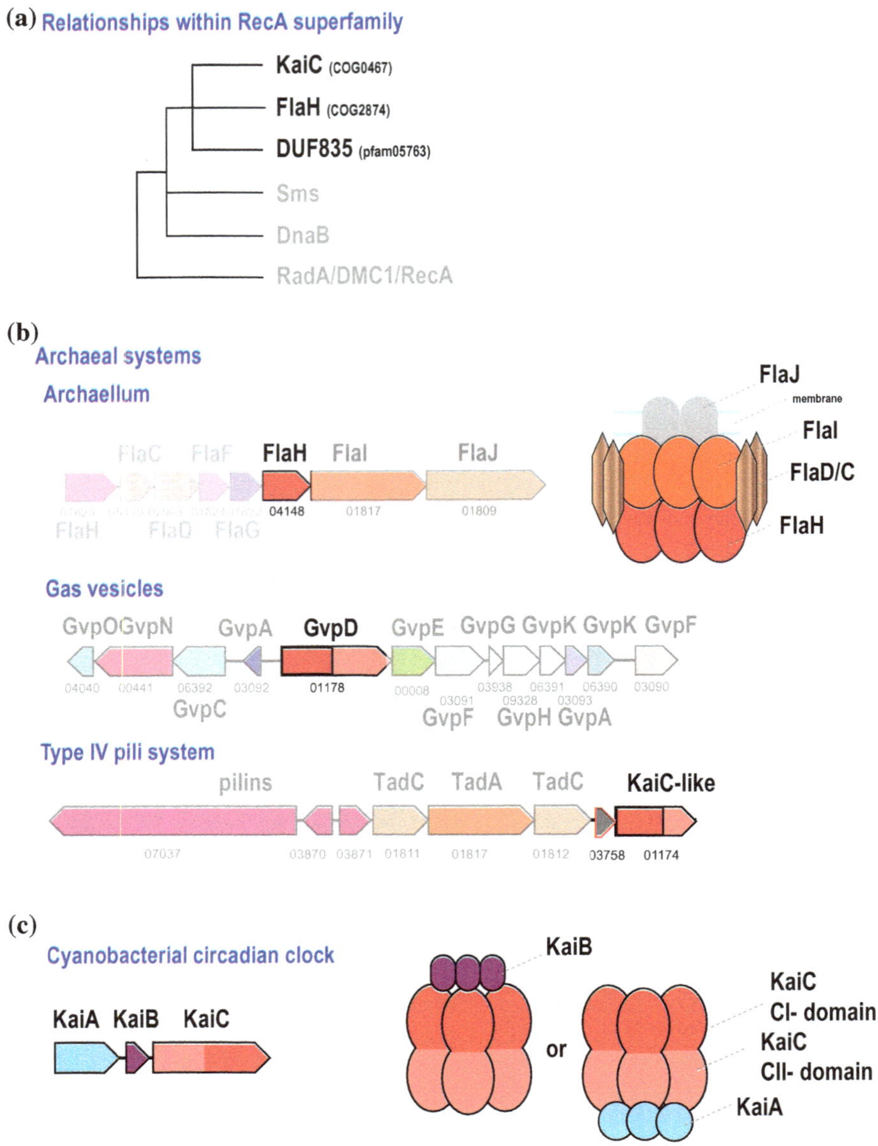

Fig. 1 Overview of the KaiC family. **a** Relationships with other RecA-like families; **b** Known archaeal systems associated with KaiC-like proteins; **c** Organization of cyanobacterial circadian clock system. Genes are shown by *arrows*. For archaeal systems, arCOG numbers are given underneath the *arrows*. Homologous genes are color-coded; KaiC family proteins are colored *red* and *pink* for the N- and C-terminal ATPase domains, respectively. Models show interactions between subunits in the respective complexes (see details and discussion in the text)

Table 2 List of arCOGs that include KaiC family members

arCOG	Annotation	Number of genomes	Number of proteins
arCOG04148	FlaH, KaiC family ATPase involved in biogenesis of archaellum	104	110
arCOG01171	RecA-superfamily ATPase implicated in signal transduction	136	400
arCOG01172	RecA-superfamily ATPase implicated in signal transduction	35	39
arCOG01173	RecA-superfamily ATPase implicated in signal transduction	41	45
arCOG01174	RecA-superfamily ATPase implicated in signal transduction	59	116
arCOG01175	RecA-superfamily ATPase implicated in signal transduction, inactivated	45	61
arCOG01176	RecA-superfamily ATPase implicated in signal transduction	11	24
arCOG01178	GvpD gas vesicle protein, contains two RecA/KaiC family ATPase domains	14	19
arCOG02453	RecA/KaiC family ATPase	27	27
arCOG02452	KaiC-like protein ATPase	50	133
arCOG02454	Rubredoxin domain fused to KaiC/Rad55 family of ATPases	2	2
arCOG01693	Inactivated KaiC-like ATPase implicated in signal transduction fused to 4VR domain	12	12
arCOG05482	RecA-superfamily ATPase implicated in signal transduction	19	20
arCOG03789	Inactivated KaiC-like ATPase, DUF835 family	12	19
arCOG03791	Inactivated KaiC-like ATPase, DUF835 family	7	7
arCOG03792	Inactivated KaiC-like ATPase, DUF835 family	18	65
arCOG03793	Inactivated KaiC-like ATPase, DUF835 family	20	30
arCOG03795	Membrane associated inactivated KaiC-like ATPase, DUF835 family	8	9
arCOG03796	Inactivated KaiC-like ATPase, DUF835 family	13	13
arCOG03797	Membrane associated inactivated KaiC-like ATPase, DUF835 family	14	35
arCOG03799	Membrane associated inactivated KaiC-like ATPase, DUF835 family	28	42
arCOG03803	Membrane associated inactivated KaiC-like ATPase, DUF835 family	17	21
arCOG03804	Membrane associated inactivated KaiC-like ATPase, DUF835 family	10	11
arCOG03805	Membrane associated inactivated KaiC-like ATPase, DUF835 family	16	40
arCOG07118	Inactivated KaiC-like ATPase	8	8

(continued)

Table 2 (continued)

arCOG	Annotation	Number of genomes	Number of proteins
arCOG03809	Inactivated KaiC-like ATPase	10	10
arCOG03808	Inactivated KaiC-like ATPase	16	30
arCOG03810	Inactivated KaiC-like ATPase	9	10

Note the data are from the latest arCOGs update (Makarova et al. 2015)

Nanoarchaea, and various KaiC families are expanded in many archaeal lineages, especially in Thermococci and Thermoproteales (Table 1). Few archaeal proteins of this superfamily have been experimentally characterized. FlaH has been shown to form a hexamer and interact with the archaellum subunit FlaX, a homolog of FlaD/C (Meshcheryakov and Wolf 2016). ATP-binding prevents this interaction and is essential for the archaellum assembly. The proposed model of interaction (Meshcheryakov and Wolf 2016) is schematically shown in Fig. 1b along with a typical organization of the archaellum locus. The KaiC-family ATPase GvpD along with the PadR-like transcriptional regulator GvpE are the main regulatory components of the Halobacteria-specific gas vesicles that increase the buoyancy of cells and enable their vertical migration in the water column in response to environmental factors, such as light, oxygen supply, temperature and salinity (Pfeifer 2015) (Fig. 1b). Several halobacterial KaiC-like proteins also have been studied with respect of their potential involvement in light-dependent gene expression (Maniscalco et al. 2014). At least four KaiC paralogs display circadian rhythm expression in *Haloferax volcanii,* and that several single- and double-gene knockouts of these genes result in disruption of light-dependent, rhythmic gene expression of the others (Maniscalco et al. 2014). Finally, sequence and comparative genomic analysis of type IV pili systems suggest that in many archaea KaiC proteins are encoded in predicted operons with other genes encoding components of these systems, suggesting that KaiC-family proteins are involved in their regulation (Makarova et al. 2016) (Fig. 1b).

Cyanobacterial circadian clock system, an ATP-dependent, post-translational molecular oscillator, is thoroughly characterized biochemically, structurally and functionally, and is described in detail in many reviews (Dvornyk et al. 2003; Nishiwaki et al. 2007; Egli and Johnson 2013; Johnson and Egli 2014; Cohen and Golden 2015; Pattanayek and Egli 2015; Shultzaberger et al. 2015; Egli 2017). Typically, the system consists of three protein components: KaiA, KaiB and KaiC (Fig. 1c). The KaiA protein forms a homodimer that interacts directly with the C-terminal ATPase domain (CII) of KaiC and promotes its phosphorylation. Structurally, KaiA is a two-domain protein, with an N-terminal four helical bundle domain and a C-terminal OmpR-like helix-turn-helix DNA-binding domain. KaiB has a thioredoxin fold and interacts with the N-terminal (CI) domain of KaiC, promoting dissociation of KaiA and dephosphorylation of KaiC. The cyanobacterium *Prochlorococcus marinus* encodes a minimal circadian system that lacks

KaiA but nevertheless shows some features of autonomous oscillator which, however, do not persist long under constant light conditions so that the system apparently requires a reset each diel cycle (Shultzaberger et al. 2015; Zinser et al. 2009). However, even when all three components are present, this is not always sufficient to reproduce all the canonical properties of circadian clock, as is the case in the purple non-sulfur bacterium *Rhodopseudomonas palustris* that only poorly maintains rhythmicity under constant conditions (Ma et al. 2016). Multiple input and output components have been shown to interact with the cyanobacterial circadian clock system, forming a complex, interconnected network. They include transcriptional regulators, receiver domains and sensory histidine kinases and light sensitive redox molecules such as quinones (Shultzaberger et al. 2015). Some of these proteins possess KaiA- or KaiB-like domains and directly interact with KaiC.

Here we present a comprehensive comparative phylogenomic analysis of the KaiC ATPase superfamily in archaea and discuss the implications for archaeal signal transduction

2 Phylogeny of the KaiC Family

KaiC homologs were identified in the set of complete archaeal and bacterial genomes that was available as of February 2016; the diverged DUF835 and FlaH subfamilies were excluded from this census. The archaeal sequences were additionally assigned to a the clusters of orthologs in the latest release of the arCOGs (Makarova et al. 2015). Despite the considerable over-representation of bacterial genomes in the database, archaeal proteins dominate the KaiC family, in agreement with the previous conclusion that this family originated in archaea (Makarova et al. 1999). A phylogenetic tree was built for a non-redundant subset of KaiC family members, with the Sms ATPase (bacterial RadA) included as the outgroup (Fig. 2a). The archaeal branches, most of them strongly supported, were designated by sequential numbers (A1–A19) or by arCOG numbers if the respective arCOGs were monophyletic. Bacterial sequences are mostly scattered over the tree, suggesting frequent HGT from archaea to bacteria. The Clade 1.1 is mostly bacterial and includes cyanobacterial KaiC components of the circadian clock (Fig. 2a). The sister clade of the cyanobacterial KaiC contains, in addition to bacterial proteins, several archaeal ones, all from different methanogens, which indicates HGT from bacteria to archaea. This branch includes experimentally studied proteins from *Rhodopseudomonas* and *Legionella* (Ma et al. 2016; Loza-Correa et al. 2014). In *Rhodopseudomonas,* involvement of the KaiC homologs in clock—like gene expression has been demonstrated, whereas in *Legionella*, these proteins are implicated in oxidative and sodium stress resistance, and do not appear to be components of an oscillator. Clade 1.1 is nested among diverse archaeal branches in Clade 1, in accord with the hypothesis that ancestral components of circadian clock were transferred from archaea to bacteria. Proteins containing two ATPase domains and those with a single ATPase domain are interspersed in Clade 1, suggesting that

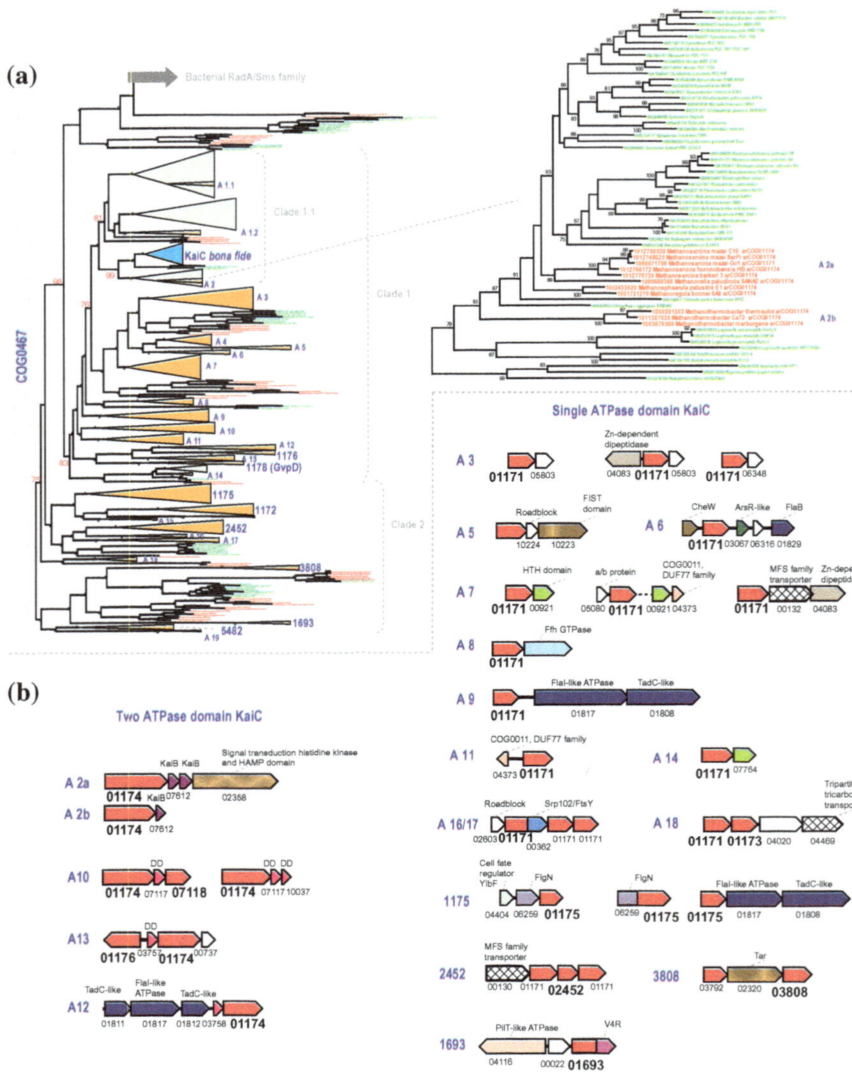

multiple gene fusions and gene fissions occurred during the evolution of this family in archaea. Furthermore, active and inactivated (as determined from the disruption of the signature motifs of the P-loop domain) ATPases are interspersed in both Clade 1 and Clade 2, suggesting multiple, independent ATPase inactivations. The archaeal branch A7 consists of KaiC-like proteins with a single ATPase domain that are well represented in both Euryarchaeota and the TACK (Thaumarchaeota, Aigarchaeota, Crenarchaeota, Korarchaeota) superphylum and thus appears to be ancestral in archaea (Table 3). The same considerations apply to branch A10, which

◄**Fig. 2 Phylogeny and conserved gene neighborhoods of the KaiC family. a** A maximum likelihood phylogenetic tree for archaeal and bacterial representatives of the KaiC protein family was constructed using the Fasttree program (Price et al. 2010). Major distinct branches are collapsed and shown by rectangles identified by numbers A1–A19 or respective arCOG numbers. Bootstrap values calculated by the same program are shown for several key nodes, and values greater than 70% are shown in *red*. The branch containing archaeal sequences most closely related to the cyanobacterial circadian clock subunit is shown on the *right*. Each sequence in the tree is described by a local numeric ID and species name. Color code: *green*, bacterial genes; *orange*, archaeal genes. **b** For each branch, a conserved gene arrangement (if detected) is shown. Genes are shown by *arrows*. arCOG numbers are indicated for each gene. Functionally linked or homologous genes are color-coded as follows: KaiC-like genes, *red*; two-component signal transduction system genes, *brown shades*; type IV pili, *dark blue*; transporters, angled grid; uncharacterized genes, *white*

includes KaiC-like proteins with two active ATPase domains. The remaining tree branches are either lineage-specific or include only a few archaeal lineages (Table 3). Thus, this analysis supports the previous conclusions that, in addition to FlaH, at least two additional KaiC families were represented in LACA (Wolf et al. 2012). The multiple long branches and inactivation of the ATPase domain imply frequent sub functionalization of the KaiC family proteins, especially in Thermococci and Thermoproteaceae, and to a lesser extent, in Aciduliprofundum and Archaeoglobi.

3 Predicted Interaction Partners of KaiC Proteins in Archaea

Analysis of conserved gene neighborhoods (Fig. 2b) and domain fusions (Fig. 3) revealed a complex and diverse set of proteins and domains that can be predicted to interact with KaiC family proteins. In particular, for the two ancestral KaiC families (see above), we can confidently predict the interacting partners. Branch A7 KaiC proteins are predicted to interact with uncharacterized proteins from arCOG00921 (COG1318). In some archaeal genomes, arCOG00921 proteins are not encoded next to A7 KaiC proteins, but genes of the two groups are always present in the same genomes, even in the smallest archaeal genome *Nanoarchaeum equitans*, suggesting that both components are involved in some essential cellular processes. A third component also could be linked to this system, namely a protein of the poorly characterized DUF77 family (pfam01910/COG0011) that is present in most archaea (arCOG04373 and appears to descend from LACA (Wolf et al. 2012). The structure for a protein from this family has been solved, revealing a ferredoxin fold, and it has been shown to forms homotetramers and bind thiamin; the expression of this gene is up-regulated in oxidizing conditions (Dermoun et al. 2010). Accordingly, it has been proposed that this protein is involved in an oxidative stress response mechanism (Wolf et al. 2012). Additionally, arCOG007764 (paralog of arCOG00921) is associated with KaiC-like ATPases of branch A14, whereas

Table 3 Description of major archaeal branches shown on the Fig. 2

Branch	Phyletic distribution	Comment
A1.1	Halobacteria only	Mostly two domain ATPase, both domains are active[a]; Apparent transfer from bacteria
A1.2	Aciduliprofundum and several Methanomicrobia genomes	Mostly two domain ATPase, both domains are active; Apparent transfer from bacteria
A2	Methanomicrobia and Methanothermobacteriales	Two active ATPase domains, most closely related to *bona fide* cyanobacterial KaiC
A3	All euryarchaeal lineages and a few thaumarchaeal lineages	Single active ATPase domain
A4	Most euryarchaeal lineages but with patchy distribution	Single active ATPase domain, fused to a large low complexity N-terminal domain, most belong to arCOG01173
A5	Patchy distribution in Archaeoglobi, Methanomicrobiales and Aciduliprofundum	Single active ATPase domain
A6	Patchy distribution in Halobacteria, Methanocella, Nitrosopumilus, present in a small genome of archaeon_GW2011_AR10	Single active ATPase domain
A7	Most archaeal lineages including Nanoarchaea	Single active ATPase domain; Possibly an ancestral group
A8	Most Thermoproteales, a few euryarchaea and in small genome of archaeon_GW2011_AR10	Single active ATPase domain
A9	Methanomicrobia and Methanothermobacteriales	Single active ATPase domain
A10	Patchy distribution in most archaeal lineages	Two active ATPase domains; Possibly an ancestral group
A11	Most TACK superphylum genomes	Single active ATPase domain
A12	Thermoproteaceae only	Two ATPase domains (second domain is inactivated and diverged)
1176	Thermoproteaceae only	Single active ATPase domain
A13	Thermoproteaceae only	Two ATPase domains (second domain is inactivated and diverged)
1178	Several Halobacteria, all Methanosaeta and all Aciduliprofundum genomes	Two ATPase domains (second domain is inactivated and diverged), in Halobacteria it is GvpD, a component and regulator of gas vesicle system
A14	Archaeoglobi only	Single active ATPase domain
1175	Patchy distribution in Archaeoglobi, Methanomicrobiales and Aciduliprofundum; present in most Halobacteria	Single inactivated ATPase domain

(continued)

Table 3 (continued)

Branch	Phyletic distribution	Comment
1172	Patchy distribution in Archaeoglobi, Methanomicrobiales and Aciduliprofundum; present in most Halobacteria	Single inactivated ATPase domain
A15	Archaeoglobi only	Single inactivated ATPase domain
2452	Patchy distribution in Archaeoglobi, Methanomicrobiales and Aciduliprofundum; present in most Halobacteria	Single inactivated ATPase domain
A16	Methanothermobacteriales	Single active ATPase domain
A17	Patchy distribution in in Methanomicrobia and Methanothermobacteriales, Thaumarchaea	Single active ATPase domain, fused to signal recognition particle receptor GTPase in most euryarchaea and to in Class III nucleotidyl cyclase in Thaumarchaea
A18	Thermococci only	Single active ATPase domain
3808	Thermococci only	Single inactivated ATPase domain
1693	Thermococci only	Single inactivated ATPase domain fused to V4R domain
5482	Thermoproteaceae only	Single active ATPase domain
A19	Several Halobacteria	Likely an active ATPase fused to metallochaperone-like domain (TRASH)

[a]ATPases are denoted active if they had intact Walker A and B motifs

arCOG04373 is also associated with KaiC-like ATPases of branch A11, reinforcing the functional link between these three protein families (Fig. 2b).

Proteins with two ATPase domains that most closely resembling the *bona fide* KaiC are typically associated with a small protein, namely, either KaiB (branch A2) or a member of a family of uncharacterized proteins (e.g. arCOGs 07117, 03757, 03758, 11224 and 10037). Proteins of the same uncharacterized family are also associated with KaiC-like ATPases of the ancestral branch A10 (Fig. 2b). The structure of one protein of this family has been solved (PDB:2p9x) and adopts a four helical bundle fold that resembles the structure of eukaryotic DEATH domain (DD) (Park et al. 2007; Kersse et al. 2011). The DEATH domains and related adaptor domains are key components of eukaryotic signal transduction pathways, particularly those involved in apoptosis, where these domains mediate connection between different components of these pathways via homotypic interactions (Park et al. 2007; Kersse et al. 2011). Exceptions are two-domain ATPases from the Halobacteria-specific branch A1.1 and from the mostly Methanomicrobial branch A 1.2, for which no small partner protein encoded in the same locus could be identified. Both of these small groups within the KaiC family seem to have been transferred relatively recently from bacteria to archaea. Taking into account the

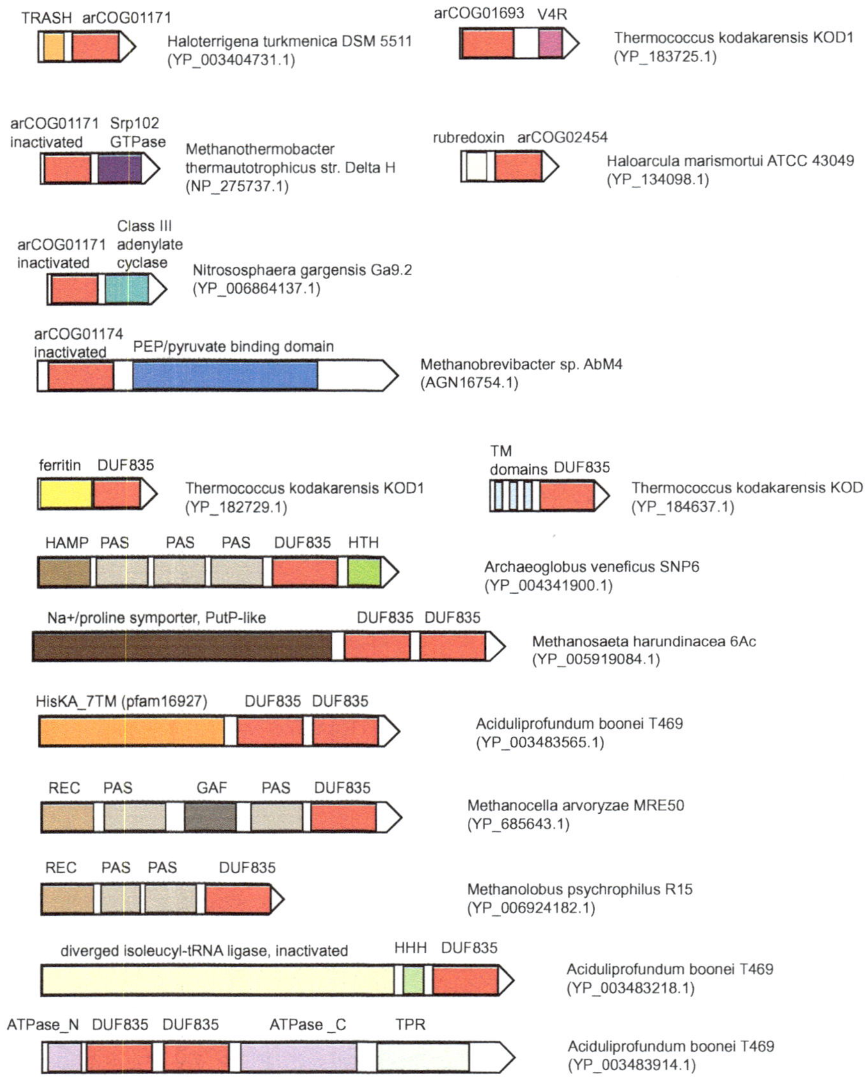

Fig. 3 KaiC protein fusions. Domains are shown by rectangles. KaiC-related domains are designated either by arCOG number or the DUF835 identifier. Genome names and respective protein IDs are indicated on the *right*. Homologous domains are color-coded. Abbreviations: HAMP, PAS, REC, HisKA_7TM, GAF are known domains shared with two-component signal transduction system; HTH, helix-turn-helix DNA-binding domain; HHH, triple helix DNA-binding domain; TRASH, metal-binding domain predicted to be involved in heavy-metal sensing; TPR, TPR repeats domain; ATPase_N—AAA ATPase N-terminal region; ATPase_C, AAA ATPase C-terminal region; TM, transmembrane segment; V4R, V4R small molecule—binding domain

shared features of the KaiB and DD-like domains, we predict that the latter also are modulators of the KaiC autophosphorylation activity.

Single-domain KaiC-like ATPases are often encoded next to their paralogs, some of which are active and others inactivated, suggesting that they might form heterodimers, recapitulating the organization of the two-domain KaiC-like ATPases.

The three most common contextual themes involving the KaiC family are: (1) type IV pili systems and other membrane-associated complexes, such as the signal recognition particle (SRP) GTPase Ffh, or a FlgN-like, flagellar biosynthesis/secretory pathway chaperone (Makarova et al. 2016); (2) signal transducing and sensory proteins that are typically associated with histidine kinases in bacteria and (3) transporters (Fig. 2b).

The fusions of KaiC with other domains are also informative, showing either the same trend as that observed for conserved neighborhoods, or suggesting involvement of KaiC-like domains in more complex signal transduction pathways (Fig. 3). Many of these fusions (e.g. to TRASH domain, rubredoxins and ferritins) point to involvement in oxidative stress.

The association with the SRP GTPase Ffh and with regulatory GTPase Srp102/FtsY indicates that KaiC-like proteins might regulate targeting of nascent secreted or membrane proteins from the ribosome to the membrane through the single recognition particle SRP (Peluso et al. 2001; Egea et al. 2008).

Inactivated KaiC-like ATPases of the DUF835 family are often found in multidomain proteins (Fig. 3). Many of these multidomain proteins contain sensory and signal-transducing domains that have been thoroughly studied in the context of bacterial two-component signal transduction systems in bacteria (Galperin 2004; Airola et al. 2013). This connection strongly suggests that DUF835 proteins are involved in signal transduction pathways. Many proteins of this family are membrane-associated, presumably interacting with other membrane proteins, some of which are fused to the DUF835 domain (e.g. Na+/proline symporter) (Table 2 and Fig. 3). Finally, fusions with other regulatory and signal transduction protein, such as AAA ATPases containing TPR repeats and cyclases, suggest that KaiC family proteins could be in involved in highly complex pathways, which include other signal transduction systems.

4 Models of KaiC–Based Signal Transduction Systems

The multiple lines of evidence discussed above indicate that KaiC family is a major hub of a versatile and complex archaeal signaling system in archaea that so far has largely escaped attention. Nevertheless, the available experimental data on circadian clock in Halobacteria (Johnson et al. 2017; Whitehead et al. 2009) and the recent progress in the study of the functions of FlaH in the archaellum (Meshcheryakov and Wolf 2016; Banerjee et al. 2013; Chaudhury et al. 2016) allow us to propose two models for the roles of KaiC-like proteins in signal transduction (Fig. 4). The first

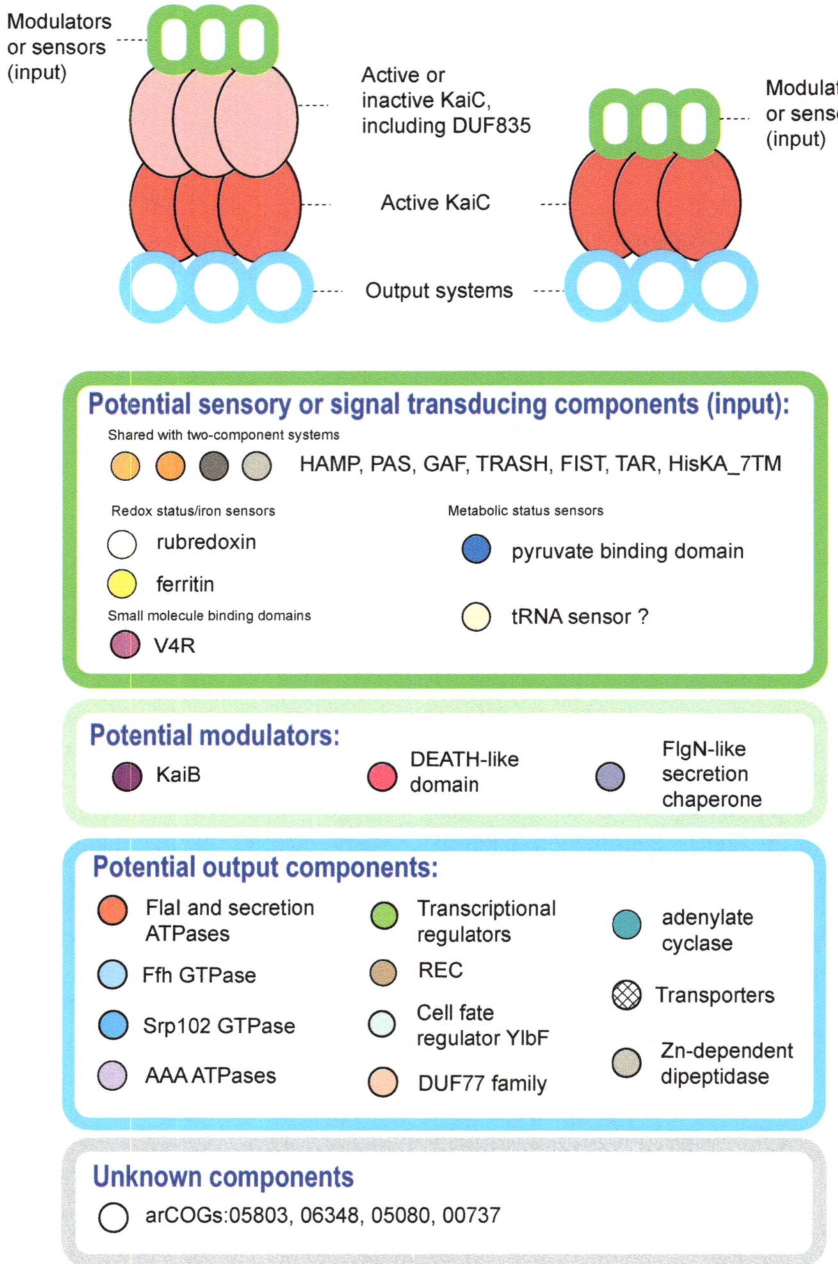

Fig. 4 Models of KaiC-based signal transduction

model is essentially identical to the circadian clock mechanism and postulates the formation of either a hexameric ring of KaiC proteins containing two ATPase domain (arCOG01174 and others) or heterohexamers in of interacting KaiC proteins each containing a single ATPase domain. Both domains can be active ATPases, or alternatively, one of the domains could be an inactivated ATPase, such as one of the multiple DUF835 domains, which would transduce the signal from an input domain to the active KaiC-like domain (Fig. 4). Each of the hexameric ATPase rings can interact with multiple partners and, as with other signal transduction systems, such partners can be roughly classified into input and output components (Fig. 4). In addition, the KaiC rings could interact with modulators of the ATPase activity, such as KaiB in the circadian clock, which might compete for binding with other output proteins. The second model postulates interaction of a single domain KaiC-like ATPase homohexamer directly with an output domain, similar to the potential interaction between FlaH and FlaI in the archaellum (Banerjee et al. 2013; Chaudhury et al. 2016) (Figs. 4 and 5). Some of the predicted components remain completely uncharacterized, and no specific functions can be predicted for them. Furthermore, the KaiC-centered signaling systems could be interconnected with other signal transduction pathways, in particular, with two-components systems via shared domains of input proteins (Fig. 4) and with Ras-like GTPases, either directly, through the interaction with Srp102/FtsY, or through Roadblock family proteins as described for both bacteria and eukaryotes (Miertzschke et al. 2011; Levine et al. 2013). The mode of signaling apparently can be modified with relative ease. For example, a two-domain KaiC-like ATPase and a DEATH-domain like protein are encoded in the type IV pili loci in Thermoproteales, whereas in euryarchaea, these loci contain a gene coding for a single domain KaiC (arCOG01175) linked to a FlgN-like secretion chaperone. Thus, two distinct models can apply to the same process of regulation of type IV pili (or archaellum) assembly in different archaea (Fig. 5).

It can be predicted that many KaiC-like proteins, despite the conservation of the Walker A and B ATPase motifs, bind ATP to transduce the signal but lack ATPase activity, similarly to FlaH (Meshcheryakov and Wolf 2016). Indeed, many of these proteins lack the pair of serine/threonine residues that are conserved among the *bona fide* KaiC and are autophosphorylated in the circadian clock system (Nishiwaki et al. 2007). However, several KaiC subfamilies, especially the two domain ATPases of arCOG01174, retain this motif or at least one of the two hydroxy amino acids, and could be active kinases.

5 Implication for Archaeal Circadian Clock

Among archaea, the diurnal gene expression has been demonstrated only in Halobacteria (Johnson et al. 2017; Whitehead et al. 2009; Edgar et al. 2012). It has been shown that KaiC-like proteins undergo cyclic expression, and deletion of most of them affected expression of the others, suggesting Halobacteria indeed execute the *bona fide* KaiC-based mechanism (Maniscalco et al. 2014). Indeed, similarly to

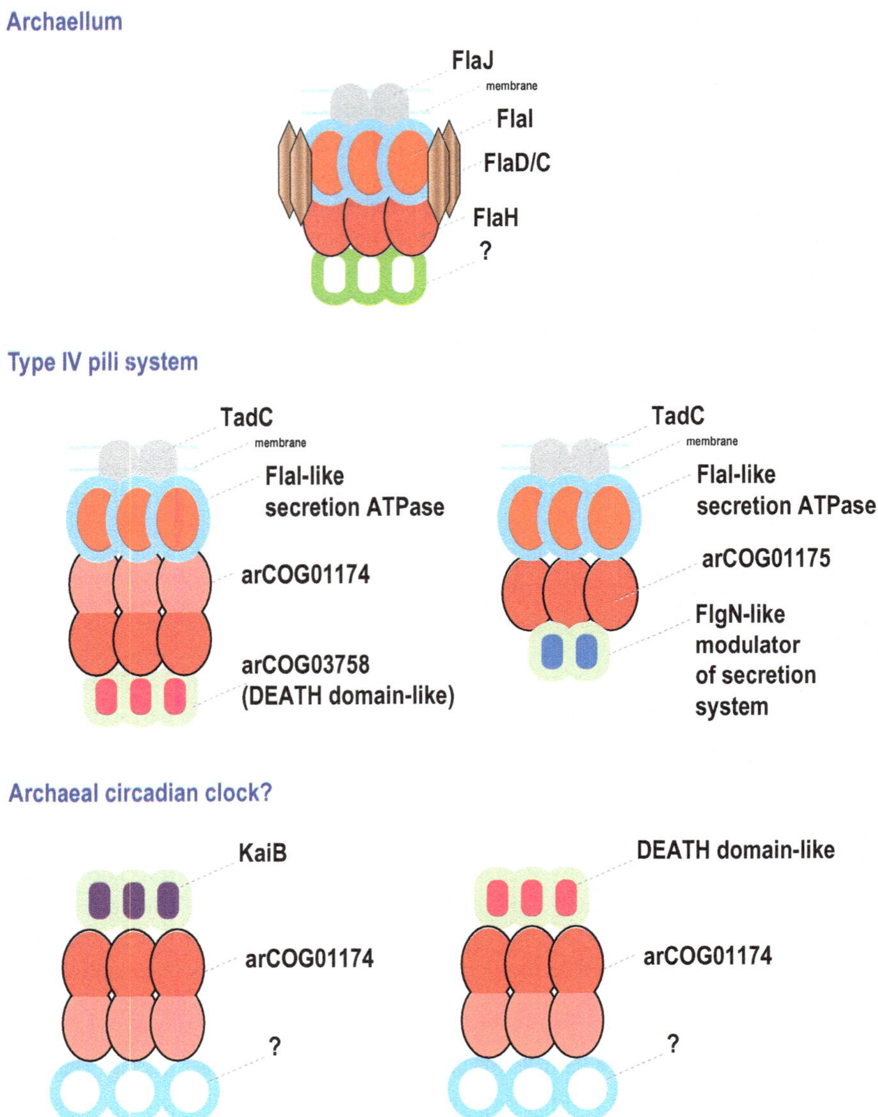

Fig. 5 Models of interactions between KaiC and membrane-associated complexes and organization of putative archaeal clock-like systems

cyanobacteria, Halobacteria adjust their metabolism to the light conditions, in particular, via rhodopsin-based proton pumps that generate a proton gradient and sensory rhodopsins that control phototaxis (Sharma et al. 2007). Halobacteria encode two-domain KaiC-like ATPases (branch 1.1 in the Fig. 2) very similar to the one from cyanobacteria, but neither KaiB, nor KaiA nor any potential analog so far have been identified in their genomes. Accordingly, the mode of KaiC action in

Halobacteria remains unclear. To the best of our knowledge, no evidence of circadian clock in any other archaea has been reported, and no rhodopsins or other light sensors have been identified.

A minimal potential circadian clock system consisting of KaiC and KaiB is present in the archaea from branch 2 (Fig. 2). However, as in the case of *Legionella*, this system could be involved in regulatory pathways distinct from the circadian clock. Considering that, apart from Halobacteria, there is seems to be no indication that archaea can sense light and modulate the metabolism accordingly, it appears unlikely that they possess clocks similar to those of photosynthetic bacteria. However, if indications of clock mechanisms in archaea (other than Halobacteria) were to be found, the best candidates would be systems from branch 2, in which a two-domain KaiC-like ATPase is associated with KaiB and from branch 10, where a two-domain KaiC-like ATPase is associated with a DEATH-like domain, a potential analog of KaiB (Figs. 2 and 5).

6 Concluding Remarks

The striking proliferation and diversification of the KaiC-like ATPase family in archaea implies that these proteins comprise the basis of diverse, unexplored signal transduction networks that appear to be specific to archaea. These signal transduction systems are likely involved in the regulations of the function of membrane-associated complexes and individual proteins, such as the archaellum, type IV pili, signal recognition particle, and transporters. Additionally, the KaiC-centered signal transduction machinery can be predicted to regulate oxidative stress response. However, it appears unlikely that archaea, apart from Halobacteria, possess cyanobacterial-type, KaiC-centered circadian clocks. The KaiC-based signaling mechanisms appear to be ancestral in archaea, with at least three KaiC paralogs projected to LACA. One ancestral subfamily that includes a protein containing an HTH domain (arCOG00921) might be involved in essential global response pathways, because it is encoded even in the minimal genome of the Nanoarchaea. The predicted KaiC-based signal transduction system appears to be interconnected with two-component signal transduction systems through inactivated ATPases of the DUF835 family that are predicted to interact with active KaiC ATPases. Clearly, the phylogenomic analysis reported here can produce only crude models of archaeal signal transduction. Nevertheless, these observations open up multiple experimental directions that can be expected to shed light on key aspects of archaeal cell biology.

Acknowledgements The authors' research is supported by intramural funds of the US Department of Health and Human Services.

References

Airola MV, Huh D, Sukomon N, Widom J, Sircar R, Borbat PP, Freed JH, Watts KJ, Crane BR (2013) Architecture of the soluble receptor Aer2 indicates an in-line mechanism for PAS and HAMP domain signaling. J Mol Biol 425(5):886–901

Axmann IM, Hertel S, Wiegard A, Dorrich AK, Wilde A (2014) Diversity of KaiC-based timing systems in marine Cyanobacteria. Mar Genomics 14:3–16

Banerjee A, Neiner T, Tripp P, Albers SV (2013) Insights into subunit interactions in the Sulfolobus acidocaldarius archaellum cytoplasmic complex. FEBS J 280(23):6141–6149

Chaudhury P, Neiner T, D'Imprima E, Banerjee A, Reindl S, Ghosh A, Arvai AS, Mills DJ, van der Does C, Tainer JA et al (2016) The nucleotide-dependent interaction of FlaH and FlaI is essential for assembly and function of the archaellum motor. Mol Microbiol 99(4):674–685

Cohen SE, Golden SS (2015) Circadian Rhythms in Cyanobacteria. Microbiol Mol Biol Rev 79 (4):373–385

Dermoun Z, Foulon A, Miller MD, Harrington DJ, Deacon AM, Sebban-Kreuzer C, Roche P, Lafitte D, Bornet O, Wilson IA et al (2010) TM0486 from the hyperthermophilic anaerobe Thermotoga maritima is a thiamin-binding protein involved in response of the cell to oxidative conditions. J Mol Biol 400(3):463–476

Deschamps P, Zivanovic Y, Moreira D, Rodriguez-Valera F, Lopez-Garcia P (2014) Pangenome evidence for extensive interdomain horizontal transfer affecting lineage core and shell genes in uncultured planktonic thaumarchaeota and euryarchaeota. Genome Biol Evol 6(7):1549–1563

Dvornyk V, Vinogradova O, Nevo E (2003) Origin and evolution of circadian clock genes in prokaryotes. Proc Natl Acad Sci U S A 100(5):2495–2500

Edgar RS, Green EW, Zhao Y, van Ooijen G, Olmedo M, Qin X, Xu Y, Pan M, Valekunja UK, Feeney KA et al (2012) Peroxiredoxins are conserved markers of circadian rhythms. Nature 485(7399):459–464

Egea PF, Tsuruta H, de Leon GP, Napetschnig J, Walter P, Stroud RM (2008) Structures of the signal recognition particle receptor from the archaeon Pyrococcus furiosus: implications for the targeting step at the membrane. PLoS ONE 3(11):e3619

Egli M (2017) Architecture and mechanism of the central gear in an ancient molecular timer. J R Soc Interface 14(128)

Egli M, Johnson CH (2013) A circadian clock nanomachine that runs without transcription or translation. Curr Opin Neurobiol 23(5):732–740

Esser D, Hoffmann L, Pham TK, Brasen C, Qiu W, Wright PC, Albers SV, Siebers B (2016) Protein phosphorylation and its role in archaeal signal transduction. FEMS Microbiol Rev 40 (5):625–647

Galperin MY (2004) Bacterial signal transduction network in a genomic perspective. Environ Microbiol 6(6):552–567

Galperin MY (2010) Diversity of structure and function of response regulator output domains. Curr Opin Microbiol 13(2):150–159

Galperin MY, Makarova KS, Wolf YI, Koonin EV (2015) Expanded microbial genome coverage and improved protein family annotation in the COG database. Nucleic Acids Res 43(Database issue):D261–269

Garushyants SK, Kazanov MD, Gelfand MS (2015) Horizontal gene transfer and genome evolution in Methanosarcina. BMC Evol Biol 15:102

Jin J, Pawson T (2012) Modular evolution of phosphorylation-based signalling systems. Philos Trans R Soc Lond B Biol Sci 367(1602):2540–2555

Johnson CH, Egli M (2014) Metabolic compensation and circadian resilience in prokaryotic cyanobacteria. Annu Rev Biochem 83:221–247

Johnson CH, Zhao C, Xu Y, Mori T (2017) Timing the day: what makes bacterial clocks tick? Nat Rev Microbiol 15(4):232–242

Kang HJ, Kubota K, Ming H, Miyazono K, Tanokura M (2009) Crystal structure of KaiC-like protein PH0186 from hyperthermophilic archaea Pyrococcus horikoshii OT3. Proteins 75 (4):1035–1039

Kersse K, Verspurten J, Vanden Berghe T, Vandenabeele P (2011) The death-fold superfamily of homotypic interaction motifs. Trends Biochem Sci 36(10):541–552

Koonin EV, Makarova KS, Aravind L (2001) Horizontal gene transfer in prokaryotes: quantification and classification. Annu Rev Microbiol 55:709–742

Koretke KK, Lupas AN, Warren PV, Rosenberg M, Brown JR (2000) Evolution of two-component signal transduction. Mol Biol Evol 17(12):1956–1970

Leipe DD, Aravind L, Grishin NV, Koonin EV (2000) The bacterial replicative helicase DnaB evolved from a RecA duplication. Genome Res 10(1):5–16

Levine TP, Daniels RD, Wong LH, Gatta AT, Gerondopoulos A, Barr FA (2013) Discovery of new Longin and Roadblock domains that form platforms for small GTPases in Ragulator and TRAPP-II. Small GTPases 4(2):62–69

Li J, Zheng X, Guo X, Qi L, Dong X (2014) Characterization of an archaeal two-component system that regulates methanogenesis in *Methanosaeta harundinacea*. PLoS ONE 9(4):e95502

Lopez-Garcia P, Zivanovic Y, Deschamps P, Moreira D (2015) Bacterial gene import and mesophilic adaptation in archaea. Nat Rev Microbiol 13(7):447–456

Loza-Correa M, Sahr T, Rolando M, Daniels C, Petit P, Skarina T, Gomez Valero L, Dervins-Ravault D, Honore N, Savchenko A et al (2014) The *Legionella* pneumophila kai operon is implicated in stress response and confers fitness in competitive environments. Environ Microbiol 16(2):359–381

Ma P, Mori T, Zhao C, Thiel T, Johnson CH (2016) Evolution of KaiC-dependent timekeepers: a Proto-circadian timing mechanism confers adaptive fitness in the purple bacterium *Rhodopseudomonas palustris*. PLoS Genet 12(3):e1005922

Makarova KS, Aravind L, Galperin MY, Grishin NV, Tatusov RL, Wolf YI, Koonin EV (1999) Comparative genomics of the Archaea (Euryarchaeota): evolution of conserved protein families, the stable core, and the variable shell. Genome Res 9(7):608–628

Makarova KS, Yutin N, Bell SD, Koonin EV (2010) Evolution of diverse cell division and vesicle formation systems in Archaea. Nat Rev Microbiol 8(10):731–741

Makarova KS, Wolf YI, Koonin EV (2015) Archaeal clusters of orthologous genes (arCOGs): an update and application for analysis of shared features between thermococcales, methanococcales, and methanobacteriales. Life (Basel) 5(1):818–840

Makarova KS, Koonin EV, Albers SV (2016) Diversity and Evolution of Type IV pili Systems in Archaea. Front Microbiol 7:667

Maniscalco M, Nannen J, Sodi V, Silver G, Lowrey PL, Bidle KA (2014) Light-dependent expression of four cryptic archaeal circadian gene homologs. Front Microbiol 5:79

McRobbie AM, Carter LG, Kerou M, Liu H, McMahon SA, Johnson KA, Oke M, Naismith JH, White MF (2009) Structural and functional characterisation of a conserved archaeal RadA paralog with antirecombinase activity. J Mol Biol 389(4):661–673

Meshcheryakov VA, Wolf M (2016) Crystal structure of the flagellar accessory protein FlaH of *Methanocaldococcus jannaschii* suggests a regulatory role in archaeal flagellum assembly. Protein Sci 25(6):1147–1155

Miertzschke M, Koerner C, Vetter IR, Keilberg D, Hot E, Leonardy S, Sogaard-Andersen L, Wittinghofer A (2011) Structural analysis of the Ras-like G protein MglA and its cognate GAP MglB and implications for bacterial polarity. EMBO J 30(20):4185–4197

Nelson-Sathi S, Dagan T, Landan G, Janssen A, Steel M, McInerney JO, Deppenmeier U, Martin WF (2012) Acquisition of 1,000 eubacterial genes physiologically transformed a methanogen at the origin of Haloarchaea. Proc Natl Acad Sci U S A 109(50):20537–20542

Nishiwaki T, Satomi Y, Kitayama Y, Terauchi K, Kiyohara R, Takao T, Kondo T (2007) A sequential program of dual phosphorylation of KaiC as a basis for circadian rhythm in cyanobacteria. EMBO J 26(17):4029–4037

Papke RT, Corral P, Ram-Mohan N, Haba RR, Sanchez-Porro C, Makkay A, Ventosa A (2015) Horizontal gene transfer, dispersal and haloarchaeal speciation. Life (Basel) 5(2):1405–1426

Park HH, Lo YC, Lin SC, Wang L, Yang JK, Wu H (2007) The death domain superfamily in intracellular signaling of apoptosis and inflammation. Annu Rev Immunol 25:561–586

Pattanayek R, Egli M (2015) Protein-protein interactions in the cyanobacterial circadian clock: structure of KaiA dimer in complex with C-terminal KaiC peptides at 2.8 A resolution. Biochemistry 54(30):4575–4578

Peluso P, Shan SO, Nock S, Herschlag D, Walter P (2001) Role of SRP RNA in the GTPase cycles of Ffh and FtsY. Biochemistry 40(50):15224–15233

Pfeifer F (2015) Haloarchaea and the formation of gas vesicles. Life (Basel) 5(1):385–402

Price MN, Dehal PS, Arkin AP (2010) FastTree 2–approximately maximum-likelihood trees for large alignments. PLoS ONE 5(3):e9490

Schlesner M, Miller A, Besir H, Aivaliotis M, Streif J, Scheffer B, Siedler F, Oesterhelt D (2012) The protein interaction network of a taxis signal transduction system in a halophilic archaeon. BMC Microbiol 12:272

Sharma AK, Walsh DA, Bapteste E, Rodriguez-Valera F, Ford Doolittle W, Papke RT (2007) Evolution of rhodopsin ion pumps in haloarchaea. BMC Evol Biol 7:79

Shultzaberger RK, Boyd JS, Diamond S, Greenspan RJ, Golden SS (2015) Giving time purpose: the synechococcus elongatus clock in a broader network context. Annu Rev Genet 49:485–505

Taylor SS, Kornev AP (2011) Protein kinases: evolution of dynamic regulatory proteins. Trends Biochem Sci 36(2):65–77

Whitehead K, Pan M, Masumura K, Bonneau R, Baliga NS (2009) Diurnally entrained anticipatory behavior in archaea. PLoS ONE 4(5):e5485

Whitworth DE, Cock PJ (2009) Evolution of prokaryotic two-component systems: insights from comparative genomics. Amino Acids 37(3):459–466

Wolf YI, Makarova KS, Yutin N, Koonin EV (2012) Updated clusters of orthologous genes for Archaea: a complex ancestor of the Archaea and the byways of horizontal gene transfer. Biol Direct 7:46

Wuichet K, Cantwell BJ, Zhulin IB (2010) Evolution and phyletic distribution of two-component signal transduction systems. Curr Opin Microbiol 13(2):219–225

Zinser ER, Lindell D, Johnson ZI, Futschik ME, Steglich C, Coleman ML, Wright MA, Rector T, Steen R, McNulty N et al (2009) Choreography of the transcriptome, photophysiology, and cell cycle of a minimal photoautotroph, *prochlorococcus*. PLoS ONE 4(4):e5135

Zschiedrich CP, Keidel V, Szurmant H (2016) Molecular mechanisms of two-component signal transduction. J Mol Biol 428(19):3752–3775

Archaea Were Trailblazers in Signaling Evolution: Protein Adaptation and Structural Fluidity as a Form of Intracellular Communication

Savannah Spradlin, Lori Cobani, Christian Brininger and Caryn Evilia

1 Introduction

Intracellular communication or cell signaling is somewhat synonymous with eukaryotes. It can be best described as elaborate, with multiple proteins interacting with each other and chemical groups exchanged, like phosphates and methyl groups. It should also be thought of as a careful balance because when signals get crossed or protein interactions goes awry, disease states, like cancers, can manifest. Bacteria, on the other hand, are not as well characterized, but several well described examples are known, like the Che system for chemotaxis and the arabinose utilization pathway (Cashin et al. 2006; Esser et al. 2016). While the work on bacterial systems is not complete, what we do know is that their signaling pathways (thus far) are much less elaborate. There are still protein-protein interactions and the exchange of chemical groups, but there doesn't seem to be as many protein "players" as there are in eukaryotes; bacterial cells are simpler and don't need as much coordinated information as eukaryotes, especially multicellular ones.

The least well characterized are the Archaea, which encompass most extremophile organisms, and little is known about their intracellular signaling pathways in general. Characterizations of archaeal genomes have revealed the components of signaling systems like those in bacteria and even some similarities to eukaryotes, however, there is still a lot to learn (Cashin et al. 2006; Esser et al. 2016).

Cell signaling pathways that involve protein-protein interactions or chemical modification involve protein structure. After all, parts of the protein must be exposed or are exposed when modifications or interactions take place. These structural elements, which sometimes have to move from one part of the protein to the other, are conserved in the adaptations found within the organism's proteins and genome. The case we will make in this review is that archaeal and/or extremophile

S. Spradlin · L. Cobani · C. Brininger · C. Evilia (✉)
Department of Chemistry, Idaho State University, Pocatello, ID 83209, USA
e-mail: evilcary@isu.edu

G. Witzany (ed.), *Biocommunication of Archaea*,
DOI 10.1007/978-3-319-65536-9_12

protein adaptations, which are more dynamic than ones from eukaryotes, could be construed as a primitive form of cell signaling. When extremophile proteins undergo changes in their environment within the cell, their proteins change shape and/or their flexibility and these changes result in differences in the transcriptional and metabolic patterns. While the same case can be made for some bacterial and eukaryotic proteins, it doesn't seem as straightforward as extremophiles, however, this is probably not unique to extremophiles or Archaea.

To understand protein adaptations and the potential cell effects in response to these adaptations, we have to subdivide archaeal extremophiles into three main subgroups: thermophiles, psychrophiles, and halophiles. Traditionally, there are four subgroups, with acidophiles being their own category, as they have their own distinct features and environments that separate them from the others (Schleper et al. 2005). However, there is a large degree of overlap between the acidophiles and thermophiles when talking about protein adaptation (Baker-Austin and Dopson 2007; Reed et al. 2013). Therefore, for the purposes of this review, the acidophiles will be grouped with the thermophiles.

These three categories show the most obvious protein adaptations to their environment but when characterizing protein adaptations in any organismal group, there are exceptions to this rule. These exceptions make forming rules about adaptation difficult, but this also backs up our assertion that protein adaptations and their structural changes within the organism might be a primitive signaling system. Proteins that aren't specifically adapted to the organism's environment might be short-lived because, not being used to the environment, they become unstable and hence, they are degraded. These same proteins might also be specifically targeted when the conditions get rough in order for the cell to re-establish control, while the adapted proteins, can repair damage or keep metabolism going because they have a range of stability in that environment. This supports the notion that extremophiles take advantage of their protein adaptations to reap the benefits of their extreme environments.

Extremophiles have had to adapt to multiple extremes because of the nature of their environments, which are geologically unstable. It could be that they have not had evolutionary time to form elaborate signaling pathways, but it could also be that they don't need them. The current research within provides examples of specific protein adaptations, how they remain stable under extreme conditions, and how this would be useful as a potential cell signal.

1.1 Protein Stability in Thermophiles

Thermophiles and hyperthermophiles both thrive at extremely hot temperatures, and while the two share similar adaptations to survive in these extremes, they differ in their temperature growth optimum. Hyperthermophiles can grow optimally up to 105 °C, whereas thermophiles are classified as growing between 50 and 0 °C. High temperatures induce molecular motions in all molecules, especially proteins, due to increased kinetic energy in these environments. Increased kinetic energy leads to greater flexibility in proteins, which is widely accepted as unfavorable in all enzymes, even though there is no generally accepted mechanism by which rigidity, and by proxy stability, is achieved (Vihinen 1987). Because of the significant kinetic energy present in such environments, thermophiles have similar adaptations to reduce protein flexibility to survive in these hot environments.

In order to function, the organisms had to adapt how they work on a fundamental level, as proteins present in mesophilic organisms are often unstable and can degrade at these extreme temperatures. At such high temperatures, proteins lacking the necessary adaptations undergo exaggerated molecular motions and unfolding, which exposes the hydrophobic core and becomes irreversible, causing aggregation (Tomazic and Klibanov 1988). Under prolonged conditions, the exaggerated molecular motions can cause proteolysis. Thermophilic proteins have several adaptations that give them the ability to retain structure and function in temperature extremes. These adaptations work to keep the cell intact and when the proteins' stability changes, either becoming "too stable" or too unfolded or flexible, the cell can change behavior to survive.

1.2 Protein Adaptations in Thermophiles

The most prominent thermophilic protein adaptations are: (1) increased number and size of hydrophobic residues which makes the protein more hydrophobic overall, but mostly giving it a large hydrophobic core, (2) increased number of disulfide bonds, which keeps the protein structure together covalently and (3) increased ionic interactions, which are more favored at high temperatures. While there are smaller adaptations that might also be present in thermophiles, these are the primary forms of adaptation that help minimize molecular motions in proteins and allow the protein to remain stable.

One of the more prominent features of thermophilic proteins is an increased hydrophobic core (Szilagyi and Zavodszky 2000; Vieille and Zeikus 2001). A good example of this is work by Tsukamoto et al., whom studied rhodopsins from thermophiles, bacteriorhodopsins from *Halobacterium salinarum*, xathorrhodopsin from Salinibacter ruber (XR). They found that thermophilic rhodopsins (TR) had alterations to both their hydrophobic core and surface amino acids. The internal residues in all rhodopsin analogs function in the proton transport pathway, and are primarily conserved between crystal structures. They observed that the TR was much more stable from pigment loss than the rhodopsins from the non-thermophiles (Tsukamoto et al. 2016). When investigating the reason for the increased stability, they found that the surface residues showed the majority of differences, with the TR containing a greater number of hydrophobic residues and hydrophobic interactions between monomers as compared to XR (Tsukamoto et al. 2016). These hydrophobic interactions served to reduce the flexibility of the protein at lower temperatures and allow optimal movement at the higher temperature at which it operates. Also, some thermostable proteins are known to deviate in their quaternary organization from their mesophilic counterparts. This allows for increased rigidity and a tighter packed hydrophobic core, which reduces the exposure of the core to the solvent (Szilagyi and Zavodszky 2000; Vieille and Zeikus 2001).

Another source of increased thermal stability comes with an increased number of disulfide bonds as compared to non-thermophile proteins. Disulfide bonds have been shown to increase stability within thermophilic proteins as well as solidify the quaternary structure (Cacciapuoti et al. 1994, 2012; Vieille and Zeikus 2001; Woycechowsky and Raines 2003; Wilkinson and Gilbert 2004). In a similar manner, salt bridging has also been shown to be a prevalent feature of thermophilic proteins when compared to the mesophilic variants (Karshikoff and Ladenstein 2001; Chan et al. 2011). The desolvation penalty and entropic costs of increased electrostatic interactions, which can be overwhelming for mesophilic organisms, can be overcome easier at the high temperatures that the thermophilic organisms reside, and thus making it a more favorable option for stabilization (Hendsch and Tidor 1994; Lee et al. 2005).

These adaptations allow for proper function at extreme temperatures, however they come at a cost of diminished functionality or stability at lower temperatures. This cost is easily avoided however, as these organisms live where the constant high temperature and high pressure is a permanent part of existence, such as a hot spring or an ocean geothermal vent. Because of their localized growth, the organisms have adapted for life at elevated temperatures, so it follows that, when the environment changes, their proteins often suffer negative effects.

Some of the above-mentioned methods of stabilization rely on increasing the rigidity of the protein, such that it is fluid at the desired temperature without too much flexibility or degradation. The balance between flexibility and rigidity at high temperatures can lead to a lower catalytic rate. An example of this is lactose dehydrogenase from *Thermotoga maritima*, which has an increased hydrophobic core and salt bridging (Peng et al. 2015). When its activity was compared to non-thermophiles, it had a much lower k_{cat} value. However, k_{cat} was seen to fall sharply for non-thermophilic strains after 40–60 °C, while the thermophilic protein k_{cat} started to climb at around 60 °C and continued climbing all the way to 100 °C. The rate determining step of lactose dehydrogenase is protein movement: the protein exists in an equilibrium between a fast reacting and slow reacting state, and the transfer between these states is the rate determining step for substrate turnover. The lower k_{cat} for the thermophilic protein at lower temperatures suggests the mechanisms which work to maintain the optimal flexibility without loss of structure in the thermophile end up causing the protein to be too rigid; this prevents the rapid transition from the slow reacting state to fast reacting state at lower temperatures, but gives the enzyme optimal activity and stability at higher temperatures (Peng et al. 2015).

1.3 Applications to Cell Signaling

When thermophiles are challenged by their environment, they need to rapidly respond in order to survive and while thermophile adaptations allow for consistent function, these adaptations also allow for rapid signaling inside the cell and an immediate response. The main challenge to a thermophile is an increase or decrease in temperature. As pointed out in many other articles, thermophilic adaptations also give thermophiles the ability to withstand other challenges, like change in pH (Baker-Austin and Dopson 2007; Sharma et al. 2012) and pressure (Boonyaratanakornkit et al. 2002; Di Giulio 2005; Hay et al. 2009), but only changes in temperature will be addressed here (Reed et al. 2013).

When thermophiles are subjected to lower temperatures, they become "frozen", which seems surreal when frozen means 50 °C. To survive the low temperature, thermophiles use their adaptations to slow down metabolism by decreasing the flexibility of their enzymes though adaptations described above, and this translates into decreases in k_{cat}. At lower temperatures, some thermophilic enzymes have an increase in binding of their substrate, even if their kinetics are slow (Peng et al. 2015). This reduction in flexibility serves two purposes: to keep the cell alive, albeit at a low metabolic rate, and to maintain integrity of the cell and its genomic materials. When thermophiles are maintained at a lower temperature, they go into senescence which can last for years, even hundreds or thousands of years.

When thermophiles are challenged by higher temperatures, proteins can denature, even with the best adaptations. While the adaptations try to beat this issue by having a more tightly packed hydrophobic core, unfolding can still occur. These unfolding events are sensed by the cell by the decrease in kinetic efficiency or decrease in binding of a target molecule to a protein. This causes an increase in the substrate and/or ligand concentration, which would lead to the cell shutting down those transport paths to prevent buildup. They would also sense the additional flexibility of its proteins by roaming chaperones and proteases, which usually have robust thermophilic adaptations. Some chaperones are heat activated and can help proteins remain stable when temperatures increase or decrease (Struvay et al. 2013; Godin-Roulling et al. 2015). Proteases, which are always on the lookout for unfolded proteins and exposed hydrophobic residues, are always active. Thermophiles (as well as other Archaea and Bacteria) have protein-based inhibitors that activate at higher temperatures to prevent proteases from degrading proteins when temperatures increase.

An example of thermophilic adaptation and cellular effect is the serine protease inhibitor, pnserpin. Pnserpin is a 44 kD protein that inhibits a variety of proteases between 20 and 100 °C. Consistently over all tested temperatures and enzymes, stoichiometry of inhibition values (the ratio of inhibitor to enzyme that causes 100% inhibition) decreased as temperature increased. This indicated a higher inhibitor activity at higher temperatures, and a corresponding lower inhibitor activity at lower temperatures. By blocking the proteases at higher temperatures, pnserpin prevents proteins from being degraded as they might start to unravel from the increased temperature. Zhang et al. found that when the temperature range was between 40 and 70 °C, the relative binding of pnserpin increased for all proteases tested; α-chymotrypsin, subtilisin, elastase, proteinase k, thrombin, and trypsin (Zhang et al. 2017). These proteases serve a variety of functions throughout the cell, and as such it can be inferred that the different binding of the inhibitor would result in differing impacts of the proteases at different temperatures. The decreased stoichiometry of inhibitions would suggest that at higher temperatures, protease are less active. A decrease in temperature would result in degradation of cellular components, depending on the protease in question.

1.4 Protein Stability in Psychrophiles

Psychrophilic extremophiles are organisms that can survive and function at extremely low temperatures—typically 20 °C or lower. Psychrophiles have been found to grow and reproduce at temperatures as low as −12 °C and maintain metabolic function at temperatures as low as −20 °C (Rivkina et al. 2000). Because many organisms can be marked as psychrotolerant and this is different from psychrophilic, it is important to distinguish between the two. Both psychrophilic and psychrotolerant organisms can survive at temperatures below 0 °C. However, psychrotolerant organisms have optimal growth above 20 °C while psychrophilic organisms have optimal growth below 15 °C (De Maayer et al. 2014). Psychrophiles, both bacterial and archaeal, face many struggles at such low temperatures. These include, but are not limited to nutrient diffusion rates, DNA supercoiling, and protein unfolding and misfolding. As such, they have undergone many adaptations (Christel et al. 2016).

1.5 Protein Adaptations in Psychrophiles

Generally, psychrophilic proteins have to be more flexible at lower temperatures than mesophiles and thermophiles due to low kinetic barriers (Najnin et al. 2016). While their adaptations seem to be the opposite of thermophiles, there are a few differences. The most prominent adaptations are such: reduced proline content, higher amounts of small amino acids (like alanine and glycine), more uncharged amino acids (like glutamine and threonine), fewer disulfide bridges, and overall fewer electrostatic interactions (hydrogen bonds, salt bridges, cation-pi interactions, aromatic-aromatic interactions) (Saunders et al. 2003; Siddiqui and Cavicchioli 2006; D'Amico et al. 2006; Piette et al. 2011; Struvay et al. 2013). Structurally, these proteins have more loops or coils, less secondary structure and a more exposed and reduced hydrophobic core. These structural adaptations lead to an overall weaker structure and fewer protein-protein interactions (Siddiqui and Cavicchioli 2006). Instead of having overall flexibility, some psychrophilic proteins have a globally rigid structure with localized flexible regions or active sites. These types of proteins can increase activity as temperature rises, and are examples of temperature induced activity (De Maayer et al. 2014; Najnin et al. 2016). The last general adaptation is one of the most impressive. Most psychrophilic proteins also have unusual folding/unfolding reversibility, meaning they can easily fold and unfold their proteins without an energetic penalty. This adaptation not only allows them a great deal of flexibility in an extreme cold environment, but would also allow them to recover from a colder or warmer challenge (Struvay et al. 2013).

An example of psychrophilic adaptations are β-galactosidases. These common and robust enzymes catalyze the breakdown of disaccharides and are present in almost all organisms, which makes comparing features of the different versions practical. The most common reaction catalyzed is the hydrolysis of the β-1,4 linkage of galactose-glucose bonds in lactose. In an analysis of β-galactosidases, the psychrophilic β-galactosidase had a strong preference towards small amino acids including glycine and alanine, especially in coil regions of the protein, and higher percentages of sheet and coil regions (Siddiqui and Cavicchioli 2006). Having larger coil regions and higher glycine and alanine content gave the protein improved flexibility and higher activity at colder temperatures, which was demonstrated by acrylamide fluorescence quenching experiments, and higher catalytic efficiency (k_{cat}/K_m) than their thermal counterparts (Georlette et al. 2003a; Siddiqui and Cavicchioli 2006; Kumar et al. 2014).

Another example of psychrophile adaptation is the protein Trigger factor (TF). TF is a large, 3 domain chaperone protein that assists protein folding and is present in Bacteria, Archaea and some Eukaryotes. It consists of an N-terminal ribosome-binding domain, a peptidyl-prolyl isomerase domain and a C-terminal domain that has an unknown function. TF exists in two forms: a monomer when bound to the ribosome and a dimer when free in the cytosol. TF is one of the first chaperonins involved with proteins synthesized by the ribosome and participates in the protein unfolding response. TF forms a stable but irreversible fold in non-psychrophiles (Robin et al. 2009; Hartl and Hayer-Hartl 2009; Hoffmann et al. 2010). However, the TF from *Pseudoalteromonas haloplanktis* (Ph TF) is a flexible enzyme and has unusually reversible folding. In work by Feller, et al., the structure and stability of TF was followed by differential scanning calorimetry and fluorescence melting curves through multiple cycles of unfolding (high temperature) and refolding (low temperature). They found that Ph TF can reform through the folding/unfolding cycles without a loss of overall structure. However, it wasn't stable at low or high temperatures. When this protein is held at high temperature (37 °C), it is 90% unfolded. When it is at lower temperatures (−4–0 °C), it also becomes unfolded (Struvay et al. 2013). Other work on TF from another psychrophile, *Psychrobacter frigidicola* (Pf TF), confirms these results. In that study, they also show that TF is flexible and does not dimerize, unlike its bacterial or other archaeal homologs (Robin et al. 2009). The flexibility, refolding and lack of protein-protein interactions have been confirmed for many other psychrophilic enzymes (Russell et al. 1998; D'Amico et al. 2001; Georlette et al. 2003b; Giaquinto et al. 2007).

1.6 Applications to Cell Signaling

When psychrophiles need to respond to changing environmental factors, it is mostly to survive temperature increases or go into senescence under extremely cold temperatures, which are similar challenges faced by thermophiles. Psychrophiles, like many other extremophiles, also have adaptations to other challenges such as changes in salinity (Dassarma et al. 2013; Karan et al. 2013). Because some characterized psychrophiles are also halophiles, as a lot of cold environments are also hypersaline, their adaptations overlap. A less common adaptation seen in psychrophiles, but seen in halophiles is the lack of lysine and arginine residues. These amino acids are disfavored in these extremophilic proteins because the long, hydrophobic chain can increase hydrophobic interactions and the positively charged amine group can encourage salt bridge formation, which neither cold or salty environments favor (Siddiqui and Cavicchioli 2006; Dassarma et al. 2013). Having multiple protein adaptations equips the cell to respond and adapt to multiple challenges quickly.

When psychrophilic proteins become too hot, they quickly unravel because they do not have many molecular interactions holding them together. This would seem to be a great disadvantage for the cell, because it could lead to protein aggregation and damage due to increased molecular motions. However, if these conditions are not prolonged, the folding/unfolding mechanism may prevent protein aggregation and damage once the temperature decreases. This allows the cell to resume the activity of the proteins. For example, the chaperone Trigger factor (TF) mentioned above, can undergo as many as 10 cycles of unfolding and refolding and resume helping other proteins refold (Robin et al. 2009; Hoffmann et al. 2010).

When psychrophilic proteins are challenged by colder temperatures, their proteins can either become more rigid, which would decrease their catalytic efficiency, or they can undergo cold-induced unfolding. Decreasing catalytic efficiency, as discussed with thermophiles, can signal the cell that conditions aren't optimal and metabolism has slowed. This signal would lead to the cell taking immediate action to counteract the effects, like downregulating metabolite transport. However, psychrophilic proteins can also undergo cold-induced unfolding, which could be a consequence of the increased flexibility, looser structure and smaller hydrophobic core (Piette et al. 2011; Struvay et al. 2013). This phenomenon, when the organism is at lower temperatures than it is adapted to, may enable water to disrupt the weak interactions maintaining the protein structure, causing unfolding. Hydration of the protein and its amino acids at low temperatures is likely the driving force of cold-induced unfolding. Additional work on other psychrophilic proteins shows a similar effect (Feller et al. 1999; Struvay et al. 2013).

This cold-based unfolding may serve psychrophiles well. By having their proteins in a state of partial denaturation, their enzymatic and binding activities come to a halt and the cell could enter senescence without any formal signaling mechanism. This also means that when temperatures warm, the cell should immediately resume activity.

1.7 Protein Stability in Halophiles

Halophilic extremophiles are characterized by the ability to grow and thrive under extremely ionic environments caused by high concentrations of inorganic ions, like sodium, potassium and other metals. These metals are often present in molarity concentrations, often at saturation or supersaturation. To combat this potential problem, some halophilic bacteria and eukaryotes prevent the entry of the inorganic salts into the cell and accumulate small organic molecules, like sugars or polyamines, to balance the osmotic pressure difference (Mevarech et al. 2000). Halophilic archaea, which can survive under higher ionic concentrations, have a different approach. They will take in the high concentrations of inorganic salt(s) to balance their cellular osmotic potential and not make many osmolytes, unlike the bacterial or eukaryotic halophiles. This strategy has its price. Proteins within the organism had to adapt to the extreme cellular environment in order to remain stable and functional.

The problem facing the cell and its proteins at high ionic conditions is a decrease in water activity. At high salt concentrations (higher than 0.1 M), water becomes less available because it is surrounded by salt in an ionic lattice (Madern et al. 2000; Mevarech et al. 2000; Soppa 2006; Kastritis et al. 2007; Tadeo et al. 2009b). The lower availability of water can cause hydrophobic amino acids in a protein to lose hydration, which strengthens the hydrophobic interactions within and between amino acids and proteins. This can lead to aggregation. In addition, due to their ionic nature, salts also interfere with electrostatic interactions between charged amino acids, which can also lead to exposure of the hydrophobic core (Tadeo et al. 2009a, b; Karan et al. 2012). Protein folding requires the proper balance between hydrophobic and electrostatic forces. Salts, especially in high concentrations, throw these forces off, which can lead to destabilization of the protein, potentially causing global unfolding or the reverse, inflexibility of the protein fold. Both of these effects can lead to aggregation and precipitation. Archaeal halophilic proteins have a number of adaptations that allow them to utilize the high concentrations of inorganic salt to stabilize their native fold. These adaptations and their corresponding effects can affect the cell's physiology and signal the cell when conditions are good or need to improve.

1.8 The Adaptations of Halophilic Proteins

Halophilic proteins have three primary adaptations: (1) increased acidic amino acid content, which is usually on the surface of the protein's structure, (2) a lower amount of hydrophobic amino acids, which forms a smaller hydrophobic core to the protein, and (3) added amino acid sequences (called "peptide insertions") that can be rich in proline resides or small amino acids, like glycine or alanine. These peptide insertions tend to be small (5–25 amino acids in length) and are generally in areas of the protein that need increased flexibility. Most halophilic proteins have one or more of these adaptations (Mevarech et al. 2000; Kastritis et al. 2007; Reed et al. 2013). When in place, these adaptations allow the cell to withstand large changes in salt concentrations. While most proteins in halophiles might have these adaptations, not all do. Those that don't could easily be turned over, as they would be destabilized and subject to degradation by proteases (Mevarech et al. 2000; Qvist et al. 2012; Longo et al. 2013).

The most "visible" adaptation in halophiles is the increased number of the acidic amino acids, aspartic and glutamic acid, which usually end up on the surface of the protein, based on structures of halophilic proteins. The surface of these proteins contains many carboxylic acid groups, which are negatively charged at physiological pH, and they can strongly coordinate metals and water. This creates shells of hydration around the protein. Cation-protein bindings appear in different crystal structures, but it is hard to distinguish between salt and water, because high resolution (less than 2.4 angstroms) is needed. Therefore, the influence of the cation-protein binding on protein's stability is still not very well known (Dym et al. 1995; Frolow et al. 1996; Richard et al. 2000; Mevarech et al. 2000; Britton et al. 2006; Zhang et al. 2013).

The halophilic proteins, compared to nonhalophilic counterparts, can also include the presence of more salt bridges. According to Dym et al., introduction of negatively charged amino acids at the N-terminus of an alpha helix increases the thermostability of malate dehydrogenase (hMDH) from *Haloarcula marismortui*. In NaCl and KCl solutions, hMDH forms hydrated salt ion network coordinated by the acidic groups on the protein surface. Nonhalophilic proteins lack this feature (Dym et al. 1995).

According to Ortega et al., an NMR analysis to study the different mechanisms contributing to protein adaptation at a per residue level, suggested that there are synergistic contributions to protein haloadaptation (Tadeo et al. 2009b; Ortega et al. 2015; Lenton et al. 2016). Cation exclusion and electrostatic repulsion between negatively charged residues destabilize the denatured state, while cation-protein interactions stabilize the folded state. Glutamic acid and aspartic acid provide polar, negatively charged carboxylate groups that are hydrated and adopt favorable conformations to interact with cations. On the contrary, lysine residues are positively charged and are depleted, because of the long, partially hydrophobic side chain. This disallows surface lysine residues from interacting with cations (Ortega et al. 2015).

The structural adaptation of halophilic proteins is also achieved through the decrease of hydrophobic contact surfaces. Most halophiles contain smaller amino acids and smaller amounts of hydrophobic amino acids. When salt concentration is higher than 0.1 M, proteins are less exposed to water because ion coordination takes precedent over water-protein interactions. The hydrophobic amino acids lose hydration and aggregate. Therefore, folding relies on the high salt concentration to strengthen the hydrophobic interactions within the protein, but repel interactions between proteins (Kastritis et al. 2007; Siglioccolo et al. 2011; Qvist et al. 2012; Zhang and Ge 2013). On the other hand, non-halophilic proteins cannot survive in high salt concentrations, because the hydrophobic and electrostatic interactions are altered, preventing the proteins from folding properly and maintaining stability (Mevarech et al. 2000; Reed et al. 2013). Siglioccolo et al. proposed that halophiles have lower hydrophobic contact in the core which may counterbalance the increased strength of hydrophobic interactions in high salt concentration. Weaker hydrophobic interactions can increase the flexibility of proteins in saline environments, since it prevents the hydrophobic core from becoming too rigid (Siglioccolo et al. 2011).

Halophilicity of a protein can be also increased by a peptide insertion in the protein, which increases protein length, keeps ions on the surface, and keeps the active site free for substrates. The inserted peptide can contain multiple acidic amino acids residues, but this is not a strict requirement. In many cases, these insertions improve the catalytic turnover of the enzyme, the flexibility and solvent-accessible surface area (Zaccai et al. 1989; Frolow et al. 1996; Taupin et al. 1997; Taupin and Leberman 1999; Marg et al. 2005; Evilia and Hou 2006). The stability of halophilic proteins also depends on the insertion and frequency of some dipeptides like DA, RA, AD, RR, AP, DD etc., while non-halophilic proteins have more LK, IL, II etc. The dipeptide composition consists mostly of acidic residues. This feature can help to distinguish between halophilic and non-halophilic proteins and aid in the classification of halophilic proteins (Zhang et al. 2013; Ortega et al. 2015).

1.9 Applications to Cell Signaling

Halophiles, like psychrophiles and thermophiles, can experience sudden environmental changes, such as an increase or decrease in salinity, that require immediate action on the part of the cell in order to survive. As the environment outside the cell changes, the intracellular environment also changes despite the efforts of the cell. For halophiles, their protein adaptations are key to surviving these changes, which could be considered more extreme than temperature changes. However, halophiles can also be affected by temperature and, therefore, can have overlapping adaptations with the psychrophiles, which was mentioned in the previous section (Karan et al. 2013; Ortega et al. 2015).

When too much salt is present, the halophile cell cannot shutdown salt import. Therefore, when salt increases on the outside, it also increases on the inside of the cell. When halophilic adapted proteins are put in too much salt, they would act "frozen", meaning through molecular crowding and ionic effects, the protein structure is not as flexible or could be considered rigid. This would serve to decrease the overall catalytic efficiency of such enzymes. Large increases in the ionic environment would affect the ability of enzymes/proteins to bind charged substrates, which might have to compete with the ions for the active site or binding pocket (Evilia and Hou 2006; Reed et al. 2013). Another potential effect of increased salt concentration would be salt-induced unfolding, which could lead to exposure of the protein's hydrophobic core. Our survey of the recent literature shows that most halophilic proteins do not have high-salt induced unfolding, which is a good thing for halophiles, as exposure of the hydrophobic core, even a weak one, would cause aggressive aggregation. High concentrations of salt exasperate the hydrophobic effect.

All of these potential protein folding effects would serve to decrease the activity of the enzymes/proteins and put the cell into a low metabolic state, which could signal senescence. Interestingly, this is why halophiles are also thought to be the oldest organisms. They go into this semi-hibernation state and can be re-awakened later by the addition of water, resulting in the dilution of the salinity. Some very old halophiles have recently been awakened and can be cultured (Fendrihan et al. 2006).

The other state for halophiles would be too little salt. Too little salt is a hard condition for halophiles to adapt to. In addition to increased flexibility of the protein due to a decrease in the hydrated ionic network around the cell, the cell also has to deal with osmotic effects with the cell membrane. If the change was gradual, the hyperflexibility of the proteins would lead to reductions in overall catalytic efficiency with the biggest effects on K_m and/or K_d for the substrate or ligand. The buildup of small molecules would be a powerful signal to the cell to indicate changing conditions. Halophiles, like other organisms, have a chaperone and protease responses to help stabilize proteins and degrade denatured proteins, but it isn't as well described as those for thermophiles or psychrophiles.

2 Conclusion

While extremophiles and Archaea have some common cell signaling systems to bacteria and eukaryotes, not many of these pathways have been described yet. Extremophiles probably use a simpler and much more direct way for intracellular communication; they use the influence of their extreme environment on their proteins' structure and flexibility. These structural changes alter the conditions in the cell and this could lead to downstream effects, like the transcription of metabolic enzymes and/or the activation of chaperones and proteases. Figure 1 summarizes these general effects that protein adaptations in extremophiles could have on cell signaling, based on our current understandings.

While this review was not all inclusive of protein adaptations in extremophiles, we intended to introduce the idea that major adaptations in proteins that result in unusual or increased stability for extreme proteins can be a simple way for the cell to sense its environment. When we see these adaptations in non-extremophiles, it could be evolution's way of giving the cell some immediate control over a particular event. For example, it has been seen in the literature that gene transfer between human pathogens and extremophiles/Archaea may have taken place (Chistoserdova et al. 1998; Faguy 2003; Tsaousis et al. 2012). To understand why a human pathogen would acquire an extremophile protein with its adaptations, it would help to first understand the signaling role that the protein may play within the

High Temperature or Low Salinity	Low Temperature or High Salinity
K_d and K_m ↑↑ k_{cat} = or ↑⁞	K_d and K_m ↓ k_{cat} ↓↓⁞
Problem: Protein folding/structure destabilized, too flexible	**Problem:** Protein folding/structure too stable/stiff, not flexible
Thermophile Adaptation: Increased hydrophobicity and electrostatic interactions help stabilize protein structure. **Halophile Adaptation**: Reduced hydrophobic core and increased acidic amino acids on surface keeps ions on the surface and the protein folded.	**Psychrophile Adaptation**: Decreased hydrophobic core and charged/electrostatic interactions keep protein structure flexible. **Halophile Adaptation**: Reduced hydrophobic core, increased acidic amino acids on surface and protein length keep ions on the surface and active site free for substrates.
Signal: Depletion of substrates for affected enzymes. The concentration decrease is "sensed" and more substrates are imported	**Signal:** Substrates bound by enzymes, but catalytically slow turnover. Import of substrates is decreased.
Chaperones are activated – increased transcription/translation and structure stabilized. Proteases activity increases.	Reduced Chaperone and protease activity
Cell increases production or import of "stabilizers" (high temperature) or increased activity of ion channels (ions in) (salinity)	Cell increases production or import of osmolytes (low temperature) or increased activity of ion channels (ions out) (salinity)

Fig. 1 A general overview of the extremophile protein adaptations and how they could relate to intracellular communication. Halophilic adaptations and their corresponding potential for signaling was put into the high temperature or low temperature category, because the effects of low and high salinity on halophilic protein structure are similar to those states in thermophiles and psychrophiles, respectively

extremophile and infer how the bacterial cell might use this information. Once we understand this, a more complete picture of how extremophiles survive and why/how pathogens adapt, might be at hand.

References

Baker-Austin C, Dopson M (2007) Life in acid: pH homeostasis in acidophiles. Trends Microbiol 15:165–171. doi:10.1016/j.tim.2007.02.005

Boonyaratanakornkit BB, Park CB, Clark DS (2002) Pressure effects on intra- and intermolecular interactions within proteins. Biochim Biophys Acta 1595:235–249

Britton KL, Baker PJ, Fisher M et al (2006) Analysis of protein solvent interactions in glucose dehydrogenase from the extreme halophile Haloferax mediterranei. Proc Natl Acad Sci USA 103:4846–4851. doi:10.1073/pnas.0508854103

Cacciapuoti G, Porcelli M, Bertoldo C et al (1994) Purification and characterization of extremely thermophilic and thermostable 5'-methylthioadenosine phosphorylase from the archaeon *Sulfolobus solfataricus*. Purine nucleoside phosphorylase activity and evidence for intersubunit disulfide bonds. J Biol Chem 269:24762–24769

Cacciapuoti G, Fuccio F, Petraccone L et al (2012) Role of disulfide bonds in conformational stability and folding of 5'-deoxy-5'-methylthioadenosine phosphorylase II from the hyperthermophilic archaeon *Sulfolobus solfataricus*. Biochim Biophys Acta 1824:1136–1143. doi:10.1016/j.bbapap.2012.06.014

Cashin P, Goldsack L, Hall D, O'Toole R (2006) Contrasting signal transduction mechanisms in bacterial and eukaryotic gene transcription. FEMS Microbiol Lett 261:155–164. doi:10.1111/j.1574-6968.2006.00295.x

Chan CH, Yu TH, Wong KB (2011) Stabilizing salt-bridge enhances protein thermostability by reducing the heat capacity change of unfolding. PLoS ONE 6:e21624. doi:10.1371/journal.pone.0021624

Chistoserdova L, Vorholt JA, Thauer RK, Lidstrom ME (1998) C1 transfer enzymes and coenzymes linking methylotrophic bacteria and methanogenic Archaea. Science 281:99–102

Christel S, Fridlund J, Watkin EL, Dopson M (2016) Acidithiobacillus ferrivorans SS3 presents little RNA transcript response related to cold stress during growth at 8 °C suggesting it is a eurypsychrophile. Extremophiles Life Extreme Cond. 20:903–913. doi:10.1007/s00792-016-0882-2

D'Amico S, Gerday C, Feller G (2001) Structural determinants of cold adaptation and stability in a large protein. J Biol Chem 276:25791–25796. doi:10.1074/jbc.M102741200

D'Amico S, Collins T, Marx JC et al (2006) Psychrophilic microorganisms: challenges for life. EMBO Rep 7:385–389. doi:10.1038/sj.embor.7400662

Dassarma S, Capes MD, Karan R, Dassarma P (2013) Amino acid substitutions in cold-adapted proteins from *Halorubrum lacusprofundi*, an extremely halophilic microbe from antarctica. PLoS ONE 8:e58587. doi:10.1371/journal.pone.0058587

De Maayer P, Anderson D, Cary C, Cowan DA (2014) Some like it cold: understanding the survival strategies of psychrophiles. EMBO Rep 15:508–517. doi:10.1002/embr.201338170

Di Giulio M (2005) A comparison of proteins from Pyrococcus furiosus and Pyrococcus abyssi: barophily in the physicochemical properties of amino acids and in the genetic code. Gene 346:1–6. doi:10.1016/j.gene.2004.10.008

Dym O, Mevarech M, Sussman JL (1995) Structural Features That Stabilize Halophilic Malate Dehydrogenase from an Archaebacterium. Science 267:1344–1346

Esser D, Hoffmann L, Pham TK et al (2016) Protein phosphorylation and its role in archaeal signal transduction. FEMS Microbiol Rev 40:625–647. doi:10.1093/femsre/fuw020

Evilia C, Hou YM (2006) Acquisition of an insertion peptide for efficient aminoacylation by a halophile tRNA synthetase. Biochemistry (Mosc) 45:6835–6845

Faguy DM (2003) Lateral gene transfer (LGT) between Archaea and *Escherichia coli* is a contributor to the emergence of novel infectious disease. BMC Infect Dis 3:13. doi:10.1186/1471-2334-3-13

Feller G, d'Amico D, Gerday C (1999) Thermodynamic stability of a cold-active alpha-amylase from the Antarctic bacterium Alteromonas haloplanctis. Biochemistry (Mosc) 38:4613–4619. doi:10.1021/bi982650+

Fendrihan S, Legat A, Pfaffenhuemer M et al (2006) Extremely halophilic archaea and the issue of long-term microbial survival. Rev Environ Sci Biotechnol Online 5:203–218. doi:10.1007/s11157-006-0007-y

Frolow F, Harel M, Sussman JL et al (1996) Insights into protein adaptation to a saturated salt environment from the crystal structure of a halophilic 2Fe-2S ferredoxin. Nat Struct Biol 3:452–458

Georlette D, Blaise V, Dohmen C et al (2003a) Cofactor binding modulates the conformational stabilities and unfolding patterns of NAD(+)-dependent DNA ligases from *Escherichia coli* and *Thermus scotoductus*. J Biol Chem 278:49945–49953. doi:10.1074/jbc.M307761200

Georlette D, Damien B, Blaise V et al (2003b) Structural and functional adaptations to extreme temperatures in psychrophilic, mesophilic, and thermophilic DNA ligases. J Biol Chem 278:37015–37023

Giaquinto L, Curmi PM, Siddiqui KS et al (2007) Structure and function of cold shock proteins in archaea. J Bacteriol 189:5738–5748. doi:10.1128/JB.00395-07

Godin-Roulling A, Schmidpeter PAM, Schmid FX, Feller G (2015) Functional adaptations of the bacterial chaperone trigger factor to extreme environmental temperatures. Environ Microbiol 17:2407–2420. doi:10.1111/1462-2920.12707

Hartl FU, Hayer-Hartl M (2009) Converging concepts of protein folding in vitro and in vivo. Nat Struct Mol Biol 16:574–581. doi:10.1038/nsmb.1591

Hay S, Evans RM, Levy C et al (2009) Are the catalytic properties of enzymes from piezophilic organisms pressure adapted? ChemBioChem 10:2348–2353. doi:10.1002/cbic.200900367

Hendsch ZS, Tidor B (1994) Do salt bridges stabilize proteins? A continuum electrostatic analysis. Protein Sci 3:211–226. doi:10.1002/pro.5560030206

Hoffmann A, Bukau B, Kramer G (2010) Structure and function of the molecular chaperone Trigger Factor. Biochim Biophys Acta BBA - Mol Cell Res 1803:650–661. doi:10.1016/j.bbamcr.2010.01.017

Karan R, Capes MD, Dassarma S (2012) Function and biotechnology of extremophilic enzymes in low water activity. Aquat Biosyst 8:4. doi:10.1186/2046-9063-8-4

Karan R, Capes MD, DasSarma P, DasSarma S (2013) Cloning, overexpression, purification, and characterization of a polyextremophilic beta-galactosidase from the Antarctic haloarchaeon *Halorubrum lacusprofundi*. BMC Biotechnol 13:3. doi:10.1186/1472-6750-13-3

Karshikoff A, Ladenstein R (2001) Ion pairs and the thermotolerance of proteins from hyperthermophiles: a "traffic rule" for hot roads. Trends Biochem Sci 26:550–556

Kastritis PL, Papandreou NC, Hamodrakas SJ (2007) Haloadaptation: insights from comparative modeling studies of halophilic archaeal DHFRs. Int J Biol Macromol 41:447–453. doi:10.1016/j.ijbiomac.2007.06.005

Kumar V, Sharma N, Bhalla TC (2014) In silico analysis of β-galactosidases primary and secondary Structure in relation to temperature adaptation. J Amino Acids 2014:475839. doi:10.1155/2014/475839

Lee CF, Makhatadze GI, Wong KB (2005) Effects of charge-to-alanine substitutions on the stability of ribosomal protein L30e from thermococcus celer. Biochemistry (Mosc) 44:16817–16825. doi:10.1021/bi0519654

Lenton S, Walsh DL, Rhys NH et al (2016) Structural evidence for solvent-stabilisation by aspartic acid as a mechanism for halophilic protein stability in high salt concentrations. Phys Chem Chem Phys PCCP 18:18054–18062. doi:10.1039/c6cp02684b

Longo LM, Lee J, Blaber M (2013) Simplified protein design biased for prebiotic amino acids yields a foldable, halophilic protein. Proc Natl Acad Sci U A 110:2135–2139. doi:10.1073/pnas.1219530110

Madern D, Ebel C, Zaccai G (2000) Halophilic adaptation of enzymes. Extremophiles 4:91–98

Marg BL, Schweimer K, Sticht H, Oesterhelt D (2005) A two-alpha-helix extra domain mediates the halophilic character of a plant-type ferredoxin from halophilic archaea. Biochemistry (Mosc) 44:29–39

Mevarech M, Frolow F, Gloss LM (2000) Halophilic enzymes: proteins with a grain of salt. Biophys Chem 86:155–164

Najnin T, Siddiqui KS, Taha T et al (2016) Characterization of a temperature-responsive two component regulatory system from the Antarctic archaeon, *Methanococcoides burtonii*. Sci Rep 6:24278. doi:10.1038/srep24278

Ortega G, Diercks T, Millet O (2015) Halophilic protein adaptation results from synergistic residue-ion interactions in the folded and unfolded states. Chem Biol 22:1597–1607. doi:10.1016/j.chembiol.2015.10.010

Peng H-L, Egawa T, Chang E et al (2015) Mechanism of thermal adaptation in the lactate dehydrogenases. J Phys Chem B 119:15256–15262. doi:10.1021/acs.jpcb.5b09909

Piette F, Struvay C, Feller G (2011) The protein folding challenge in psychrophiles: facts and current issues. Environ Microbiol 13:1924–1933. doi:10.1111/j.1462-2920.2011.02436.x

Qvist J, Ortega G, Tadeo X et al (2012) Hydration dynamics of a halophilic protein in folded and unfolded states. J Phys Chem B 116:3436–3444. doi:10.1021/jp3000569

Reed CJ, Lewis H, Trejo E et al (2013) Protein adaptations in archaeal extremophiles. Archaea Vanc BC 2013:373275. doi:10.1155/2013/373275

Richard SB, Madern D, Garcin E, Zaccai G (2000) Halophilic adaptation: novel solvent protein interactions observed in the 2.9 and 2.6 A resolution structures of the wild type and a mutant of malate dehydrogenase from *Haloarcula marismortui*. Biochemistry (Mosc) 39:992–1000

Rivkina EM, Friedmann EI, McKay CP, Gilichinsky DA (2000) Metabolic activity of permafrost bacteria below the freezing point. Appl Environ Microbiol 66:3230–3233

Robin S, Togashi DM, Ryder AG, Wall JG (2009) Trigger factor from the psychrophilic bacterium *Psychrobacter frigidicola* is a monomeric chaperone. J Bacteriol 191:1162–1168. doi:10.1128/JB.01137-08

Russell RJ, Gerike U, Danson MJ et al (1998) Structural adaptations of the cold-active citrate synthase from an Antarctic bacterium. Structure 6:351–361

Saunders NF, Thomas T, Curmi PM et al (2003) Mechanisms of thermal adaptation revealed from the genomes of the Antarctic Archaea *Methanogenium frigidum* and *Methanococcoides burtonii*. Genome Res 13:1580–1588. doi:10.1101/gr.1180903

Schleper C, Jurgens G, Jonuscheit M (2005) Genomic studies of uncultivated archaea. Nat Rev Microbiol 3:479–488. doi:10.1038/nrmicro1159

Sharma A, Kawarabayasi Y, Satyanarayana T (2012) Acidophilic bacteria and archaea: acid stable biocatalysts and their potential applications. Extremophiles 16:1–19. doi:10.1007/s00792-011-0402-3

Siddiqui KS, Cavicchioli R (2006) Cold-adapted enzymes. Annu Rev Biochem 75:403–433. doi:10.1146/annurev.biochem.75.103004.142723

Siglioccolo A, Paiardini A, Piscitelli M, Pascarella S (2011) Structural adaptation of extreme halophilic proteins through decrease of conserved hydrophobic contact surface. BMC Struct Biol 11:50. doi:10.1186/1472-6807-11-50

Soppa J (2006) From genomes to function: haloarchaea as model organisms. Microbiology 152:585–590

Struvay C, Negro S, Matagne A, Feller G (2013) Energetics of protein stability at extreme environmental temperatures in bacterial trigger factors. Biochemistry (Mosc) 52:2982–2990. doi:10.1021/bi4002387

Szilagyi A, Zavodszky P (2000) Structural differences between mesophilic, moderately thermophilic and extremely thermophilic protein subunits: results of a comprehensive survey. Structure 8:493–504

Tadeo X, Lopez-Mendez B, Castano D et al (2009a) Protein stabilization and the Hofmeister effect: the role of hydrophobic solvation. Biophys J 97:2595–2603. doi:10.1016/j.bpj.2009.08.029

Tadeo X, Lopez-Mendez B, Trigueros T et al (2009b) Structural basis for the aminoacid composition of proteins from halophilic archea. PLoS Biol 7:e1000257. doi:10.1371/journal.pbio.1000257

Taupin CM, Leberman R (1999) Archaeabacterial seryl-tRNA synthetases: adaptation to extreme environments and evolutionary analysis. J Mol Evol 48:408–420

Taupin CM, Hartlein M, Leberman R (1997) Seryl-tRNA synthetase from the extreme halophile *Haloarcula marismortui*–isolation, characterization and sequencing of the gene and its expression in *Escherichia coli*. Eur J Biochem 243:141–150

Tomazic SJ, Klibanov AM (1988) Mechanisms of irreversible thermal inactivation of Bacillus alpha-amylases. J Biol Chem 263:3086–3091

Tsaousis AD, Ollagnier de Choudens S, Gentekaki E et al (2012) Evolution of Fe/S cluster biogenesis in the anaerobic parasite Blastocystis. Proc Natl Acad Sci U S A 109:10426–10431. doi:10.1073/pnas.1116067109

Tsukamoto T, Mizutani K, Hasegawa T et al (2016) X-ray crystallographic structure of thermophilic rhodopsin: implications for high thermal stability and optogenetic function. J Biol Chem 291:12223–12232. doi:10.1074/jbc.M116.719815

Vieille C, Zeikus GJ (2001) Hyperthermophilic enzymes: sources, uses, and molecular mechanisms for thermostability. Microbiol Mol Biol Rev 65:1–43. doi:10.1128/MMBR.65.1.1-43.2001

Vihinen M (1987) Relationship of protein flexibility to thermostability. Protein Eng 1:477–480

Wilkinson B, Gilbert HF (2004) Protein disulfide isomerase. Biochim Biophys Acta 1699:35–44. doi:10.1016/j.bbapap.2004.02.017

Woycechowsky KJ, Raines RT (2003) The CXC motif: a functional mimic of protein disulfide isomerase. Biochemistry (Mosc) 42:5387–5394. doi:10.1021/bi026993q

Zaccai G, Cendrin F, Haik Y et al (1989) Stabilization of halophilic malate dehydrogenase. J Mol Biol 208:491–500

Zhang G, Ge H (2013) Protein hypersaline adaptation: insight from amino acids with machine learning algorithms. Protein J 32:239–245. doi:10.1007/s10930-013-9484-3

Zhang G, Huihua G, Yi L (2013) Stability of halophilic proteins: from dipeptide attributes to discrimination classifier. Int J Biol Macromol 53:1–6. doi:10.1016/j.ijbiomac.2012.10.031

Zhang H, Fei R, Xue B et al (2017) Pnserpin: a novel serine protease inhibitor from extremophile *Pyrobaculum neutrophilum*. Int J Mol Sci. doi:10.3390/ijms18010113

Protein Phosphorylation-Dephosphorylation and Signal Processing in the Archaea

Peter J. Kennelly

1 Introduction: "The Phone Is Ringing"

What do you think of when someone mentions the word "communication"? An 18th century individual might visualize a letter being carried vast distances across land and sea by wagon, railway, ship, and the footfalls of the local postman. A 19th century individual might envision miles of wires carrying dots and dashes hammered out using a telegraph key. A 20th century individual might picture miles of wire supported by a veritable forest of wooden poles enabling them to carry on a conversation with someone many miles distant while each of them sits in their home. Today, most people would undoubtedly picture the personal electronic device that keeps them connected at all times and in all places via voice mail, email, or text messages. While no doubt there would be exceptions to the rule, most of us tend to think of communication mainly in terms of the means of conveyance: transmitters, receivers, wires, satellites, etc.

However, a complete picture of communication must also include an understanding of the factors that triggered and shaped each transmission. What were the underlying factors that initiated the act of dialing the phone and selection of a specific "number"? What happens after the recipient picks up our metaphorical receiver? Is their response automatic and invariant (e.g. answering machine), or does it differ depending upon circumstances (e.g. human operator)? Is there a mechanism for addressing the simultaneous receipt of messages that, in isolation, would stimulate antagonistic responses? The aim of this chapter is to discuss our current, albeit fragmentary, appreciation of the intracellular apparatus responsible for the initiation of and reaction to inter organismal signals.

P.J. Kennelly (✉)
Department of Biochemistry, Virginia Polytechnic Institute and State University, Blacksburg, VA 24061, USA
e-mail: pjkennel@vt.edu

G. Witzany (ed.), *Biocommunication of Archaea*,
DOI 10.1007/978-3-319-65536-9_13

2 From Reflex to Information Processing

2.1 In the Beginning: Sensor-Response

Communication is, at its heart, a sensor-response process. What distinguishes recruitment, a form of "communication", from chemotaxis, a generic sensor-response process, are the characteristics of the signal. These, in turn, are largely contextual in nature: the source of the signal, whether it is a dedicated informational entity or is "moonlighting" as such, and perhaps the nature of the circumstances that triggered its dispatch. The bottom line is that the evolution of archaeal and other biologic communication mechanisms is inextricably tied to the development of molecular sensor-response processes in general.

The ability to sense and respond to changing circumstances constituted a watershed in the evolution of living organisms. It is easy to imagine the incredible competitive advantage that was conferred on the first organism capable of coordinating carbohydrate and amino acid metabolism or to chemotax toward an essential and/or scarce nutrient, however crude and "accidental" the original mechanism might have been. Early sensor-response systems undoubtedly relied heavily on internal signals, such as the concentrations of key metabolites, as their source of sensory inputs. While the levels of pathway end products or key metabolites such as 5′-AMP or citrate provided only indirect, derivative information regarding a cell's extracellular milieu, their levels nonetheless reflected to a meaningful degree the abundance and composition of the available nutrient pool. Moreover, the ability to coordinate two or more metabolic processes in and of itself, by enhancing the efficiency with which available resources were utilized, enhanced the fitness of the affected organism.

While directly monitoring extracellular inputs using extracellular receptors is far and away superior to inferring external circumstances from internal cues, the latter likely emerged first simply because the necessary components were readily available. The concentrations of adenine nucleotides, in particular 5′-AMP, provided an immediate readout of the overall energy status of the cell, while the level of acetyl-CoA served as a gauge of available metabolic fuel. Perhaps more importantly, the metabolic enzymes encoded by primitive genomes provided a set of modular ligand binding domains that could be combined through gene duplication and gene fusion events to generate allosterically-regulated proteins.

It is hard to envision equally rapid and facile pathways for the emergence of transmembrane receptors. Some external signals, such as light, certain hydrophobic organic molecules, and diatomic gases, were sufficiently membrane permeable to permit monitoring using intracellular receptors. However, while some proteins undoubtedly proved capable of autonomously inserting themselves into or across a lipid membrane; the proliferation of transmembrane receptors remained contingent on the development of efficient mechanisms for the secretion and membrane insertion of proteins. Once this capitalization cost had been paid, the types of transmembrane signal transduction systems that we associate with hormone action, chemotaxis, neural responses, olfaction, and other forms of sensing and communication rapidly proliferated, overtaking and in some cases replacing allosteric monitoring of indicator metabolites.

2.2 *Reflex Versus Calculation*

If the ability to sense and swim towards a single nutrient confers a competitive advantage, then adding more chemoreceptors should increase that advantage. If coordinating flux through glycolysis and the tricarboxylic acid cycle is good, then coordinating flux through the pyrimidine and purine biosynthetic pathways should be even better. However, there is a limit to what can be achieved by simple addition. The first sensor-response apparatuses were simple, and therefore reactive or reflexive in nature. Receipt of a signal elicited a hard-wired response at all times and under all circumstances. Over time, as exemplified by the manner in which methylation of chemotaxis receptors continually recalibrated the sensitivity of the sensor while moving along a gradient or the elegantly-layered web of feedback events that regulated flux and distribution within the multi-branched pathway by which the aromatic amino acids phenylalanine, tyrosine and tryptophan are synthesized, individual sensor-response systems could expand their hard-wired response repertoire beyond a binary on-off paradigm to a more graduated one. However, as long as each module functioned in isolation, the system remained reflexive in nature with no means of coordination among different molecular processes and little capacity to integrate and weigh multiple inputs.

The diauxic growth exhibited by many microorganisms when cultured in the presence of multiple potential carbon sources illustrates the difference in performance between a set of semi-autonomous sensor-response subsystems and a fully integrated network (Fig. 1). Initially, the microorganism metabolizes only one of the available substrates, generally that matching the default configuration of its metabolic pathways, until this energy source is exhausted. The organism then enters a stationary period, the diauxic lag, while it awaits production of the gene products needed to begin processing the second, alternate carbon source. A fully integrated sensor response system, it can be argued, would eliminate the lag phase by either anticipating the need to switch carbon sources, or configuring metabolism to process the components in parallel rather than sequentially. This is not to argue that the latter solutions would be superior to that which has sufficed over millennia of evolution—indeed the artificial nature of laboratory culture conditions may exacerbate the duration of the lag period.

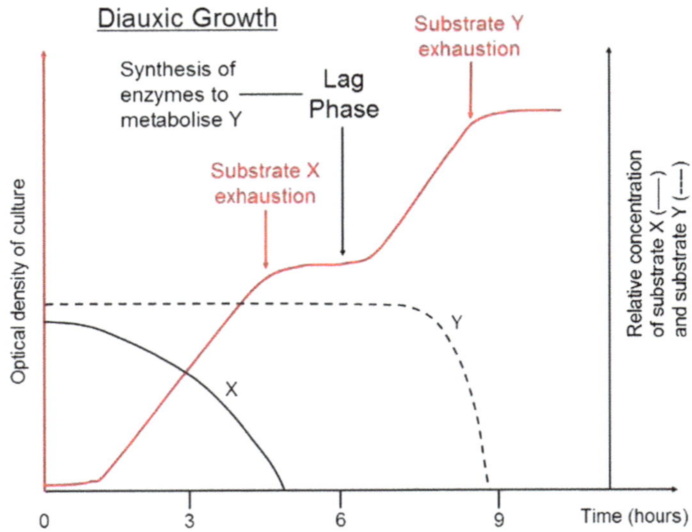

Fig. 1 Diauxic growth of a bacterium on media containing limited amounts of two carbon sources

2.3 Connectivity Is Key

The key factor in the transition living cells from reflexive reactors to calculating adaptors, from symbiotic individuals to components of multicellular organisms, can be summed up in a single word: connectivity (Beltrao et al. 2013; Bray 1995, 2003). If a cell is to weigh multiple inputs, both external and internal, in an integrated manner and generate a calculated, comprehensive, and nuanced response; the sensor-response pathways targeted by each signal must be physically and functionally interconnected. Incorporation into a larger network, in turn, transforms individual macromolecular on-off switches into binary elements of an information processing network without altering their essential chemical nature. In other words, the ability to formulate and respond to signals in a sophisticated manner—communication—arose, not from the emergence of some new macromolecular paradigm such as the proposed evolutionary linkage between protein-tyrosine phosphorylation multicellular eukaryotes (Jin and Pawson 2012; Hunter 2014), but from increases the number and architectural complexity with which extant macromolecular bits became interconnected.

2.4 Bits for an Organic Microprocessor

Like electronic computers, biological information processing is founded upon the assembly of binary elements, or bits, capable of oscillating back and forth between two physically distinct states (Bray 1995; Marks 1996; Kennelly 1999; Landry et al. 2014). Moreover, these states must be consequential, not simply formal. The properties affected must be capable of influencing other elements in the network. While cellular sensor-response pathways can and do employ a diverse mosaic of different biomolecular switches and bits; not all were created equal with respect to their suitability as foundational elements for integrated bio-organic information processing and response networks. For example, while partial proteolysis acts as a very effective trigger for the activation of many digestive enzymes as well as the blood clotting and complement cascades found in plasma, entropic factors render it an operationally irreversible process in vivo, thus rendering zymogen activation incapable of acting as a bioorganic bit.

The binding of dissociable allosteric effectors such as second messengers or feedback metabolites is an inherently reversible, and thus naturally binary, mechanism for transitioning macromolecules between functional states. However, while it is relatively easy for one allosteric ligand to target multiple proteins, it is difficult to diversify the timing and duration of their ligand-induced transitions since the status of each target is directly tied to its concentration. While coordinate regulation of multiple targets is quite feasible, in the absence of any additional regulatory intervention, transitioning occurs across the target population in a concerted "all or none" manner. Perhaps more importantly, while some proteins are capable of responding to multiple allosteric effectors, such as phosphofructokinase and isocitrate dehydrogenase, this method of accommodating multiple inputs has enjoyed only limited application. This is presumably because the protein engineering requirements of productively linking multiple three-dimensionally complex allosteric domains to some catalytic or other functional domain—either through gene fusion or formation of hetero-oligomers—is quite formidable. As a consequence, covalent modification eventually surpassed allosteric regulation as organo-molecular information processing/communication networks grew in size and capability.

Proteins undergo numerous covalent modifications throughout their life cycle. Many of these represent structural alterations required to transform a genetically encoded polypeptide into a functional protein: glycosylation, proteolytic processing of transiently needed leader sequences, the addition of functionally essential prosthetic groups such as lipoic acid (pyruvate dehydrogenase complex), biotin (carboxylases), and Fe-S centers. These modifications are distinguished by their obligate nature, stoichiometric level, early or even co-translational occurrence, and stability over time rather than their chemical nature. Regulatory covalent modifications are characterized by their temporal and stoichiometric dynamism and linkage to a specific trigger. Thus, partial proteolysis occurs as both a common processing event and a trigger for unleashing the latent catalytic activity of zymogens such as trypsinogen, chymotrypsinogen, and s majority of the components of the blood clotting and complement cascades.

The lack of an entropically-favorable means for restoring an activated zymogen to its dormant, preprotein state renders it operationally irreversible. Hence, while partial proteolysis can and is used as an effective regulatory trigger, it lacks the "reversibility" essential for serving as a bioorganic bit. So, what other characteristics are necessary or desirable in a staple building block for cellular information processing and communication networks? One is that the site of modification or acceptor site requires little associated structural overhead. This is important for permitting a suitably diverse spectrum of proteins to be linked into the network to accommodate the widest possible set of inputs and cellular processes. Second, this facilitates the introduction of multiple modification sites into a target, an important means of generating proteins capable of serving as foci for signal convergence: a key pre-requisite for processing (Beltrao et al. 2013).

2.5 *Phosphorylation as the Lynchpin of Molecular Information Processing Networks*

Linking together the mechanistically diverse cellular systems responsible for sensing and responding to intra- and extra-cellular signals into an information processing and communication network requires a very special molecular glue. The agent for accomplishing this must be a potent and relatively non-specific agent for perturbing protein conformation. In order to do so, it must be able to act through space rather than through covalent bonds, as the latter would impose substantial constraints on the proteins susceptible to modulation and narrow the menu of potential regulatory covalent modification sites within them. The principal biophysical forces responsible for determining and stabilizing a protein's three-dimensional conformation are charge-charge interactions, hydrogen bonding, hydrophobic interactions, and steric hindrance. It therefore follows that a universal bioorganic bit must possess relatively extreme characteristics with regard to one or more of these forces.

Among the various chemical moieties utilized for the reversible covalent modification of proteins, such as methyl groups, glutathione, ubitquitin, small ubiquitin-like modifier (SUMO), acetyl groups, etc., it is the phosphoryl group that possesses the most potent suite of biophysical attributes (Westheimer 1987). When linked to a carboxyl, amino, imino, or hydroxyl group; the phosphoryl group remains a polyprotic acid with two dissociable hydrogens. The pKa's for dissociation of the first and second protons from phosphoesters are typically around 0 and 6 (Vogel 1989); which means at physiological pH a phosphoserine side chain would bear a charge of at least −1 and perhaps −2. Moreover, the region of the protein to which the phosphate is bound can shift pKa_2 to a more basic or acidic value (Jaquet et al. 1993; Singer and Forman-Kay 1997). The potential for a net change in charge of −2 (or even −3 in the case of phosphohistidine, phospholysine, and phosphoarginine) on the surface of a protein is greater than can be achieved by the methylation of glutamate side chains, −1 to 0, or via the acetylation of lysine (+1 to 0).

Many researchers attempt to mimic the phosphorylated state of some proteins by utilizing site-directed mutagenesis to replace a phosphoacceptor serine, threonine, or even tyrosine residue with aspartic or glutamic acid; since dissociation of the proton from their carboxyl side chain (pKa 3–5) will introduce a charge of −1. While sometimes successful, other times it is not. This is not unexpected, given that the two-component system works by phosphorylating the carboxylate group of a glutamate residue on the response regulator, indicating that the modulatory effects of protein phosphorylation are not simply or perhaps always about charge. The phosphoryl group also avidly forms salt bridges with the protonated guanidino group of arginine, the common denominator between the two being the dispersal of their respective charges over multiple atoms via resonance (Johnson and Lewis 2001). Phosphoryl groups also possess a high propensity to form an extensive network hydrogen bonds—either with other groups on the protein or surrounding water molecules. Hence introduction of a phosphoryl group increases the hydrophilicity of the surrounding region, while a shell of adhering water molecules magnifies its physical dimensions and steric impact. Perhaps the strongest evidence for the ability of phosphoryl groups to perturb protein structure is the recent observation that in many instances where regulation of a protein by phosphorylation has been conserved, the specific site(s) at which the phosphoryl groups attach oftentimes have not (Gnad et al. 2010; Landry et al. 2014).

2.6 *Multisite Phosphorylation*

The cumulative impact of its high charge density, extreme hydrophilicity, exceptional hydrogen bonding capacity, and steric enlargement by hydrating water molecules renders phosphorylation the most potent and versatile perturber of protein conformation among nature's arsenal of covalent modifiers (Kennelly 1999). However, protein phosphorylation-dephosphorylation also exemplifies another key characteristic of a general or all-purpose building block for cellular information processing and communication networks: the capacity to introduce multiple input sites into a single protein in order to generate points of signal convergence and integration (Roach 1991; Salazar and Hofer 2003; Yang et al. 2004; Doan et al. 2006; Park et al. 2006). Protein phosphorylation-dephosphorylation generates convergence nodes in three ways.

First, given their minimal structural requirements, introduction of multiple phosphorylation sites into a protein is not just feasible, but extremely common (Yachie et al. 2009). Second, multiple protein kinases oftentimes can phosphorylate the same acceptor site on a protein (Pearson and Kemp 1991). Third, there is no obligate relationship between the protein kinases that introduce a phosphoryl group into a protein and the protein phosphatases that remove them. Each discriminates their targets using different cues. While multiple proteins may be phosphorylated by a single protein kinase, their subsequent dephosphorylation may be carried out by different protein phosphatases. Since these kinases and phosphatases are likely responding to different internal and/or external inputs, including other protein kinases and protein phosphatases, one can easily imagine how this can be exploited for signal integration and information processing as well as temporal differentiation in the regulation of target phosphoproteins (Marks 1996; Kennelly 1999).

Much of this versatility arose as a natural consequence of the basic character of protein phosphorylation-dephosphorylation. Once the evolutionary capitalization cost had been paid, the emergence of dedicated, independent regulatory protein kinases and protein phosphatases, the development of interconnected networks became, relatively speaking, straightforward (Alon 2003; Beltrao et al. 2013; Derouiche et al. 2016; Jin and Pawson 2012; Yachie et al. 2009). However, while protein-serine/threonine/tyrosine phosphorylation relied on separate, autonomous protein kinases and protein phosphatases to catalyze the phosphorylation and dephosphorylation of target proteins, the bacterial two-component system embedded these functions in a single polypeptide, the response regulator protein. Not only is the response regulator the target of phosphorylation, it is the catalyst for its phosphorylation as well.

In two-component signaling, the role of histidine kinase is not to catalyze the transfer of phosphate from ATP to the response regulator (Zschiedrich et al. 2016). Rather, the autophosphorylated histidine residue serves as a specialized phosphodonor substrate for the intrinsic protein kinase activity in its cognate response regulator. Thus, signaling via the two-component paradigm is modulated by controlling the availability of the substrate phosphoprotein. In addition, many response regulators serve as their own autophosphatases. Whatever the initial advantages of this arrangement, its inherent restrictiveness severely limited the ability to develop the capacity to network beyond other two-component systems (Jung et al. 2012).

3 Protein-Ser/Thr/Tyr Phosphorylation and Multicellular Life

3.1 Eukaryotic Protein Phosphorylation

Early on, regulatory protein-ser/thr/tyr phosphorylation-dephosphorylation was closely associated with the cell-cell communication needs of "higher" organisms composed of multiple, specialized cell types. The first protein-ser/thr phosphorylation-mediated signal transduction cascade, the "fight or flight response", outlined how the hormone epinephrine, acting through its receptor, mobilized glycogen bound glucose through the phosphorylation of glycogen synthase and glycogen phosphorylase. Further studies uncovered details of the role of protein phosphorylation-dephosphorylation in the mediation of gluconeogenesis and glycolysis by insulin and glucagon, the contraction of striated muscle tissue in response to neuronal signals, and the stimulation of mitogenesis by peptide hormones. These first signal transduction cascades functioned as the intracellular extension of the neuroendocrine systems in animals, an association that suggested that protein phosphorylation-dephosphorylation had emerged specifically to meet the communication needs of multicellular, eukaryotic organisms (Kennelly and Potts 1996).

The quantum leap in cellular communication and coordination associated with multicellularity also offered a convincing rationale for why an organism would meet the high initial capitalization costs of protein phosphorylation-dephosphorylation, i.e. the development of unique, dedicated protein kinases and protein phosphatases. Why would a microbe pay this when allosterism appeared to be perfectly capable of meeting its much more limited needs? Early studies using radiolabeling and electrophoresis revealed that, while mammalian cells contained hundreds and perhaps thousands of phosphoproteins, bacteria such as *Escherichia coli* and *Salmonella typhimurium* could muster barely a dozen each (Manai and Cozzone 1979; Ames and Nikaido 1981). Ironically, when the first phosphorylation-based sensor-response pathways were unambiguously identified in *Bacteria*, they took the form of two-component systems whose chemistry, catalysts, and mechanism radically differed from existing eukaryotic models (Gross et al. 1989; Igo et al. 1990).

3.2 *Discovery of "Eukaryotic" Protein Phosphorylation Components in the* Archaea

Given the prevailing consensus, it was therefore unexpected and intriguing when, in the early 1980s, Spudich and coworkers discovered that the photosynthetic prokaryote *Halobacterium* contained multiple proteins containing phosphate groups that were covalently linked via phosphoester bonds. (Spudich and Stoeckenius 1980; Spudich and Spudich 1981). Shortly thereafter Skorko (1984) identified [^{32}P] phosphoserine and [^{32}P]phosphothreonine in acid hydrolysates of phosphoproteins isolated from extracts of the archaeon *Sulfolobus acidocaldarius* following incubation with radiolabeled ATP. This challenge to the domain-exclusivity of the eukaryotic protein phosphorylation-dephosphorylation paradigm was further undermined by the discovery that several archaeons, including *Sulfolobus solfataricus* (Leng et al. 1995), *Methanosarcina thermophila* TM-1 (Solow et al. 1997), and *Pyrodictium abyssi* TAG11 (Mai et al. 1998) contained homologs of one of the two major families of protein-serine/threonine phosphatases present in eukaryotes, the PPP-family (Barford 1996). Moreover, these were not phylogenetic strays acquired by horizontal gene transfer. The common progenitor of these protein-serine/threonine phosphatases was present in the last common ancestor of the *Archaea* and *Eucarya*, prior to the emergence of eukaryotes as an autonomous phylogenetic domain (Beltrao et al. 2013).

3.3 The Demise of the "Eukaryotic" Monopoly

The advent of genome sequencing soon revealed that, in addition to the PPP-family protein-serine/threonine phosphatases, homologs of the so-called eukaryotic protein kinase superfamily (Smith and King 1995) as well as the PPM-family protein-serine/threonine phosphatases and both the conventional and low molecular weight [LMW] protein-tyrosine phosphatases could be found encoded in the genomes of the *Archaea* (Shi et al. 1998). The sole absentee was the Cd25 dual-specificity protein phosphatase that helps regulate the eukaryotic cell cycle. As the number of genome sequences grew, it became possible to trace the evolutionary history of the various families of protein kinases and protein phosphatases encountered in nature (Beltrao et al. 2013).

These analyses confirmed that the histidine kinases and response regulators of the two-component system were bacterial in origin and were likely acquired by eukaryotes through endosymbiosis (Koretke et al. 2000; Galperin 2004). The progenitors of the "eukaryotic" protein kinase [ePK] superfamily were the "atypical" ePKs of the piD261/Bud32 family and the RIO family, from which the typical ePKs, a subgroup encompassing a large majority, i.e. 478 of 518, of the known or deduced protein-serine/threonine/tyrosine kinases encoded by the human genome (Manning et al. 2002), emerged. Atypical ePKs were thus part of the original equipment of both the *Archaea* and *Eucarya*, and found their way into bacterial organisms by horizontal gene transfer (Leonard et al. 1998; Ponting et al. 1999). Similarly, the PPP-family protein-serine/threonine phosphatases and conventional protein tyrosine phosphatases [PTPs] were both present in the last common ancestor of the *Archaea* and *Eucarya*, while the PPM family protein-serine/threonine phosphatases originated in the *Eucarya* and acquired by the *Bacteria*, as well as a handful of the *Archaea*, by lateral gene transfer (Ponting et al. 1999).

4 Taking Stock. Archaeal Protein Kinases, Protein Phosphatases, and Phosphoproteins

4.1 The Archaeal Kinome

Virtually all archaeons contain at least one open reading frame encoding a deduced, atypical ePK (Kennelly 2014; Esser et al. 2016). However, a clear dichotomy exists between the suite of protein kinases present in members of the euyarchaea and those present in the crenarchaea. Crenarchaeons contain only deduced or known ePKs, whereas many of the euryarchaea possess large suites of two-component histidine kinases that were acquired along with their cognate response regulators from the *Bacteria* (Briegel et al. 2015). The co-opting of the two-component sensor-response modules may explain, to some degree, why the crenarchaeons listed in Table 1 encode an average of slightly more than five potential ePKs, whereas the

Table 1 Distribution of ORFs encoding known or deduced protein kinases and protein phosphatases among selected members of the euryarchaea [E] and crenarchaea [C]

Organism	C/E	Genome (Mbp)	ORFs	HisKs	ePKs	PPP	PPM	cPTP	LMW PTP
Aeropyrum pernix K1	C	1.67	1752	0	1	1	0	0	0
Archaeoglobus fulgidus	E	2.18	2486	15	4	1	0	0	1
Caldiarchaeum subterraneum	C	1.68	1742	0	3	1	0	1	0
Halobacterium NRC-1	E	2.57	2674	14	0	0	0	0	0
Haloferax volcanii DS2	E	4.01	4130	24	1	3	0	0	0
Haloquadratum walsbyi DSM16790	E	3.18	2914	10	4	1	0	1	0
Methanobacterium thermoautotrophicum	E	1.75	1921	16	3	0	0	1	1
Methanocaldococcus janaschii	E	1.74	1823	0	4	0	0	3	0
Methanohalophilus mahii DSM5219	E	2.01	2032	17	3	0	1	0	0
Pyrobaculum calidifontis JCM 11548	C	2.01	2226	0	5	1	0	1	0
Pyrococcus horikoshii OT3	E	1.80	2000	1	4	0	0	1	0
Staphylothermus marinus F1	C	1.57	1655	0	6	1	0	1	0
Sulfolobus acidocaldarius	C	2.23	2330	0	11	1	0	1	0
Sulfolobus islandicus REY15A	C	2.52	2819	0	8	1	0	1	0
Sulfolobus solfataricus P2	C	2.99	3034	0	11	1	0	1	0
Thermococcus kodakarensis KOD1	E	2.09	2358	1	3	0	0	1	0
Thermofilum pendens	C	1.81	1947	0	3	1	0	1	0
Thermoplasma acidophilum	E	1.56	1532	0	1	0	0	0	0
Thermoplasma volcanium	E	1.58	1550	0	3	1	1	0	0

Table 2 Evidence for physiologic impacts or triggers of protein-S/T/Y phosphorylation in the *archaea*

Archaeon	Functional effect or trigger	Reference
Halobacterium halobium	Light-dark adaptation	Spudich and Stoeckenius (1980)
Sulfolobus acidocaldarius	Phosphate starvation	Osorio and Jerez (1996)
Sulfolobus acidocaldarius	Archaellum biosynthesis	Haurat et al. (2017)
Sulfolobus acidocaldarius	Motility	Reimann et al. (2013)
Sulfolobus acidocaldarius	Gene expression	Reimann et al .(2013)
Sulfolobus solfataricus	D-gluconate dehydratase activity (phosphatases inactivate in vitro)	Kim and Lee (2005)
Sulfolobus solfataricus	Phosphohexomutase activity (phosphorylation on Ser309 inactivates)	Ray et al. (2005)
Sulfolobus solfataricus	Carbon source adaptation (glucose versus tryptone)	Esser et al. (2012)
Sulfolobus solfataricus	Oxidative stress (activation of SsoPK4 by oxidized CoA)	Ray et al. (2015)
Thermococcus kodakaraensis KOD1	Phenylalanine-tRNA synthetase (substrate trapping with inactive form of Tk-PTP and probing with anti-PY antibodies)	Jeon et al. (2002)
Thermococcus kodakaraensis KOD1	Phosphomannomutase (substrate trapping with inactive form of Tk-PTP and probing with anti-PY antibodies)	Jeon et al. 2002
Thermococcus kodakaraensis KOD1	RNA terminal phosphate cyclase (substrate trapping with inactive form of Tk-PTP and probing with anti-PY antibodies)	Jeon et al. (2002)
Thermoplasma acidophilum	2-Keto-3 deoxy-gluconate kinase activity (inactivated by phosphatases in vitro)	Jung and Lee (2005)

euryarchaeons encode an average of three and *Halobacteirum* NRC-1 none at all. Also, while euryarchaea contain only ancient, atypical ePKs, several of the euryarchaea possess typical ePKs in addition to their atypical ePKs of the piD262/BUD32 and RIO families (Esser et al. 2016).

One can speculate that the "resistance" of the crenarchaea to acquisition of two-component histidine kinases and response regulators reflects lifestyle rather than some inherent genetic trait. The crenarchaea are rich in hyperthermophiles that live at temperatures where the much greater lability to hydrolysis of phosphohistidine and phosphoglutamate may necessitate the exclusive use of phosphoesters (Buss and Stull 1983; Daniel and Cowan 2000). It therefore may be notable that the thermophilic euryarchaeons in Table 1 possessed no, or at most one, deduced histidine kinase while their non-thermophilic counterparts possessed from ten to twenty-four.

We know relatively little regarding the physiologic functions of archaeal ePKs (Table 2). Two members of the RIO family of ePKs from *Archaeoglobus fulgidus*,

Rio1 and Rio2, have been expressed and demonstrated to phosphorylate themselves on serine and to phosphorylate exogenous proteins such as myelin basic protein (Rio1) or myelin basic protein, casein, and histone H1 (Rio2) on yet to be identified amino acids in vitro (Kiburu and LaRonde-LeBlanc 2012; LaRonde-LeBlanc and Wlodawer 2004; LaRonde-LeBlanc et al. 2005a, b). In eukaryotes, the RIO kinases are critical for growth and ribosome biogenesis. However, no information on the function of archaeal RIOs has come to light.

The largest number of ePKs present in a single archaeon is eleven, in *Sulfolobus solfataricus*. Thus far, four of these deduced protein kinases have been expressed and observed to phosphorylate exogenous proteins on serine and threonine residues in vitro (Haile and Kennelly 2011; Lower and Kennelly 2003; Lower et al. 2004; Ray et al. 2015). These include the piD261/Bud32 homologue SsoPK5 (Haile et al. 2011) and SsoPK4 (Ray et al. 2015), one of three typical ePKs encoded by its genome. All of three of these deduced typical ePKs show recognizable homology to the eucaryal ePKs responsible for phosphorylating eukaryotic initiation factor-2α [eIF2α], which halts global protein synthesis in response to stress (Donnelly et al. 2013).

PH0152, an eIF2α kinase homolog from *Pyrococcus horikoshii* OT3, was the first typical ePK from an archaeon to be expressed and characterized. Tahara and coworkers (2004) observed that recombinantly expressed PH0152 phosphorylated both eIF2α and its archaeal homolog, aIF2α, on the same conserved serine residue, or its archaeal equivalent, respectively, that is targeted in eukaryotes. More recently an eIF2α kinase homolog from *S. solfataricus,* SsoPK4, was expressed and characterized. Like PH0152, it exhibited protein-serine/threonine kinase activity toward its cognate aIF2α as well as several other exogenous phosphoproteins. It autophosphorylated itself, in trans, on multiple residues. Intriguingly, it's K_m for substrates was decreased several-fold in the presence of oxidized Coenzyme A, i.e. two molecules of CoASH linked by an S–S bond (Ray et al. 2015). Since archaeons lack a dedicated protective thiol such as glutathione, when subject to oxidative stress CoASH moonlights as an ROS scavenger, forming CoA disulfide. The dimeric disulfide, in turn, recruits two SsoPK4 molecules; facilitating the authophosphorylation and activation of its protein kinase activity when this microaerophile is subject to oxidative stress to form a simple and economical sensor-response pathway (Ray et al. 2015).

More recently, Albers and colleagues have engaged in an elegant series of biochemical and molecular genetic studies mapping the role of protein phosphorylation in regulatory network responsible for controlling the flagellum, also known as the archaellum, in the crenarchaeon *Sulfolobus acidocaldarius*. Genome mining suggested that *Sulfolobales* genomes contain a conserved archaellum regulatory network gene cluster, or Arn. Deletion of genes encoding either of two Arn proteins, ArnA and ArnB, repressed the expression of archaellum genes (Reimann et al. 2012). Both ArnA and ArnB could be phosphorylated by recombinant Saci1193 and Saci1694, homologs of a typical and atypical ABC-1 type ePK encoded by the *S. acidocaldarius* genome (Reimann et al 2012). As both ArnA and ArnB could be dephosphorylated by the sole PPP-family protein-ser/thr phosphatase in this hyperthermophile, Saci-PP2A, the genes encoding Saci-PP2A or the sole protein-tyrosine phosphatase homolog, the dual-specific Saci-PTP, were deleted. Loss of Saci-PP2A resulted in hypermotile behavior when mutant *S. acidocaldarius* were deprived of nutrients that was accompanied by a reduction in growth rate and alterations in the pattern of gene expression (Reimann et al. 2013).

Subsequent studies revealed that deletion of the genes encoding either Saci1193 or two other deduced ePKs, the RIO kinase homolog Saci0965 and typical ePK Saci1181, resulted in a significant reduction in cell motility (Hoffmann et al. 2016; Haurat et al 2017). Conversely, deletion of the gene encoding the atypical ABC-1 family ePK, Saci1694, yielded a hypermotile phenotype (Hoffmann et al. 2016). In addition, Hoffmann and coworkers (2016) verified that Saci1193 and Saci1694 phosphorylated their targets on serine and threonine residues.

4.2 *The Archaeal Phosphatome*

As can be seen in Table 1, the set of archaeal phosphatomes appears to be even less orderly than the archaeal kinomes. While the number of ePKs in the archaeal kinome was relatively modest, the phosphatome generally consists of a single PPP protein-ser/thr phosphatase and a single conventional PTP [cPTP]. However, this is not universal. Some archaeons harbor LMW PTPs instead of cPTPs, or no PTPs at all. In these latter organisms, a PPM-family protein-ser/thr phosphatase homolog is often present.

All four of the deduced archaeal PPP-family protein phosphatases expressed and tested to date have been observed to be serine/threonine specific in vitro, just like their eucaryal homologs: PP1-arch1 from *S. solfataricus* (Leng et al. 1995), P1-arch2 from *M. thermophila* TM-1 (Solow et al. 1997), Py-PP1 from *P. abyssi* TAG11 (Mai et al. 1998), and Saci-PP2A from *S. acidocaldarius* (Reimann et al. 2013). Thus far, only two archaeal cPTPs have been rigorously tested for protein-ser/thr/tyr phosphatase activity, Tk-PTP from *T. kodakaraensis* KOD11 (Jeon et al. 2002) and Saci-PTP from *S. acidocaldarius* (Reimann et al. 2013). Both were found to possess dual specific, i.e. both serine/threonine- and tyrosine-specific protein phosphatase activity in vitro. The sole archaeal PPM examined to date, Tvn-PPM from *T. volcanium*, was found unexpectedly to possess protein-tyrosine phosphatase activity in addition to its predicted protein-ser/thr phosphatase activity (Dahche et al. 2009).

From a genomic perspective, the ratio of known or deduced ePKs to protein-ser/thr/tyr phosphatases is not unusual. In eukaryotes, the ratio of kinases to phosphatases varies from 3:1 to 5:1 (Table 1). For many years the presence of PTP homologs in the *Archaea* was quite puzzling. No canonical protein-tyr kinases, either eucaryal or bacterial, were evident in archaeal genomes (Esser et al. 2016) while attempts to detect protein-tyrosine kinase activity in cell extracts were also unsuccessful. The lack of *bona fide* protein-tyr kinases, protein-tyr phosphatases, or tyrosine-phosphorylated proteins in fungi such as yeast had long been cited as proof that protein-tyrosine phosphorylation evolved late in evolutionary time, coincident with the emergence of "higher" eukaryotes (Jin and Pawson 2012; Hunter 2014). Even though proteins immunoreactive toward anti-PTyr antibodies had been detected in several archaeons (Smith et al. 1997), until the determination of the *S. solfataricus* phosphoproteome provided unambiguous chemical evidence for vigorous trafficking in protein-tyrosine phosphorylation (Esser et al. 2012), it was difficult to rationalize how it could have skipped over an entire eukaryotic genus if present in the common archaeal-eucaryal ancestor.

4.3 Archaeal Phosphorylomes

To date, only four archaeons have been subject to comprehensive proteomic analysis of their global phosphoprotein population, or phosphorylome: *H. salinarium* (Aivaliotis et al. 2009), *S. solfataricus* (Esser et al. 2012), *S. acidocaldarius* (Reimann et al. 2013), and *Methanohalophilus postcalensis* (Wu et al. 2016). It is apparent that, as with the distribution of protein kinases, the members of the Archaea are somewhat schizophrenic with respect to their phosphoprotein content (Table 3). The relative abundance of phosphoproteins in the euryarchaeons *H. salinarium* and *Methanohalophilus postcalensis* was relatively low, encompassing just a few percent of total ORFs, more comparable to bacterial organisms than to eukaryotes for which, in most cases, a third or more of the proteins encoded by their genomes are phosphorylated on serine, threonine, or tyrosine residues.

As might be expected for the two archaeons encoding the highest number of known or deduced ePKs among the *Archaea,* the crenarchaeons *S. acidocaldarius* and *S. solfataricus* contained the largest population of phospho-ser/thr/tyr-containing proteins in both absolute and relative terms. These proportions, 35 and 18%, respectively, were comparable to the fraction of eucaryal proteins subject to protein-ser/thr/tyr phosphorylation. While the number of archaeal phosphoproteomes is too small in number to draw any definitive conclusions, as was observed for the ePKs, a dichotomy appears to exist between the *Euryarchaea* and *Crenarchaea* with regards to protein phosphorylation. Euryarchaeal kinomes and phosphoproteomes bear a striking resemblance, at least at a gross level, to those found in *Bacteria*, while the crenarchaea are much more eukaryote-like.

The roster of phosphoproteins in *S. acidocaldarius* and *S. solfataricus* embraces the full structural spectrum (Esser et al. 2012; Reimann et al. 2013). Thus, it would appear that, in these arhcaeons, protein-S/T/Y phosphorylation may help control and coordinate any and all cellular functions. However, beyond the control of archaellum synthesis and function as described above, available data on the functional role of protein phosphorylation-dephosphorylation in the *Archaea* is extremely limited (Table 2).

Perhaps the most surprising revelation provided by the current library of archaeal phosphoprotomes is the high abundance of phosohotyrosine. Whereas only 1–2% of the phosohorylation sites in eucaryal proteins employ tyrosine as their

Table 3 Comparison of archaeal phosphoproteomes with representatives from the *bacteria* [B] and *eucarya* [E]

Organism	Domain	ORFs	P-S/T/Y-proteins (%)	P-S/T/Y-sites	PS:PT:PY	ePKs	HisKs	PPPs and PPMs	PTPs and DSPs	Phosphoproteome Reference
Halobacterium salinarium	A	2722	69 (2.5%)	81	86:12:1	0	14	0	0	Aivaliotis et al. (2009)
Methanohalophilus postcalensis	A	2186	127 (6%)	217	64:19:27	4	23	1	1	Wu et al. (2016)
Sulfolobus solfataricus	A	2977	540 (18%)	1318	26:21:54	11	0	1	1	Esser et al. (2012)
Sulfolobus acidocaldarius	A	2292	801 (35%)	1206	33:28:36	11	0	1	1	Reimann et al. (2013)
Bacillus subtilus (sporulating)	B	4100	155 (3.8%)	124	48:34:19	2	36	6	4	Rosenberg et al. (2015)
Escherichia coli	B	4289	108 (2.5%)	150	76:17:7	3	30	3	3	Soares et al. (2013)
Synechocystis sp. PCC 6803	B	3725	188 (5.1%)	242	nd	9	47	6	3	Spat et al. (2015)
Candida albicans	E	4419	2896 (66%)	15906	80:18:2	91	3	16	18	Wilger et al. (2015)
Saccharomyces cerevisiae	E	6604	2374 (36%)	9783	81:17:2	113	1	18	13	Amoutzias et al. (2012)
Caenorhabditis elegans	E	19099	2373 (12%)	6780	80:18:2	411	0	65	109	Plowman et al. (1999), Zielinska et al. (2009)
Homo sapiens	E	20412	7832 (38%)	38,229	84:16:1	518	0	35	108	Sharma et al. (2014)

phosphoacceptor residue, the proportion in *S.solfataricus* rises to an unprecedented 54%, more than half. Although the proportion is lower in *S. acidcaldarius* and *M. postcalensis*, tyrosine accounts for more than one-quarter to one-third of the phosphoacceptor sites in these archaeons. Since phosphotyrosine is even more resistant to hydrolysis that phosphoserine or phosphothreonine (Buss and Stull 1983), one can rationalize its abundance in the hyperthermophilic *Sulfolobales*. However, *M. postcalensis* is a mesophile.

Assuming one can dismiss *H. salinarium*, which has lost all of its ePKs, as an aberration, it is clear that the accepted evolutionary history of protein-tyrosine phosphorylation (Jin and Pawson 2012; Hunter 2014) will need to be rewritten. On the other hand, given the abundance of protein-tyrosine phosphorylation in *S. acidcaldarius* and *S. solfataricus*, one cannot help but wonder why, following years of genomic and biochemical study, no source of protein-tyrosine kinase activity has been uncovered in these organisms (Kennelly 2014; Hoffmann et al. 2016). Esser et al (2016) report that the *Sulfolobales* lack recognizable homologues of either eukaryotic protein-tyrosine kinases or bacterial BY protein-tyrosine kinases. In view of this disconnect, as well as the disparity in the fraction of their genome devoted to protein kinases and protein phosphatases relative to eukaryotes (0.1–0.6% vs. >2%), one cannot help but speculate that the *Archaea* harbor some novel, yet to be recognized catalyst for protein-tyrosine phosphorylation (Kennelly 2014).

5 Do the Archaea Interpret and Broadcast Information Using Integrated Protein Phosphorylation Networks?

Eukaryotic cells utilize pervasive and highly integrated biomolecular information processing networks to communicate with one another whose central core is dominated by thousands of ser/thr/tyr phosphorylated proteins and the hundreds of protein kinases and phosphatases responsible for modulating their state of modification. Indeed, roughly one-third of all eukaryotic proteins are targeted for phosphorylation and nearly 2.5% of the entire human genome encodes protein-ser/thr/tyr kinases and phosphatases.

At the other end of the spectrum, while the *Bacteria* contain many relatively sophisticated regulatory subsystems dedicated to the control of central carbon metabolism, coordination of carbon and nitrogen metabolism, guiding cellular locomotion, and adapting metabolism to existing carbon sources. However, it does not appear that they have achieved the type of comprehensive integration and global information processing characteristic of the *Eucarya*, or have managed to do so with minimal reliance on protein-ser/thr/tyr phosphorylation (Galperin 2005). As can be seen in Tables 3, the proportion of bacterial proteins that belong to their phosphorylome is roughly an order of magnitude lower than is typical for the *Eucarya*, 2–5% versus 20–50%, while their genomes encode only a handful of protein-ser/thr/tyr kinases.

The *Archaea* offer a paradoxical profile. In terms of the number and nature of their protein kinases, protein phosphatases, and phosphoproteins many euryarchaea differ little from bacterial organisms, suggesting that they possess relatively limited information processing capacity. The crenarchaotes *S. acidocaldarius* and *S. solfataricus,* on the other hand, possess an abundance of ser/thr/tyr-phosphorylated proteins, enough that, if properly interlinked, they could form the foundation for a very sophisticated and capable information processing and communication network. However, this wealth of ser/thr/tyr-phosphorylated proteins is not matched with a comparable abundance of know or deduced protein-ser/thr/tyr kinases or phosphatases—totals that are much more comparable to the *Bacteria* and euryarchaea.

This dichotomy begs the question as to whether the *Archaea,* and in particular the crenarchaea, harbor as yet unrecognized sources of protein-ser/thr and in particular protein-tyr kinase and phosphatase activity to support their burgeoning phosphoprotein populations. Whereas in Table 3 the ratio of phosphoproteins to protein-ser/thr/tyr kinases in the *Eucarya* ranged from 6:1 for *C. elegans* to 32:1 for *C. albicans*, with an average of ≈19:1, the corresponding ratios for *S. acidocaldarius* and *S. solfataricus* were 49:1 and 73:1, respectively. It is tempting to speculate that members of the crenarachaea have achieved a threshold of information processing power comparable to unicellular eukaryotes, however many, many gaps need to filled in before speculation might turn into fact.

References

Aivaliotis M, Macek B, Gnad F, Reichelt P, Mann M, Oesterhelt D (2009) Ser/Thr/Tyr protein phosphorylation in the archaeon *Halobacterium salinarium*—a representative of the third domain of life. PLoS ONE 4:e4777

Alon U (2003) Biological networks: the tinkerer as an engineer. Science 301:1866–1867

Ames GF-L, Nikaido K (1981) Phosphate-containing proteins of *Salmonella typhimurium* and *Escherichia coli.* Analysis by a new two-dimensional gel system. Eur J Biochem 115:525–531

Amoutzias GD, He Y, Lilley KS, Van de Peer Y, Oliver SG (2012) Evaluation and properties of budding yeast phosphoproteome. Mol Cell Proteomics 11(M111):009555

Barford D (1996) Molecular mechanisms of the protein-serine/threonine phosphatases. Trends Biochem Sci 21:407–412

Beltrao P, Bork P, Krogan NJ, van Noort V (2013) Evlution and functional cross-talk of protein post-translational modifications. Mol Systems Biol 9:714

Bray D (1995) Protein molecules as computational elements in living cells. Nature 376:307–312

Bray D (2003) Molecular networks: the top-down view. Science 301:1864–1865

Briegel A, Ortega DR, Huang A, Oikonomou CM, Gunsalus RP, Jensen GJ (2015) Structural conservation of chemotaxis machinery across Archaea and Bacteria. Envron Microbiol Rep 7:414–419

Buss JE, Stull JT (1983) Measurement of chemical phosphate in proteins. Methods Enzymol 99:7–14

Chu H-M, Wang AH-J (2007) Enzyme-substrate interactions revealed by crystal structures of the archaeal Sulfolobus PTP-fold phosphatase and its phosphopeptide complexes. Proteins 66:996–1003

Dahche H, Abdullah A, Potters MB, Kennelly PJ (2009) A PPM-family protein phosphatase from *Thermoplasma volcanium* hydrolyzes protein-bound phosphotyrosine. Extremophiles 13:371–377

Daniel RM, Cowan DA (2000) Biomolecular stability and life at high temperatures. Cell Mol Life Sci 57:250–264

Derouiche A, Shi L, Kalantari A, Mijakovic I (2016) Evolution and tinkering: what do a protein kinase, a transcriptional regulator, and chromosome segregation/cell division proteins have in common? Curr Genet 62:67–70

Doan T, Mendez A, Detwiler PB, Chen J, Rieke F (2006) Multiple phosphorylation sites confer reproducibility of the rod's single-photon responses. Science 313:530–533

Donnelly N, Gorman AM, Gupta S, Samali A (2013) The eIF2α kinases: their structures and functions. Cell Mol Life Sci 70:3493–3511

Esser D, Pham TK, Reimann J, Albers SV, Seibers B, Wright PC (2012) Change of carbon source causes dramatic effects in the phosphoproteome of the archaeon *Sulfolobus solfataricus*. J Proteome Res 11:4823–4833

Esser D, Hoffmann L, Pham TK, Brasen C, Qiu W, Wright PC, Albers S-V, Siebers B (2016) Protein phosphorylation and its role in archaeal signal transduction. FEMS Microbiol Rev. doi:10.1093/femsre/fuw020

Galperin MY (2004) Bacterial signal transduction network in a genomic perspective. Environ Microbiol 6:552–567

Galperin MY (2005) A census of membrane-bound and intracellular signal transduction proteins in bacteria: bacterial IQ, extroverts and introverts. BMC Microbiol 5:35

Gnad F, Forner F, Zielinska DF, Birney E, Gunawarden J, Mann M (2010) Evolutioonary constraints of phosphorylation in eukaryotes, prokaryotes, and mitochondria. Mol Cell Proteomics 9:2642–2653

Gross R, Arico B, Rappuoli R (1989) Families of bacterial signal-transducing proteins. Mol Microbiol 3:1661–1667

Haile JD, Kennelly PJ (2011) The activity of an ancient atypical protein kinase is stimulated by ADP-ribose in vitro. Arch Biochem Biophys 511:56–63

Haurat MF, Figueiredo AS, Hoffmann L, Li L, Herr K, Wlson AJ, Beeby M, Schaber J, Albers S-V (2017) ArnS, a kinase involved in starvation-induced archaellum expression. Mol Microbiol 103:181–194

Hoffmann L, Schummer A, Reimann J, Haurat MF, Wilson AJ, Beeby M, Warscheid B, Albers S-V (2016) Expanding the archaellum regulatory network—the eukaryotic protein kinases ArnC and ArnD influence motility of *Sulfolobus acidocaldarius*. Microbiol Open 6:e00414

Hunter, T (2014) The genesis of tyrosine phosphorylation. In: Schlessinger J, Lemmon MA (eds) Cold Spring Harb Perspec Biol 6:a020644

Igo MM, Slauch JM, Silhavy TJ (1990) Signal transduction in bacteria: kinases that control gene expression. New Biol 2:5–9

Jaquet K, Korte K, Schnackerz K, Vyska K, Heilmeyer LMG Jr (1993) Characterization of the cardiac troponin I phosphorylation domain by ^{31}P nuclear magnetic resonance spectroscopy. Biochemistry 32:13873–13878

Jeon S-J, Fujiwara S, Takagi M, Tanaka T, Imanaka T (2002) Tk-PTP, a protein-tyrosine phosphatase from the hyperthermophilic archaeon *Thermococcus kodakaraensis KOD1*: enzymatic characteristics and identification of its substrate proteins. Biochem Biophys Res Commun 295:508–514

Jin J, Pawson T (2012) Modular evolution of phosphorylation-based signalling systems. Philos Trans R Soc Lond B Biol Sci 367:2540–2555

Johnson LN, Lewis RJ (2001) Structural basis for control by phosphorylation. Chem Rev 101:2209–2242
Jung JH, Lee SB (2005) Identification and characterization of *Thermoplasma acidophilum* 2-keto-3-deoxy-D-gluconate kinase: a new class of sugar kinases. Biotechnol Bioprocess Eng 10:535–539
Jung K, Fried L, Behr S, Heermann R (2012) Histidine kinases and response regulators in networks. Curr Opin Microbiol 15:118–124
Kennelly PJ (1999) Bits for an organic microprocessor. In: Sitaramaya A (ed) Introduction to cellular signal transduction. Birkhauser, Boston, pp 235–263
Kennelly PJ (2014) Protein serine/threonine/tyrosine phosphorylation in the archaea. J Biol Chem 289:9480–9487
Kennelly PJ, Potts M (1996) Fancy meeting you here! A fresh look at "prokaryotic" protein phosphorylation. J Bacteriol 178:4759–4764
Kiburu IN, LaRonde-LeBlanc N (2012) Interaction of Rio1 kinase with toyocamycin reveals conformational switch that controls oligomeric state and catalytic activity. PLoS ONE 7: e37371
Kim S, Lee SB (2005) Identification and characterization of Sulfolobus solfataricus D-gluconate dehydratase: a key enzyme in the non-phosphorylated Entner-Doudoroff pathway. Biochem J 386:271–280
Koretke KK, Lupas AN, Warren PV, Rosenberg M, Brown JR (2000) Evolution of two-component signal transduction. Mol Biol Evol 17:1956–1970
Landry CR, Freschi L, Zarin T, Moses AM (2014) Turnover of protein phosphorylation evolving under stabilizing selection. Front Genet 5:245
LaRonde-LeBlanc N, Guszczynski T, Copeland T, Wlodawer A (2005a) Autophosphorylation of *Archaeoglobus fulgidus* Rio2 and crystal structures of its nucleotide-metal complexes. FEBS J 272:2800–2810
LaRonde-LeBlanc N, Guszczynski T, Copeland T, Wlodawer A (2005b) Structure and activity of the atypical serine kinase Rio1. FEBS J 272:3698–3713
Laronde-LeBlanc N, Wlodawer A (2004) Crystal structure of *A. fulgidus* Rio2 defines a new family of serine protein kinases. Structure 12:1585–1594
Leng J, Cameron AJ, Buckel S, Kennelly PJ (1995) Isolation and cloning of a protein-serine/threonine phosphatase from an archaeon. J Bacteriol 177:6510–6517
Leonard CJ, Aravind L, Koonin EV (1998) Novel families of putative protein kinases in bacteria and archaea: evolution of the "eukaryotic" protein kinase superfamily. Genome Res 8:1038–1047
Lower BH, Kennelly PJ (2003) Open reading frame Sso2387 from the archaeon *Sulfolobus solfataricus* encodes a polypeptide with protein-serine kinase activity. J Bacteriol 185:3436–3445
Lower BH, Potters MB, Kennelly PJ (2004) A phosphoprotein from the archaeon *Sulfolobus solfataricus* with protein-serine kinase activity. J Bacteriol 186:463–472
Mai B, Frey G, Swanson RV, Mathur EJ, Stetter KO (1998) Molecular cloning and functional expression of a protein-serine/threonine phosphatase from the hyperthermophilic archaeon *Pyrodictium abyssi* TAG11. J Bacteriol 180:4030–4035
Manai M, Cozzone AJ (1979) Analysis of the protein-kinase activity of *Escherichia coli* cells. Biochem Biophys Res Commun 91:819–826
Manning G, Whyte DB, Matrinez R, Hunter T, Sudarsanam S (2002) The protein kinase complement of the human genome. Science 298:1912–1934
Marks F (1996) The brain of the cell. In: Marks F (ed) Protein phosphorylation. VCH Publishers Inc, New York, pp 1–35
Osorio G, Jerez CA (1996) Adaptive response of the archaeon *Sulfolobus acidocaldarius* BC65 to phosphate starvation. Microbiology 142:1531–1536
Park K-S, Mohapatra DP, Misonou H, Trimmer JS (2006) Graded regulation of the Kv2.1 potassium channel by variable phosphorylation. Science 313:976–978

Pearson RB, Kemp BE (1991) Protein kinase phosphorylation site sequences and consensus specificity motifs: tabulations. Methods Enzymol 200:62–81

Plowman GD, Sudaransam S, Bingham J, Whyte D, Hunter T (1999) The protein kinases of *Caenorhabditis elegans*: a model for signal transduction in multicellular organisms. Proc Natl Acad Sci U S A 96:13603–13610

Ponting CP, Aravind L, Schultz J, Bork P, Koonin EV (1999) Eukaryotic signaling domain homologues in archaea and bacteria: ancient ancestry and horizontal gene transfer. J Mol Biol 289:729–745

Ray WK, Keith SM, DeSantis AM, Hunt JP, Larson TJ, Helm RF, Kennelly PJ (2005) A phosphohexomutase from the archaeon *Sulfolobus solfataricus* is covalently modified by phosphorylation on serine-309. J Bacteriol 187:4270–4275

Ray WK, Potters MB, Haile JD, Kennelly PJ (2015) Activation of SsoPK4, an archaeal eIF2α kinase homolog, by oxidized coenzyme A. Proteomes 3:89–116

Reimann J, Lassak K, Khadouma S, Ettema TJG, Yang N, Driessen AJM, Kling A, Albewrs S-V (2012) Regulation of archaella expression by the FHA and von Willebrand domain-containing proteins ArnA and ArnB in *Sulfolobus acidocaldarius*. Mol Microbiol 86:24–36

Reimann J, Esser D, Orell A, Amman F, Pham TK, Noirel J, Lindas A-C, Bernander R, Wright PC, Siebers B, Albers S-V (2013) Archaeal signal transduction: Impact of protein phosphatase deletions on cell size, motility, and energy metabolism in *Sulfolobus acidocaldarius*. Mol Cell Proteomics 12:3908–3923

Roach PJ (1991) Multisite and hierarchal protein phosphorylation. J Biol Chem 266:14139–14142

Rosenberg A, Soufi B, Ravikumar V, Soares NC, Krug K, Smith Y, Macek B, Ben-Yehuda S (2015) Phosphoproteome dynamics mediate revival of bacterial spores. BMC Biol 13:76

Salazar C, Hofer T (2003) Allosteric regulation of the transcription factor NFAT1 by multiple phosphorylation sites: a mathematical analysis. J Mol Biol 327:31–45

Sharma K, D'Souza RCJ, Tyanova S, Schaab C, Wisniewski JR, Cox J, Mann M (2014) Ultradeep human phosphoproteome reveals a distinct regulatory nature of Tyr and Ser/Thr-based signaling. Cell Rep 8:1583–1594

Shi L, Potts M, Kennelly PJ (1998) The serine, threonine, and/or tyrosine-specific protein kinases and protein phosphatases of prokaryotic organisms. A family portrait. FEMS Microbiol Rev 22:229–253

Singer AU, Forman-Kay JD (1997) pH titration studies of an SH2 domain-phosphopeptide complex: unusual histidine and phosphate pK_a values. Protein Sci 6:1910–1919

Skorko R (1984) Protein phosphorylation in the archaebacterium *Sulfolobus acidocaldarius*. Eur J Biochem 145:617–622

Smith RF, King KY (1995) Identification of a eukaryotic-like protein kinase gene in archaebacteria. Prot Sci 4:126–129

Smith SC, McCartney B, Kennelly PJ, Potts M (1997) Protein-tyrosine phosphorylation in the archaea. J Bacteriol 179:2418–2420

Soares NC, Spat P, Krug K, Macek B (2013) Global dynamics of the *Escherichia coli* proteome and phosphoproteome during growth in minimal media. J Proteome Res 12:2611–2621

Solow B, Young JC, Kennelly PJ (1997) Gene cloning and expression and characterization of a toxin-sensitive protein phosphatase from the methanogenic archaeon *Methanosarcina thermophila* TM-1. J Bacteriol 179:5072–5075

Spat P, Macek B, Forchhammer K (2015) Phosphoproteome of the cyanobacterium *Synechocystis* sp. PCC 6803 and its dynamics during nitrogen starvation. Front Microbiol 6:248

Spudich EN, Spudich JL (1981) Photosensitive phosphoproteins in halobacteria: regulatory coupling of transmembrane proton flux and protein dephosphorylation. J Cell Biol 91:895–900

Spudich JL, Stoeckenius W (1980) Light-regulated retinal-dependent reversible phosphorylation of Halobacterium proteins. J Biol Chem 255:5501–5503

Tahara M, Ohsawa A, Saito S, Kimura M (2004) In vitro phosphorylation of initiation factor 2α (aIF2α) from hyperthermophilic archaeon *Pyrococcus horikoshii* OT3. J Biochem 135:479–485

Vogel HJ (1989) Phosphorous-31 nuclear magnetic resonance of phosphproteins. Meth Enzymol 177:263–282

Westheimer FH (1987) Why nature chose phosphates. Science 235:1173–1178

Wilger SD, Liu Z, Olarte RA, Adamo ME, Stajich JE, Myers LC, Kettenbach AN, Hogan DA (2015) Analysis of the *Candida albicans* phosphoproteome. Eukaryot Cell 14:474–485

Wu W-L, Lai S-J, Yang J-T, Chern J, Liang S-Y, Chou C-C, Kuo C-H, Lai M-C, Wu S-H (2016) Phosphoproteomic analysis of *Methanohalophilus portucalensis* FTF1T identified the role of protein phosphorylation in methanogenesis and osmoregulaton. Sci Rep 6:29013

Yachie N, Saito R, Sugahara J, Tomita M, Ishihama Y (2009) In silico analysis of phosphoproteome data suggest a rich-get-richer process of phosphosite accumulation over evolution. Mol Cell Proteomics 8:1061–1071

Yang L, MacLellan WR, Han Z, Weiss JN, Qu Z (2004) Multisite phosphorylation and network dynamics of cyclin-dependent kinase signaling in the eukaryotic cell cycle. Biophys J 86:3432–3443

Zielinska DF, Gnad F, Jedrusik-Bode JR, Mann M (2009) *Caenorhabditis elegans* has a phosphoproteome typical for metazoans that is enriched in developmental and sex determination proteins. J Proteome Res 8:4039–4049

Zschiedrich CP, Keidel V, Szurmant H (2016) Molecular mechanisms of two-component signal transduction. J Mol Biol 428:3752–3775

Secondary Metabolites in Archaea and Extreme Environments

Shengqin Wang and Zuhong Lu

1 Introduction

Archaea are often viewed as living in extreme environments such as high salt niches, extreme temperature solutions, as well as acid and alkaline conditions (Javaux 2006; Rampelotto 2013), where cannot be tolerated by other forms of life. They should evolve a series of systems to sense and adapt to environmental stress in living such conditions. Interestingly, there are still many other microorganisms living around, and archaea need the ability to fight against other microbiology. For example, a lot of CRISPR spacers have been detected in the archaea genome, which means that archaea still live with phages in the extreme environment (Jaubert et al. 2013).

It is well known that secondary metabolites in bacterial and plant are bioactive compounds and serve survival functions for the organisms producing them (Demain and Fang 2001). These secondary metabolites are not the necessary production in living and development, but they are thought to be important components of the innate immune system against other organisms and play essential roles in improving tolerance against environmental stress (Piasecka et al. 2015). These secondary metabolites exhibit many antimicrobial mechanisms including specific receptors, membrane components and cytoplasmic targets (Wallace 2004; Mohan et al. 2016).

Compared with bacteria or plants, secondary metabolites in archaea have been far less researched (Charlesworth and Burns 2015). Currently, some researches

S. Wang (✉)
College of Life and Environmental Science, Wenzhou University,
Wenzhou 325035, China
e-mail: mysango@163.com

Z. Lu
State Key Lab of Bioelectronics, School of Biological Science
and Medical Engineering, Southeast University, Nanjing 210096, China

G. Witzany (ed.), *Biocommunication of Archaea*,
DOI 10.1007/978-3-319-65536-9_14

Table 1 Recurrent secondary metabolites detected in archaea genomes by antismash

Types	# of this kind of secondary metabolites
Nonribosomal peptides	7
Bacteriocin	59
Cyanobactin	3
Lantipeptide	3
Lassopeptide	2
Terpene	76
Ectoine	6
Siderophore	5

have been established to explain their potential microbial mechanism or interaction in archaea. One of the roles of these moleculars should be important components of the defense system to fight against microbial infection like that in bacteria. However, the detail description of the secondary metabolites interaction in archaea is still lacking. In this chapter, we listed the variety of molecules which detected in more than one complete archaea genomes by antismash (Table 1) (Weber et al. 2015), and discuss their potential role in the biocommunication of archaea under extreme environments.

2 Secondary Metabolites Serve as Defense Molecules Against Other Organisms

Nonribosomal peptides are synthesized by nonribosomal peptide synthetases usually produced by microorganisms, and present extremely broad range of biological activities, such as toxins or pigments (Finking and Marahiel 2004). Many nonribosomal peptides antibiotic are produced by bacteria, including penicillin and vancomycin (Felnagle et al. 2008). Genomic mining has revealed the wide distribution of nonribosomal peptides across the three domains of life, and three nonribosomal peptides have been found in two archaea genera (Wang et al. 2014). Insteresting, both of these two archaea genera live in the temperature range from 20 °C to 45 °C, a proper temperature for other microorganism, suggesting nonribosomal peptides in these strains may implement a function like that in bacteria.

Bacteriocin is antibiotic peptides produced by microorganisms to inhibit the growth of other bacterial strains living in a similar environment. It has been found that bacteriocin is widely distributed in archaea, such as haloarchaea and Sulfolobus genus (Torreblanca et al. 1994; O'Connor and Shand 2002; Ellen et al. 2011). Typically, the bacteriocins are classified by size, and the bacteriocins from haloarchaea can be divided into two classed: microcins (less than 20 kDa in size) and colicin-like proteins (20–90 kDa in size). They are hydrophobic and robust, withstanding heat, which assist them to implement the function in extreme environment. The antimicrobial activity of these peptides is also different, and some bacteriocins having a narrow range of activity from affecting only close relative microorganisms but others can inhibit the growth of another phylum of archaea, such as *Sulfolobus solfataricus* (Haseltine et al. 2001). Though these bacteriocins can inhibit some archaea, there is still not proved evidence that bacteriocins from archaea can implement the inhibition of bacteria, although some archaea have been reported capable of inhibiting the bacteria (Shand and Leyva 2007). In addition, cyanobactin is a class of low-molecular-weight cyclic peptides prevalent in planktonic bloom-forming cyanobacteria (Leikoski et al. 2009, 2010; Sivonen et al. 2010). Interestingly, Cyanobactin was detected in three *Methanosarcina barkeri* genome sequences with the genome mining approaches. It was suggested that these peptides were probably acquired through horizontal gene transfer (Martins and Vasconcelos 2015).

Lantipeptide is the most widely distributed ribosomally synthesized and post-translationally modified peptides (RiPPS) formed by dehydration of Serine/Threonine residues and subsequent addition of cysteine residues to the resulting dehydro amino acid meso-lanthionine with a serial reaction of post translation modification. It has been reported as the most widely distributed RiPPS, and also found in Archaea (Skinnider et al. 2016). Lasso peptide, is also a RiPP subfamily. Recently, by using pattern-matching algorithm and repeating search the query set contained lasso peptide cyclases, novel lasso peptide gene clusters were detected across some archaeal genomes (Maksimov et al. 2012; Tietz et al. 2017). Both Lantipeptide and Lasso peptide should be important components of the defense system in archaea.

Terpenes are a major class of biology compounds usually found in plant. It is reported that the terpenes can protect the plant that produces them by deterring herbivores and by attracting predators and parasites of herbivores with a strong odor (Pichersky et al. 2006). In a recent study, hundreds of terpenes have been detected in bacteria, suggesting this kind of secondary metabolites should have more widely distribute than before(Yamada et al. 2015; Dickschat 2016). Currently, there is no experimental evidence found in archaea, perhaps because we still can not find the proper condition to express this kind of substance in the natural environment.

3 Secondary Metabolites Serve as a Protective Substance in Extreme Environment

Ectoine is usually found in high concentrations in halophilic microorganisms and serves as a protective substance to help organisms that produced them survive extreme osmotic stress by acting as an osmolyte. Evidence of ectoine biosynthesis gene cluster in archaea has been detected by genome mining of the Thaumarchaeon (Walker et al. 2010), and then has been proved by functionally expressed in response to high salinity (Widderich et al. 2016).

It is very little known about assimilation of iron in archaea. Iron is critical to biological processes like respiration or DNA synthesis. The previous study in bacteria shows that the siderophores are produced to combat low iron stress. The solubility of iron is expected to be very low in many natural environment and the siderophores can overcome the insolubility and immobility of ferric ions through chelation. The evidence of siderophores produced from halophilic archaea was studied by chemical assays test in Indian haloarchaeal strains (Dave et al. 2006). These archaea are living in extreme haloalkaliphiles with a minimum salt requirement of 20% and optimum pH 8.0–9.0, where the solubility of iron is very low.

4 Conclusions

For living in extreme environments, secondary metabolites can help archaea that produced them flight against other microorganism living around, and serve as a protective substance to gain more resource from extreme environment. Therefore, the secondary metabolites can act as an alternative defense mechanism to archaea living in extreme environments.

References

Charlesworth JC, Burns BP (2015) Untapped resources: biotechnological potential of peptides and secondary metabolites in archaea. Archaea 2015:1–7. doi:10.1155/2015/282035

Dave BP, Anshuman K, Hajela P (2006) Siderophores of halophilic archaea and their chemical characterization. Indian J Exp Biol 44:340–344

Demain AL, Fang A (2001) The natural functions of secondary metabolites. In: History of modern biotechnology I. Springer, Berlin, pp 1–39

Dickschat JS (2016) Bacterial terpene cyclases. Nat Prod Rep 33:87–110. doi:10.1039/c5np00102a

Ellen AF, Rohulya OV, Fusetti F et al (2011) The sulfolobicin genes of *Sulfolobus acidocaldarius* encode novel antimicrobial proteins. J Bacteriol 193:4380–4387. doi:10.1128/JB.05028-11

Felnagle EA, Jackson EE, Chan YA et al (2008) Nonribosomal peptide synthetases involved in the production of medically relevant natural products. Mol Pharm 5:191–211. doi:10.1021/mp700137g

Finking R, Marahiel MA (2004) Biosynthesis of nonribosomal peptides. Ann Rev Microbiol 58:453–488. doi:10.1146/annurev.micro.58.030603.123615

Haseltine C, Hill T, Montalvo-Rodriguez R et al (2001) Secreted euryarchaeal microhalocins kill hyperthermophilic crenarchaea. J Bacteriol 183:287–291. doi:10.1128/JB.183.1.287-291.2001

Jaubert C, Danioux C, Oberto J et al (2013) Genomics and genetics of *Sulfolobus islandicus* LAL14/1, a model hyperthermophilic archaeon. Open Biol 3:130010. doi:10.1098/rsob.130010

Javaux EJ (2006) Extreme life on Earth–past, present and possibly beyond. Res Microbiol 157:37–48. doi:10.1016/j.resmic.2005.07.008

Leikoski N, Fewer DP, Jokela J et al (2010) Highly diverse cyanobactins in strains of the genus Anabaena. Appl Environ Microbiol 76:701–709. doi:10.1128/AEM.01061-09

Leikoski N, Fewer DP, Sivonen K (2009) Widespread occurrence and lateral transfer of the cyanobactin biosynthesis gene cluster in cyanobacteria. Appl Environ Microbiol 75:853–857. doi:10.1128/AEM.02134-08

Maksimov MO, Pelczer I, Link AJ (2012) Precursor-centric genome-mining approach for lasso peptide discovery. Proc Natl Acad Sci U S A 109:15223–15228. doi:10.1073/pnas.1208978109

Martins J, Vasconcelos V (2015) Cyanobactins from cyanobacteria: current genetic and chemical state of knowledge. Mar Drugs 13:6910–6946. doi:10.3390/md13116910

Mohan G, Thipparamalai Thangappanpillai AK, Ramasamy B (2016) Antimicrobial activities of secondary metabolites and phylogenetic study of sponge endosymbiotic bacteria, *Bacillus* sp. at Agatti Island, Lakshadweep Archipelago. Biotechnol Rep (Amst) 11:44–52. doi:10.1016/j.btre.2016.06.001

O'Connor EM, Shand RF (2002) Halocins and sulfolobicins: the emerging story of archaeal protein and peptide antibiotics. J Ind Microbiol Biotechnol 28:23–31. doi:10.1038/sj/jim/7000190

Piasecka A, Jedrzejczak-Rey N, Bednarek P (2015) Secondary metabolites in plant innate immunity: conserved function of divergent chemicals. New Phytol 206:948–964. doi:10.1111/nph.13325

Pichersky E, Noel JP, Dudareva N (2006) Biosynthesis of plant volatiles: nature's diversity and ingenuity. Science 311:808–811. doi:10.1126/science.1118510

Rampelotto PH (2013) Extremophiles and extreme environments. Life (Basel) 3:482–485. doi:10.3390/life3030482

Shand RF, Leyva KJ (2007) Peptide and Protein Antibiotics from the domain archaea: halocins and Sulfolobicins. Bacteriocins. Springer, Berlin, pp 93–109

Sivonen K, Leikoski N, Fewer DP, Jokela J (2010) Cyanobactins-ribosomal cyclic peptides produced by cyanobacteria. Appl Microbiol Biotechnol 86:1213–1225. doi:10.1007/s00253-010-2482-x

Skinnider MA, Johnston CW, Edgar RE et al (2016) Genomic charting of ribosomally synthesized natural product chemical space facilitates targeted mining. Proc Natl Acad Sci 113:E6343–E6351. doi:10.1073/pnas.1609014113

Tietz JI, Schwalen CJ, Patel PS et al (2017) A new genome-mining tool redefines the lasso peptide biosynthetic landscape. Nat Chem Biol 13:470–478. doi:10.1038/nchembio.2319

Torreblanca M, Meseguer I, Ventosa A (1994) Production of halocin is a practically universal feature of archaeal halophilic rods. Lett Appl Microbiol 19:201–205. doi:10.1111/j.1472-765X.1994.tb00943.x

Walker CB, de la Torre JR, Klotz MG et al (2010) Nitrosopumilus maritimus genome reveals unique mechanisms for nitrification and autotrophy in globally distributed marine crenarchaea. Proc Natl Acad Sci 107:8818–8823. doi:10.1073/pnas.0913533107

Wallace RJ (2004) Antimicrobial properties of plant secondary metabolites. Proc Nutr Soc 63:621–629

Wang H, Fewer DP, Holm L et al (2014) Atlas of nonribosomal peptide and polyketide biosynthetic pathways reveals common occurrence of nonmodular enzymes. Proc Natl Acad Sci 111:9259–9264. doi:10.1073/pnas.1401734111

Weber T, Blin K, Duddela S et al (2015) antiSMASH 3.0-a comprehensive resource for the genome mining of biosynthetic gene clusters. Nucleic Acids Res 43:W237–43. doi:10.1093/nar/gkv437

Widderich N, Czech L, Elling FJ et al (2016) Strangers in the archaeal world: osmostress-responsive biosynthesis of ectoine and hydroxyectoine by the marine thaumarchaeon *Nitrosopumilus maritimus*. Environ Microbiol 18:1227–1248. doi:10.1111/1462-2920.13156

Yamada Y, Kuzuyama T, Komatsu M et al (2015) Terpene synthases are widely distributed in bacteria. Proc Natl Acad Sci 112:857–862. doi:10.1073/pnas.1422108112

Evolutionary Interaction Between Archaeal-Eukaryal Cell Lineages and Viruses

Masaharu Takemura and Tomohiro Mochizuki

1 Introduction

Phylogenetically, archaea is a relative of us eukarya. Since its original recognition in 1977 (Woese and Fox 1977) archaea have been well known to be extremophiles, and have been considered to be divided into two major phyla: Euryarchaeota and Crenarchaeota. Recent culture-independent metagenome based phylogenetic studies are starting to provide a more complex view of the domain. Including the non-extremophilic archaeons, the domain can be divided into several groups named Euryarchaeota, and the TACK (Thaum-, Aig-, Cren-, and Korarcheota) superphylum. More recently, several new archaeal lineages have been proposed, named Lokiarchaeota, Thorarchaeota, Odinarchaeota, Heimdallarchaeota, and another "Asgard" superphylum (Woese et al. 1990; Preston et al. 1996; Brochier-Armanet et al. 2008; Guy and Ettema 2011; Spang et al. 2015; Seitz et al. 2016; Zaremba-Niedzwiedzka et al. 2017). Nowadays eukarya are thought to have arisen from one of these archaeal cell lineages around 2500–2000 mya. Symbiosis is a state of contiguousness in which organisms of different species provide mutual benefits to one another, thereby facilitating their mutual existence. Eukarya is thought to have arisen from the permanent fusion between bacteria and archaea, and subsequent additional incorporation of bacteria into the host cell resulting in the evolution of mitochondria and chloroplasts (Margulis 1993; Doolittle and Brown 1994; Margulis 1996; Martin and Müller 1998; Archibald 2014). One aspect of

M. Takemura (✉)
Laboratory of Biology, Department of Liberal Arts, Faculty of Science,
Tokyo University of Science, 1-3 Kagurazaka, Shinjuku-ku, Tokyo 162-8601, Japan
e-mail: takemura@rs.kagu.tus.ac.jp

T. Mochizuki (✉)
Earth-Life Science Institute, Tokyo Institute of Technology, 1-12-20 Ookayama,
Meguro-ku, Tokyo 152-8550, Japan
e-mail: tomo.mochiviridae@elsi.jp

G. Witzany (ed.), *Biocommunication of Archaea*,
DOI 10.1007/978-3-319-65536-9_15

eukaryal cellular evolution over which there is little consensus on is the origin of the nucleus.

During symbiotic contacts between these bacterial ancestors and archaeal-eukaryal ancestors, horizontal gene transfer (HGT) (or lateral gene transfer: LGT) is thought to have occurred between these organisms (Takemura et al. 2015). The most famous example is the gene coding the mitochondrial DNA replication protein, DNA polymerase γ, which is located in the host genome and not in the mitochondrial genome, which should be its "homeground" (Kaniak-Golik and Skoneczna 2015). HGT is now known to have contributed to the molecular evolution of various organisms, in many cases being mediated by numerous viruses, as "gene robbers" or "gene suppliers." Thus it is reasonable to expect that the evolution of archaeal-eukaryal cell lineages may had been affected by such events, possibly being mediated by some archaeal or eukaryal viruses.

Recently, new viruses infecting archaeal cells or eukaryal cells have been isolated from various environments worldwide. The archaeal viruses, which have mainly been isolated from hyperthermophilic Crenarchaeota and halophilic Euryarchaeota host strains, are known to possess amazingly diverse virion morphotypes (for reviews, Pina et al. 2011; Snyder et al. 2015; Prangishvili et al. 2016). So-called "giant viruses", initially isolated from acanthamoeba cells in 2003, have since been discovered in many environments and have alerted us to the fact that a very large attractive world of such previously unexpected viruses exist on modern earth (La Scola et al. 2003; Abergel et al. 2015).

In this review, we would like to give outlines of the archaea, archaeal viruses, eukaryal giant viruses, and their possible evolutional interactions, focusing on HGT events which occurred in their genes such as the B-family DNA polymerases. First we will give an introduction to the archaeal domain and archaeal viruses. Then we will mention "giant viruses," which are thought to have evolved in contact with archaeal-eukaryal cell lineages, and finally we will describe the evolutionary relationship between archaeal-eukaryal cell lineages and viruses, including the perspectives of the eukaryogenesis hypothesis.

2 Diversity of Archaea

2.1 Traditional Archaeal Classification

Traditionally, many cultivated archaeal organisms had been isolated from extreme environments, such as salt saturated lakes, hot springs, deep sea hydrothermal vents, or oxygen depleted sediments. Many still perceive archaea as extremophiles. First classification within the domain archaea was proposed by Carl Woese himself, dividing the domain into two "kingdoms" (or later "phyla"), Euryarchaeota and Crenarchaeota (Woese et al. 1990). Euryarchaeota consists of a diverse group of organisms, ranging from halophiles, methanogens, and to hyperthermophiles.

Crenarchaeota mainly consists of hyperthermophilic archaea. Later on, the use of culture independent microbial population analysis, such as 16S rRNA targeted PCR analysis, also revealed the existence of archaeal lineages which do not fall into these two kingdoms. The phylum Korarchaeota was the first of its kind, identified from 16S based analysis on geothermal environments, such as Yellowstone National Park (Barns et al. 1996). Although no pure culture has been obtained, its whole genome sequence has been achieved using enrichment cultures, and revealed that they have thin filamentous cell morphology (Elkins et al. 2008).

Another group of archaea, *Cenarchaeum symbiosum,* was detected as a symbiont of a sea sponge, and was originally reported as a phychrophilic Crenarchaeote (Preston et al. 1996). Their relatives, referred to as marine Crenarchaea, were found to be ubiquitously present in marine waters. The first cultivable isolate, *Nitrosopumilus maritimus*, was obtained from an aquarium sediment, and it was shown that it is a chemolithoautotroph which oxidizes ammonia to nitrate (Könneke et al. 2005). The whole genome sequence information of the marine sponge symbiont *C. symbiosum* obtained later (Hallam et al. 2006) revealed that what had been referred to as the mesophilic marine Crenarchaea was in fact not a member of the phylum Crenarchaeota, but formed the distinct phylum Thaumarchaeota (Brochier-Armanet et al. 2008).

2.2 *Expanding New Archaeal Phyla in the NGS Era*

The recent technological advances in next generation DNA sequencing techniques boosted our understanding of uncultivable archaea from various environments. The full genome sequence of Candidatus "*Caldiarchaeum subterraneum*" was recovered from a geothermal water stream of a subsurface gold mine (Nunoura et al. 2011). The genome sequence information suggested that Ca. "*C. subterraneum*" should either be placed at the ancestral position of other Thamuarchaeota, or form an independent phylum of its own with the proposed name of "Aigarchaeota".

Since the time when our understanding of microbial diversity was biased toward cultured organisms and the phylogeny of the archaeal domain was primarily divided between Euryarchaeota and Crenarcheaota, which one of the two served as the potential ancestor of the eukaryal domain was always under debate. While members of the Crenarchaeota were exclusively limited to hyperthermophiles, the Euryarchaeota phylum showed much higher ecological diversity, and some euryarchaeotes in non-thermal environments could be seen as having more similarities with eukaryotes. Euryarchaeota were known to have histone proteins, which were lacking in Crenarchaeota, except for the order Thermoproteales. On the other hand, crenarchaeal cell division mechanisms were closer to eukaryotes. While Euryarchaeotes uses the FtsZ ring, which is also used in bacteria, Crenarchaeotes use the ESCRT system, which is used in eukaryotic endosome division. In fact, in the very early days, Crenarchaeota phylum as we know them today were also suggested to be much more closely related to eukaryotes rather than to the

Euryarchaeota phylum. It was primarily based on electron microscopic observation of the superficial structure of the ribosome, and the term "Eocyte" was given to what we refer to today as Crenarchaeota (Lake et al. 1984).

2.3 TACK, Asgard Superphylum, and the Eocyte Theory

The discovery of Korarchaeota, Thaumarchaeota and Aigarchaeota in recent years is shedding light on this Eocyte hypothesis. These three phyla, together with Crenarcheota, are known to share several key physiological properties, such as cytokinesis, membrane remodeling, cell shape determination and protein recycling. Functional and comparative genomics studies have shown that significant similarities shared among these four phyla, proposed to be termed "TACK superphylum" (Guy and Ettema 2011), or "Proteoarchaeota" kingdom (Petitjean et al. 2015), are also shared with eukaryotes.

Culture independent metagenomic excavation of the potential eukaryotic ancestors has continued in recent years. In 2015, a reconstruction of a near full-length archaeal genome was reported from a deep sea sediment sample taken near the Loki's Castle hydrothermal vent site in the Mid-Atlantic Ridge (Spang et al. 2015). The genome was a bona fide archaeon, described as the new phylum "Lokiarchaeota", which showed monophyletic grouping with eukarya. The over 5 Mbp genome of Lokiarchaeota harbored genes involved in cytoskeleton dynamics, which may enable phagocytosis—the event expected to have occurred with alphaproteobacteria in acquisition of the mitochondria. The discovery of Lokiarchaeota was followed by additional proposal of the new archaeal phyla Thorarchaeota (Seitz et al. 2016), Odinarchaeota, and Heimdallarchaeota, and the "Asgard" superphylum was suggested to group all these metagenome based archaeal phyla (Zaremba-Niedzwiedzka et al. 2017). The authors claim that Asgard archaea seem to be even closer ancestors of the eukaryotic domain than the TACK archaea.

In any case, the recent metagenome based discoveries point to the direction that there seems to be a significant evolutionary difference between Euryarchaeota and the rest of the archaea, and that eukaryotes arose from the latter—what had originally been hypothesized thirty years ago as the Eocyte theory. Culture independent metagenome reconstruction always harbors the potential of artificial assembling errors. However, as more data accumulates, we should be able to get closer to identifying the potential evolutionary path. It seems that we are still quite far from identifying what the more evolutionarily ancient Last Archaeal Common Ancestor (LACA) or the Last Universal Common Ancestor (LUCA) was like. Although we are undoubtedly approaching towards understanding of the Last Eukaryal Common Ancestor (LECA).

3 Archaeal Viruses

3.1 Surprisingly Diverse Archaeal Viruses

Viruses are found from all three domains of life. Studies on viruses infecting eukarya and bacteria have been done for over a century, and under medical, agricultural or bacteriological purpose, several thousand strains have already been isolated by numerous laboratories (King et al. 2012). Recent progress of the study of "giant viruses," which infect eukaryal cells as described in Chap 4, has opened up a new field of virology.

Studies on viruses infecting the third domain of life, archaea, have been done by a very small community in the past two to three decades, and just roughly over a hundred strains have been isolated (for reviews, Pina et al. 2011; Snyder et al. 2015; Prangishvili et al. 2016). Viruses infecting bacteria (also referred to as phages) are known to be dominated by head-tail morphotypes, which consist of double-stranded (ds-) DNA viruses of the families *Myoviridae*, *Siphoviridae* and *Podoviridae*, which together are classified into the order Caudovirales (Fig. 1a). Bacterial viruses with other morphotypes, such as icosahedral or helical/linear, as well as other genotypes, such as single-stranded (ss-) DNA or ds-/ssRNA, exist in relatively smaller proportions. On the other hand, the viruses of the other prokaryotic counterpart, which have mainly been isolated from hyperthermophilic Crenarchaeota and halophilic Euryarchaeota host strains, are known to possess amazingly diverse virion morphotypes. It includes lemon/spindle—(*Fuselloviridae*) (Martin et al. 1984; Wood et al. 1989; Bath and Dyall-Smith 1998), tailed spindle—(*Bicaudaviridae*) (Häring et al. 2005a), droplet—(*Guttaviridae*) (Arnold et al. 2000a), bottle—(*Ampullaviridae*) (Häring et al. 2005b) and coil/spring-shaped (*Spiraviridae*) (Mochizuki et al. 2012) viruses (Fig. 1b). Many of these morphotypes were never before observed in nature. In addition, common virion morphotypes such as helical/linear—(*Rudiviridae*, *Lipothrixviridae*, and recently proposed "*Tristromaviridae*") (Janekovic et al. 1983; Zillig et al. 1993; Arnold et al. 2000b; Rensen et al. 2016), polyhedral (*Turriviridae*, *Sphaerolipoviridae*) (Rice et al. 2001; Bamford et al. 2005), globular (*Globuloviridae*) (Häring et al. 2004), bacilliform (*Clavaviridae*) (Mochizuki et al. 2010), pleomorphic (*Pleolipoviridae*) (Pietilä et al. 2009), as well as head-tailed Caudovirales viruses have been isolated (Fig. 1b). While the highly studied bacterial viruses are currently classified into 10 taxonomical families, archaeal viruses are classified into 16 families (Fig. 2). New viruses which correspond to new viral families are continuously being isolated from the archaea, but it is extremely rare in the bacterial domain. However, the cultivated archaeal viruses today are all exclusively DNA viruses. They are mostly dsDNA viruses, and only halophilic *Pleolipoviridae* (Pietilä et al. 2009) and hyperthermophilic *Spiraviridae* (Mochizuki et al. 2012) viruses have ssDNA genomes. An RNA virus has yet to be found from the archaeal domain. Moreover, no RNA virus has been found from thermophilic bacteria either. As our understanding of viruses in extreme conditions is still just the tip of the iceberg, we have yet to know whether or not an RNA virus truly exists in the archaeal domain or in any geothermal environment. However, the very fragile RNA molecule is thought to have played a critical role in the origin of life, known as the RNA-world theory, in an ancient geothermal environment. If extant RNA viruses do not truly exist in the modern analog of such an environment, it may require us to reexamine the theories on the origin and evolution of life.

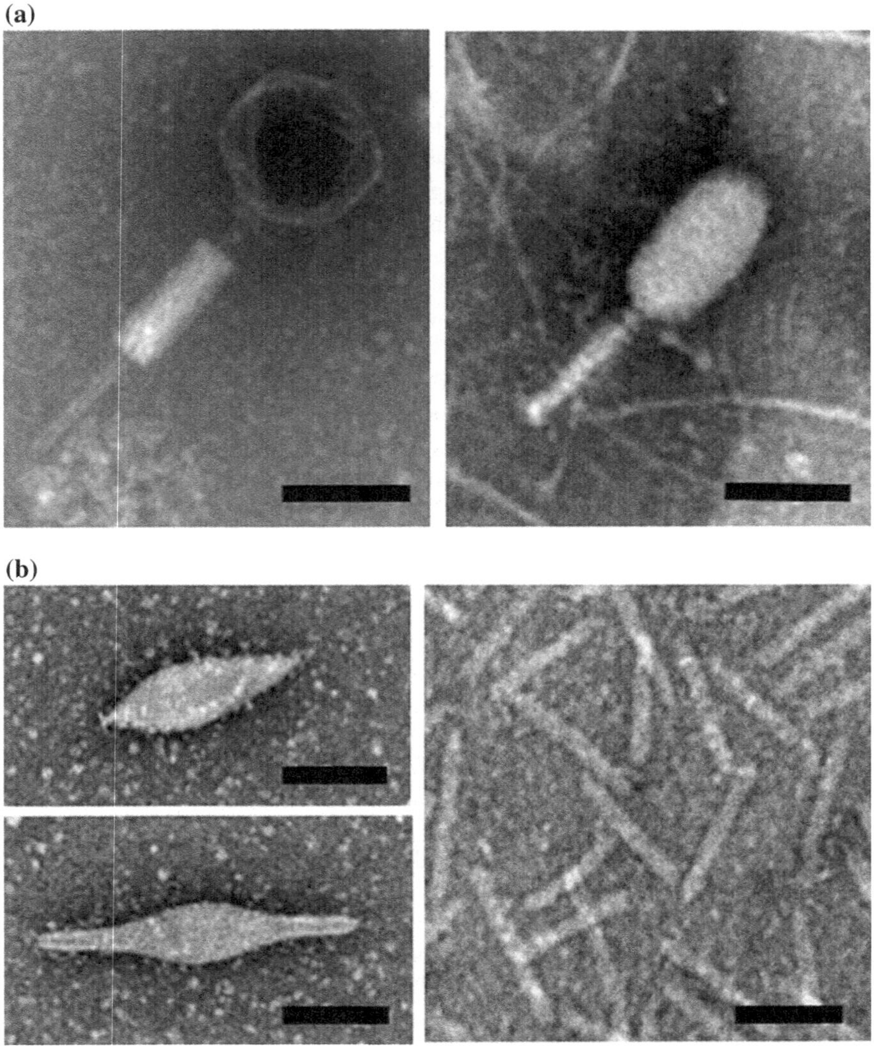

Fig. 1 Transmission electron micrograph of prokaryotic viruses. **a** Bacterial head-tailed viruses of the order Caudovirales. **b** Hyperthermophilic archaeal spindle-shaped virus APSV1 with tail-less (*top left*) and elongated form (*bottom left*), and bacilliform APBV1, a member of the family *Clavaviridae* (*right*). All scale bars, 100 nm

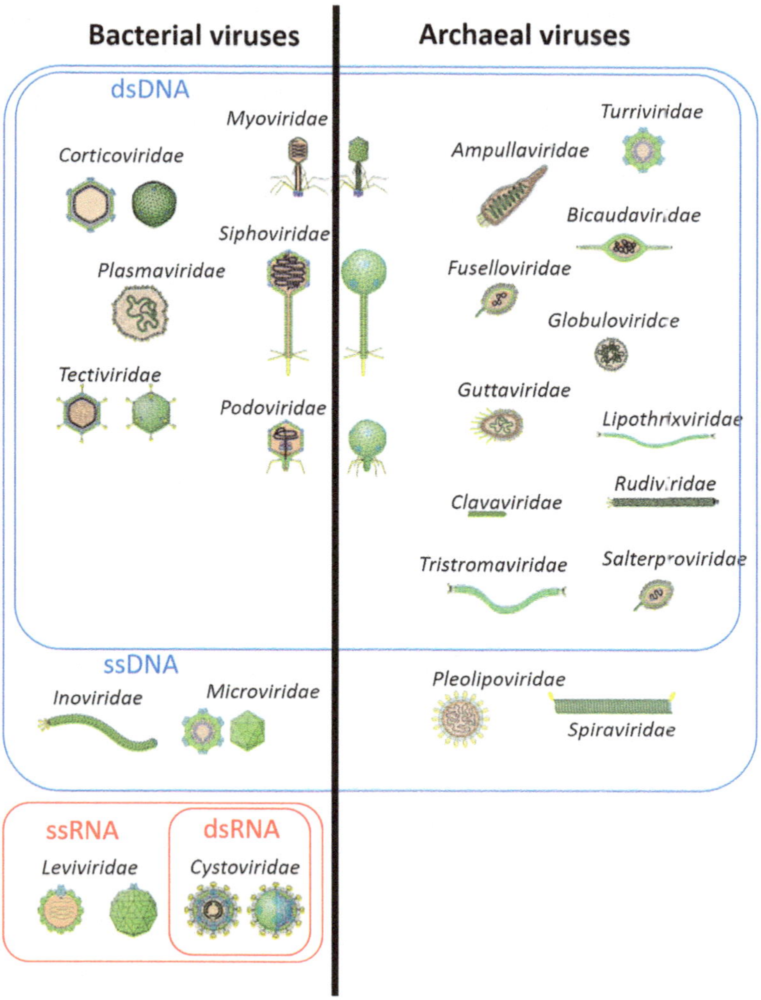

Fig. 2 Graphical representation of diversity of bacterial (*left*) and archaeal (*right*) viruses. DNA viruses are enclosed in blue line(s), and RNA viruses in red. Single-stranded (ss-) viruses are enclosed within a single line, and double-stranded (ds-) viruses in double lines. Images adopted from ViralZone with permission, and with slight modifications. Illustrations: ViralZone, Swiss Institute of Bioinformatics

Most genes encoded within the archaeal viral genomes have no homologous sequence in the public databases, except for viruses of the same family (Prangishvili et al. 2006). Thus, the genomic annotation of the predicted genes results in a very high percentage of "hypothetical unknown protein". While many bacterial or eukaryal viruses encode detectable DNA polymerase or RNA polymerase genes which are involved in their genome replication, such genes are rarely detected from archaeal viral genome sequences. It may be that they are simply borrowing the polymerases from the host system, but the possibility that they encode novel types of polymerases may also be possible.

3.2 *Possible Evolutionary Implication Witnessed by Different Virospheres*

The remarkable morphological and sequential diversity observed in archaeal viruses is a clear indication that archaeal viruses constitute a completely different virosphere, and that an evolutionary implication must lie behind it. Currently no solid explanation is given, but an interesting hypothesis involving the viruses and divergence of the two domains from the last universal common ancestor (LUCA) has recently been suggested (Forterre and Prangishvili 2009; Prangishvili 2013). They assume that the ancestral LUCAs were prone to viral infections by various capsid morphotypes (Fig. 3). To prevent frequent viral infection, one group invented the cell wall and later constituted the bacterial domain. In order to deal with the physical barrier, their viruses invented the head-tail system, which physically uses the tail structure to penetrate and inject the genome through the physical barrier and into the host cell. On the other hand, another group glycosylated their cell surface proteins to escape from virus-cell recognition, which later constituted the archaeal domain (Fig. 3). Archaeal viruses also responded by glycosylating

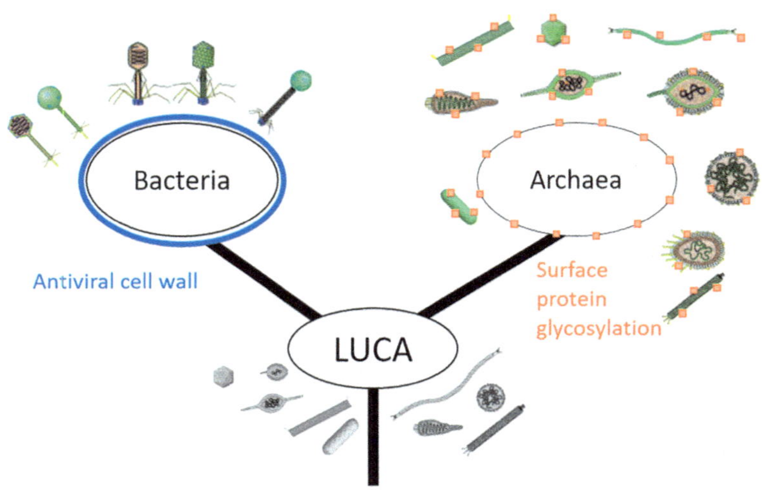

Fig. 3 Graphical description of the previously suggested evolutionary scenario on virus-driven divergence of archaeal and bacterial domains. *Large circles* indicate representative cellular organisms, *blue circle* indicate the bacterial cell wall, *orange squares* represent glycosylation of surface proteins. *Black and white* viruses represent theoretical viruses of the LUCA (Last Universal Common Ancestor) era. Virus images adopted from ViralZone with permission, with slight modifications. Illustrations: ViralZone, Swiss Institute of Bioinformatics

their capsid proteins. Intense post translation glycosylation seen in many archaeal and archaeal viral proteins is rarely found in bacteria or bacterial viruses.

As our understanding of the LUCA is still very poor, how archaea, bacteria, and furthermore eukarya diversified remains to be a major evolutionary mystery. But as presented above, the idea of incorporating their respective virosphere may facilitate our understanding of bacterial and archaeal-eukaryal cellular evolution as well.

4 Giant Viruses and Evolution of Eukarya

4.1 *Giant Viruses*

In 2003, the first documented so-called "giant virus" was reported. La Scola et al. published a one-page paper in Science describing the discovery of *Acanthamoeba polyphaga mimivirus* (APMV), from the water of a cooling tower of a hospital in Bradford, The United Kingdom (La Scola et al. 2003). This virus was previously thought to be a contamination of an unknown bacterial species, named *Bradfordcoccus*. Various observations, such as the absence of rRNA genes, the presence of an eclipse phase in acanthamoeba cells, and their icosahedral morphological properties revealed by transmission electron microscope observations indicated, and later concluded, that this *bradfordcoccus* was in fact a virus of unbelievably huge size. The virion of APMV was 400 nm or larger, by far the biggest virion size ever discovered (Fig. 4). Its whole genome was sequenced in 2004, and it was reported to be even larger than 1 Mb (1,181,404 bp), with 1262 open reading frames (ORFs) (Raoult et al. 2004). The genome size and the number of encoded ORFs were the largest among all previously known viruses. APMV and its relatives, the *Mimiviridae* family (named after their mimicking of bacteria), are now known to have long surface fibrils layers, so the virion size including the surface fiber layer is 750 nm in diameter (La Scola et al. 2003; Raoult et al. 2004; Abergel et al. 2015).

More than a decade after the initial surprising discovery, many giant viruses have been found and isolated from various environments including sea water, fresh water, and several living things, and they are now classified into several groups (Table 1). The members of *Mimiviridae* are continuing to expand (Abergel et al. 2015). To date, over 100 isolates have been reported, including *Acanthamoeba castellanii mamavirus*, *Samba virus*, *Kroon virus*, *Mimivirus Bombay*, *Mimivirus shirakomae*, *Acanthamoeba polyphaga moumouvirus*, *Megavirus chilensis*, and *Cafeteria roenbergensis virus* (La Scola et al. 2008; Fischer et al. 2010; Arslan et al. 2011; Yoosuf et al. 2012; Colson et al. 2012; Desnues et al. 2012; Ngounga et al. 2013; Campos et al. 2014; Assis et al. 2015; Boratto et al. 2015; Andrade et al. 2015; Chatterjee et al. 2016; Takemura et al. 2016). The second group of giant viruses have amphora-shaped particles of around 1 μm in diameter, which include the family *Pandoraviridae* (*Pandoravirus salinus*, *Pandoravirus dulcis*, and

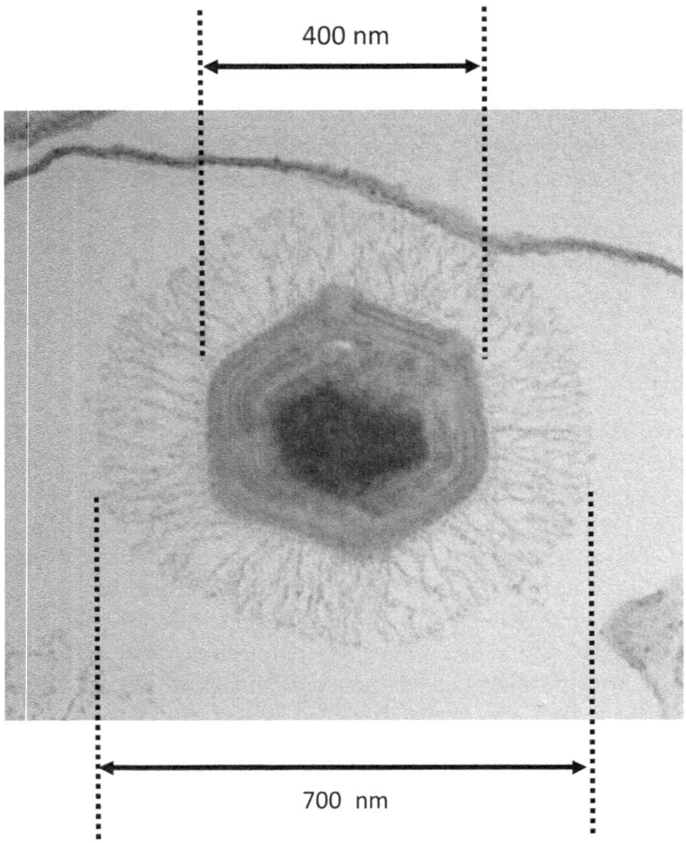

Fig. 4 TEM image of *Mimivirus kasaii* particle in acanthamoeba cell. *Mimivirus* virions possess an icosahedral capsid 400–500 nm in diameter. Capsid surface is covered by a fibril layer with 150 nm thickness. Inner membrane exists inside capsid layers, and a core including genomic DNA is inside the inner membrane. *Mimivirus kasaii* was isolated from the water sample from a Japanese river mouth (Takemura et al. 2016)

Pandoravirus inopinatum) and *Pithovirus sibericum*, *Cedratvirus*, and *Mollivirus sibericum* (Philippe et al. 2013; Legendre et al. 2014, 2015; Andreani et al. 2016), of which the latter has not been assigned a taxonomical family. The third type of giant viruses is *Faustovirus*, which is closely related to *Asfarviridae*, one of a big group named nucleo-cytoplasmic large DNA viruses (NCLDV) (Reteno et al. 2015). And the fourth group of giant viruses is *Marseilleviridae*, with a particle size of around 200 nm, which is smaller than the other giant viruses, and a genome size of 300–400 kbp, including *Marseillevirus*, *Lausannevirus*, *Melbournevirus*, *Tunisvirus*, *Insectomime virus*, *Brazilian marseillevirus*, *Golden marseillevirus*, and *Tokyovirus* (Boyer et al. 2009; Thomas et al. 2011; Colson et al. 2013; Doutre et al. 2014; dos Santos et al. 2016; Takemura 2016a, 2016b). Based on shared genes among them, these giant viruses are known to belong to NCLDV, which also

Table 1 Representative giant viruses

Family or genus	Virus name	Genome size (bp)
Pandoravirus	*Pandoravirus salinus*	2,473 870
	Pandoravirus dulcis	1,908 524
Mimiviridae	*Megavirus chilensis*	1,259 197
	Acanthamoeba polyphaga mimivirus	1,181 549
	Acanthamoeba polyphaga moumouvirus	1,021 348
Mollivirus	*Mollivirus sibericum*	651.523
Pithovirus	*Pithovirus sibericum*	610.033
Marseilleviridae	*Tokyovirus*	372.707
	Marseillevirus	368.454

includes *Poxviridae*, *Iridoviridae*, *Phycodnaviridae*, *Ascoviridae*, and *Asfarviridae* (Iyer et al. 2001, 2006).

4.2 B-Family DNA Polymerases

Eukaryotes possess four B-family DNA polymerases, DNA polymerase α, δ, ε, and ζ (Wang 1991; Hübscher et al. 2002; Burgers 2009). DNA polymerase α is thought to be involved in the synthesis of RNA/DNA primers in both leading and lagging strand synthesis. DNA polymerase δ is involved in the synthesis of Okazaki fragment in the leading strand, and DNA polymerase ε is involved in the synthesis of the long leading strand (Takemura 2011). These eukaryal B-family DNA polymerases are thought to have evolved from one of the ancestral DNA polymerases. The "duplication hypothesis" has mainly been considered to explain the evolution of eukaryal B-family DNA polymerases, in which it is proposed that DNA polymerase α and δ have evolved from ancestral DNA polymerases, presumably the DNA polymerase ε (Edgell et al. 1998). Nowadays DNA polymerase ε is known to be a "chimera" of two archaeal DNA polymerases, one belonging to the B-family and the other to the D-family (Tahirov et al. 2009), and the other three (α, δ, and ζ) are thought to have arisen from another archaeal B-family DNA polymerase (Tahirov et al. 2009). On the other hand, taking into account the B-family DNA polymerases of NCLDVs including the giant viruses, an alternative evolutionary scenario of these DNA polymerases could be hypothesized (Takemura 2001; Takemura et al. 2015).

Previous phylogenetic analyses of the B-family DNA polymerases of giant viruses, archaea, and eukarya suggested a relationship between the NCLDVs and the evolution of archaeal-eukaryal cell lineage (Claverie 2006; Takemura et al. 2015). Phylogenetic trees constructed using amino acid sequences of the B-family DNA polymerase suggested that DNA polymerases of NCLDVs belong to phylogenetic clusters separate from those of archaea and eukaryal DNA polymerases (Monier et al. 2008; Takemura et al. 2015).

4.3 Evolutionary Relationship Between Viruses and Cells

Takemura et al. have reported that in all the unrooted phylogenetic trees, which were reconstructed using PhyML, MEGA6, MrBayes, and PhyloBayes, the eukaryal B-family DNA polymerases were always grouped into four clans, corresponding to DNA polymerases α, δ, ε, and ζ groups, respectively (Takemura et al. 2015). In these trees, DNA polymerases δ and ζ were more closely related to each other than to DNA polymerases ε and DNA polymerase α respectively (Fig. 5). On the other hand, in some of these phylogenetic analyses, B-family DNA polymerases of NCLDVs were divided into four clans: *Poxviridae*, *Pithovirus*, Chlorella viruses and *Pandoraviruses*, and other NCLDVs including *Mimiviridae*. Interestingly, in all phylogenetic trees, eukaryal DNA polymerase δ was grouped with a major clan of *Mimiviridae* DNA polymerases, with DNA polymerase ζ clan being placed outside this DNA polymerase δ-*Mimiviridae* group (Takemura et al. 2015) (Fig. 5). Although the analyses data lacked eukaryal DNA polymerase ε, *Pandoravirus* DNA polymerases, and *Pithovirus* DNA polymerase, similar results had already been reported previously (Monier et al. 2008). On the other hand, the position of

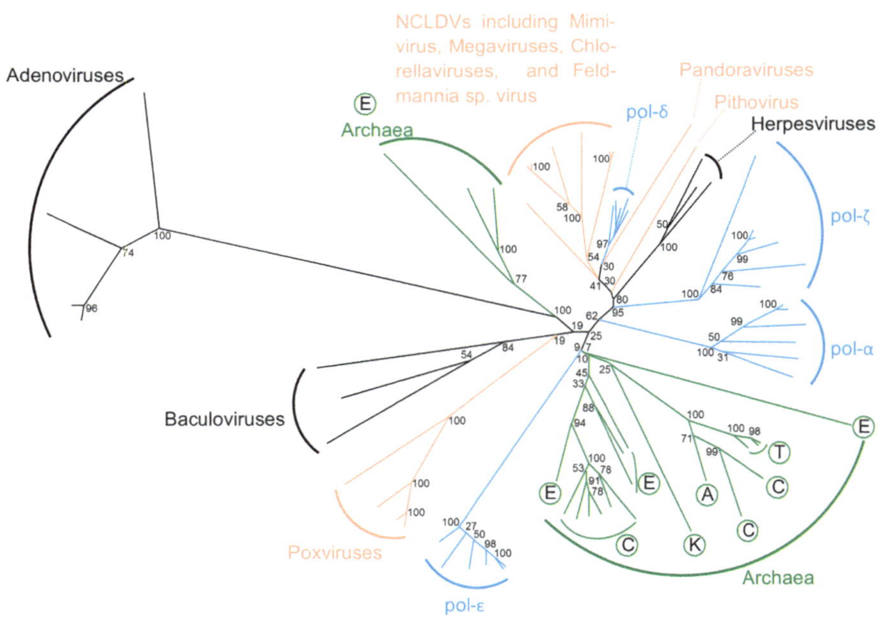

Fig. 5 Phylogenetic tree of B-family DNA polymerases of archaea, eukaryotes, and NCLDVs. Unrooted MEGA6 phylogenetic tree was built based on alignment derived from the full-length alignment of B-family DNA polymerases where any columns containing gaps were discarded (Tamura et al. 2013; Takemura et al. 2015). Numbers at the branch points indicate bootstrap values. Abbreviations are: pol-α, eukaryotic DNA polymerase α; pol-δ, eukaryotic DNA polymerase δ; pol-ε, eukaryotic DNA polymerase ε, pol-ζ, eukaryotic DNA polymerase ζ; E, euryarchaeota; T, thaumarchaeota; A, aigarchaeota; C, crenarchaeota; and K, korarchaeota

Poxviridae DNA polymerases was always far from other NCLDVs, and was outside the clan containing DNA polymerases δ and ζ (Takemura et al. 2015) (Fig. 5). Eukaryal DNA polymerase ε was positioned close to the archaeal clan, suggesting DNA polymerase ε was the most ancestral among the four eukaryotic B-family DNA polymerases (Edgell et al. 1998; Tahirov et al. 2009; Takemura et al. 2016) (Fig. 5). These phylogenetic analyses suggest that the different NCLDV lineages such as *Poxviridae* and *Mimiviridae* were involved in the molecular evolution of different eukaryal B-family DNA polymerases in archaeal-eukaryal cell lineages. It is reasonable to expect that these interactions between viruses and archaeal-eukaryal lineages should have been mediated by HGT.

4.4 *Inferred Scenario of Evolution of B-Family DNA Polymerases*

According to the phylogenetic analyses described above, here we suggest the inferred evolutionary scenario of B-family DNA polymerases, and further archaeal-eukaryal cell lineages and some giant viruses (Takemura et al. 2015). The last common ancestor of the archaeal-eukaryal cell lineage (AAE) and (or) NCLDV ancestor retained an ancestral DNA polymerase ε (Fig. 6). Putative HGT occurred between these ancestors, although the direction of HGT, whether from cells to viruses or vice versa, is unknown (Fig. 6). An ancestral DNA polymerase α evolved from the ancestral DNA polymerase ε in AAE or viruses, and HGT occurred (Fig. 6). This NCLDV lineages evolved into the *Poxviridae* family, and archael-eukaryal cells with two types of B-family DNA polymerase have evolved into archaea, with or without the loss of the other type (Fig. 6). The ancestral DNA polymerase ζ has then evolved putatively in the viral lineage from ancestral NCLDV possessing ancestral DNA polymerase α-like DNA polymerase, and had transferred to eukaryal cell lineage by HGT (Fig. 6). Continuously, DNA polymerase δ has evolved from ancestral DNA polymerase ζ putatively in the viral lineage, and also transferred to eukaryal cells by HGT. This NCLDV lineages have evolved into *Mimiviridae* and other NCLDVs including *Pandoraviruses* and *Pithovirus*, and cells having four types of B-family DNA polymerase have evolved into modern eukarya (Fig. 6).

This scenario, however, does not take into account archaeal diversity as described in Chap. 2. It is not fully elucidated and which archaeal group could truly be the ancestor of our eukaryal domain is still under debate.

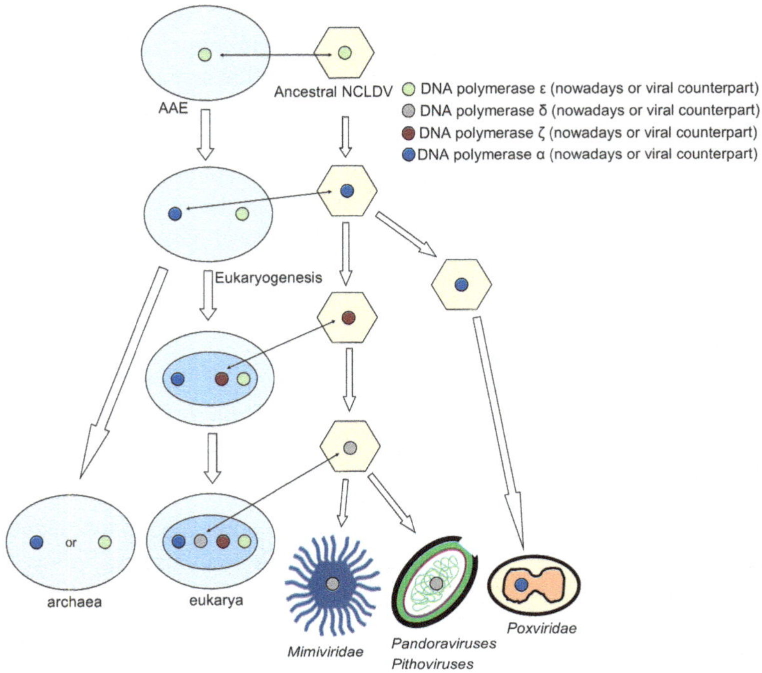

Fig. 6 Hypothetical scenario of the evolution of B-family DNA polymerases. Black bi-directional arrows indicate HGT. Open big arrows respectively indicate evolution of each living organism or virus. Small colored circles indicate the different eukaryotic DNA polymerases or their viral and archaeal counterparts. Abbreviations are: AAE, ancestral archaeal-eukaryotic cell lineage

5 Eukaryogenesis in Archaea

5.1 Eukaryogenesis Hypothesis

As described previously, eukaryotes have arisen from the ancestral archaeal cells in which the nucleus, and membrane-coated giant components (organelles), have emerged via an unknown mechanism. Several researchers have advanced persuasive hypotheses related to the origin of the nucleus, such as the invagination model, hydrogen hypothesis, syntrophic hypothesis, vesicle model, and planctomycete model (Stanier 1970; Cavalier-Smith 1983; Moreira and López-García 1998; Horiike et al. 2001; Embley and Martin 2006; Fuerst and Sagulenko 2012; Martin 2005; Cavalier-Smith 2010). The common feature of such models is that the endomembrane system must have emerged prior to the formation of the nuclear membrane. In the invagination model, the endomembrane system evolved by invagination of the cell membrane, followed by the emergence of the nuclear membranes. In the syntrophic symbiosis model, the endomembrane system emerged via the formation of a consortium of several bacteria and archaea from their cell membranes, and subsequently the nuclear membrane was formed. Regardless of whether the nucleus emerged by invagination, vesicle, or symbiotic models, several questions remain: When and how did the intracellular innovation start? Are there any signs of a discrete evolutionary event? How were ribosomes excluded from the nucleus?

5.2 *Complicated DNA Viruses in Eukaryogenesis*

The hypotheses that Takemura and Bell proposed independently were more controversial than these other hypotheses (Takemura 2001; Bell 2001). Both researchers proposed that the eukaryotic nucleus originated from DNA-virus ancestors that invaded eukaryal ancestors, based on phylogenetic analyses of eukaryal B-family DNA polymerases and those of DNA viruses, and of other genes involved in gene expression. Bell designated his own hypothesis, the "viral eukaryogenesis (VE) hypothesis" (Bell 2001). Although Takemura had described "undigested inner virion membranes at numerous sites resulting in the formation of a prototype nucleus dramatically smaller than today's" in his hypothesis (Takemura 2001), Takemura's theory has recently been reinforced by the discovery of several giant viruses and the observation of virion factories, which are cytoplasmic compartments formed by giant viruses in infected cells. Thus, we provide an update and a reconstruction of Takemura's previous hypothesis.

In the previous hypothesis, Takemura mentioned that a poxvirus-like ancestor could provide clues of eukaryogenesis because *Poxviridae* viruses are very complicated DNA viruses, and are able to replicate their own genomes in the cytoplasm of the host cells independent of the host nucleus, suggesting that these viruses can infect cells "without a nucleus." (Takemura 2001).

5.3 *Virion Factories of Giant Viruses*

After publication of papers by Takemura and Bell (Takemura 2001; Bell 2001), it was reported that in infections by vaccinia virus, one of the *Poxviridae*, replication of DNA occurs at endoplasmic reticulum (ER)-enclosed cytoplasmic sites (Tolonen et al. 2001). For at least 3 h post-infection, the replication sites of the vaccinia virus in HeLa cell cytoplasm are entirely surrounded by ER membranes, which the authors called "mini-nuclei." These mini-nuclei are thought to be virion factories, which produce numerous new virions of vaccinia virus in HeLa cells. More interestingly, it was also reported that HeLa cell ribosomes are mostly on the side of the ER membrane facing the cytoplasm, not on the side of the virion factory (Tolonen et al. 2001).

Recently *Mimiviridae* family has been found to construct large virion factories in their infected host cell cytoplasm, emerging around 3 h post infection. Between 2 and 8 h post infection, the virion factory grows in size as large as the host cell nucleus, and continues to produce *Mimivirus* virions until 24 h post infection (Suzan-Monti et al. 2007; Kuznetsov et al. 2013; Mutsafi et al. 2014) (Fig. 7). In early stage of the Mimivirus virion factory formation, ER membranes in the amoebal cytoplasm seem to engulf the replicating *Mimivirus* genomic DNA, resulting in an early virion factory surrounded by ER membranes—as with the "mini-nuclei" of *vaccinia viruses* (Mutsafi et al. 2014). On the other hand, in the matured phase of virion factory, ER membranes do not seem to constantly surround the virion factory without a gap. Instead, a large amount of fragmented membranes, presumably derived from host ER membranes, start to surround the virion factory, forming the inner membranes of *Mimivirus* virions.

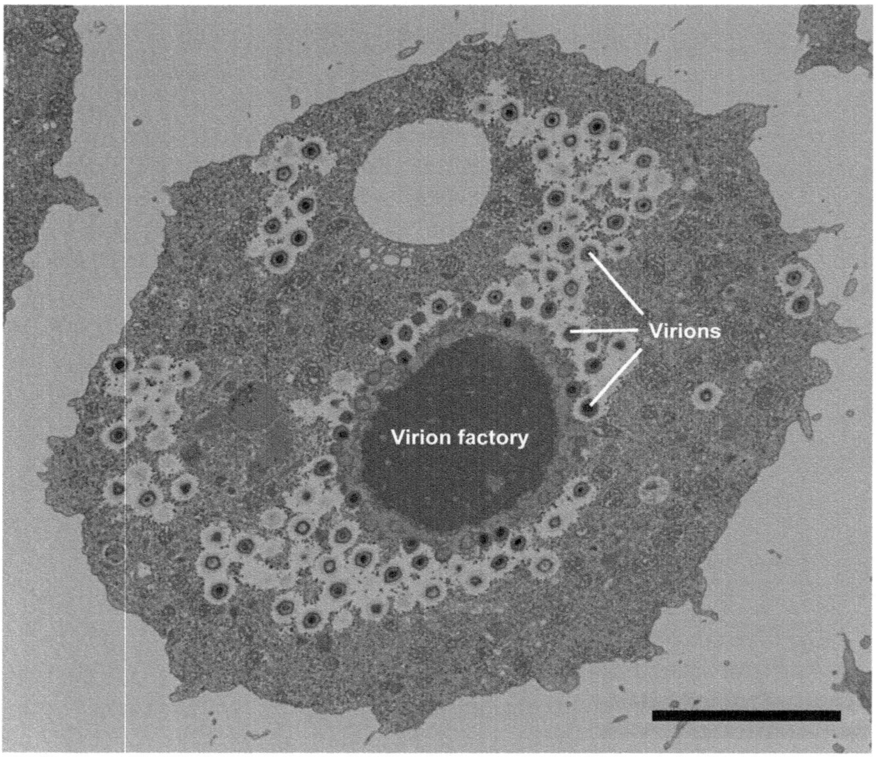

Fig. 7 Virion factory of *Mimivirus shirakomae*. *Acanthamoeba castellanii* cells infected by *Mimivirus shirakomae* (8 h post infection) are shown. Cell nucleus does not appear in this section. The scale bar represents 4 μm. *Mimivirus shirakomae* was isolated from the water sample from Shirakoma pond located in Nagano prefecture of Japan (Takemura et al. 2016)

5.4 *Putative Contribution of Virion Factory to Eukaryogenesis*

These interesting features of virion factories of vaccinia virus and Mimivirus have allowed us to pose an interesting scenario for eukaryogenesis (Fig. 8), as reported previously (Forterre and Gaïa 2015). An archaeal host was "invaded" by symbiotic association with alpha-proteobacteria, which became the mitochondria (Martin and Koonin 2006). The archaeal host was also infected with ancestral NCLDVs at that time (Fig. 8). Evolution of phagocytosis then occurred, creating a permanent endomembrane as in the process described previously (Cavalier-Smith 2010). The occasional or eventual lysis of the newly formed mitochondrion, which was derived from an alpha-proteobacterium, led to HGT from the symbiont to the host as described previously (Martin and Koonin 2006), and the genomes of both symbiont and the host were mixed (Lake et al. 1999) (Fig. 8). Candidate genes that may have arisen in eukaryotes in this way include the following: group II introns (Martin and Koonin 2006); the *FtsZ* gene, which gave rise to tubulin genes by gene duplication (Cavalier-Smith 2010); genes encoding ribosome export factors (Ohyanagi et al. 2008); genes encoding ribosomal processing in the present nucleolus (Staub et al. 2004); and genes for lipid metabolism, which brought ester-linked lipid membranes to the host-retained ether-linked lipid membrane. Infection with NCLDV might have triggered the fusion of ER cisternae because it can instigate the formation of a virion factory surrounded by ER membranes (Tolonen et al. 2001; Schramm and Locker 2005) (Fig. 8). Furthermore, viral genomes would have eventually mixed with the host genomes as well. During eukaryogenesis, the DNA-virus system of mRNA capping and linear genome with short tandem repeats (STR) at both ends (Bell 2001, 2009) may have been imported into the newly formed "nucleus," and mature ribosomes, which are active in translation, would have been excluded from the nucleus as described previously (Tolonen et al. 2001) (Fig. 8). This compartmentalization of the nucleus and the cytoplasm would effectively separate the mRNA splicing and translation steps, which occur at different speeds (Martin and Koonin 2006). Eukaryotic ribosomes are assembled in the nucleolus, and exported to the cytoplasm through nuclear pores. This suggests that the mechanism of eukaryotic ribosomal assembly might be different from that of prokaryotes (Hage and Tollervey 2004). Possibly, the ribosomal subunits of early eukaryotes, shortly after the emergence of the nucleus, were of a similar size to prokaryotic ribosomes and could thus pass through the early nuclear pores (Staub et al. 2004). Alternatively, early eukarya may have had a simpler mechanism of ribosomal assembly and transport to the cytoplasm.

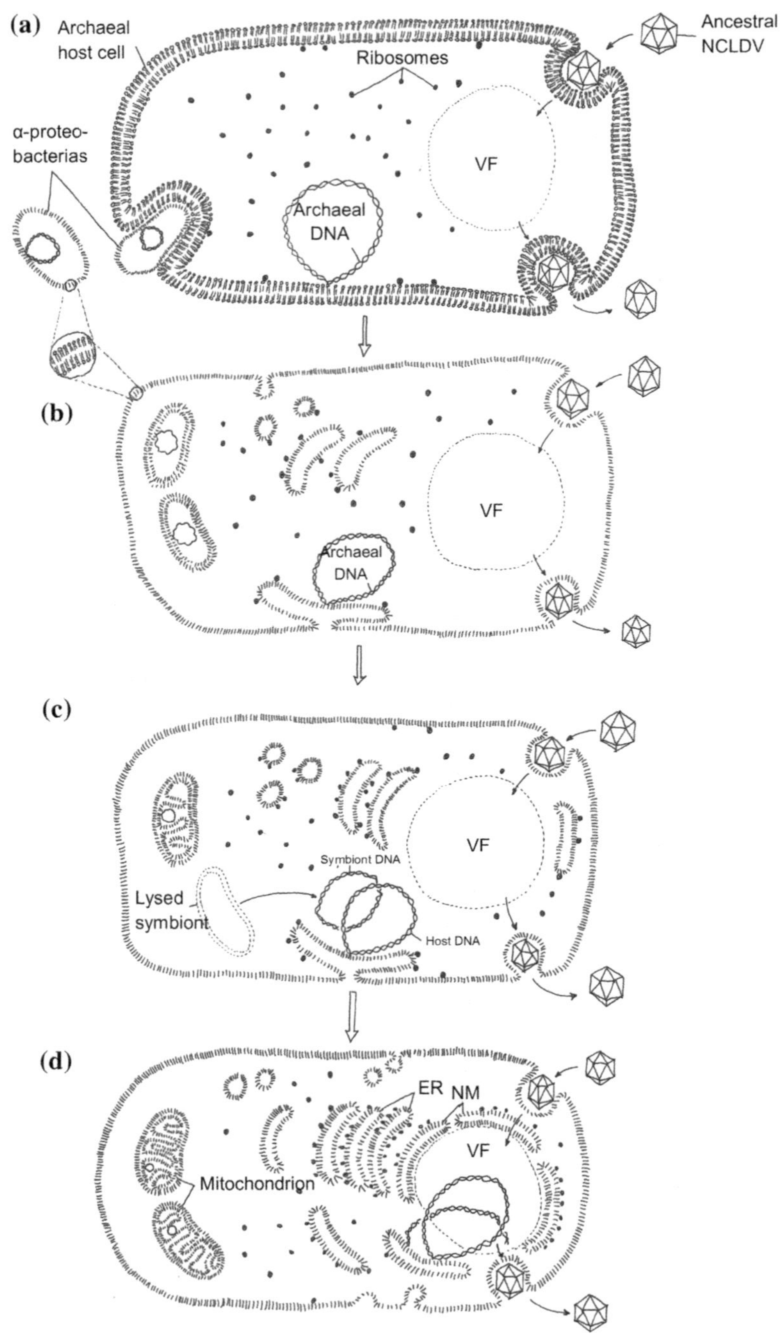
(a)
Archaeal host cell
Ancestral NCLDV
Ribosomes
α-proteo-bacterias
VF
Archaeal DNA
(b)
VF
Archaeal DNA
(c)
VF
Symbiont DNA
Lysed symbiont
Host DNA
(d)
ER
NM
VF
Mitochondrion

◂**Fig. 8** Hypothetical scenario of the eukaryogenesis based on virion factory. Virion factory as the origin of the nucleus. **a** An archaeal host was "invaded" by symbiotic association with alpha-proteobacteria, and was also infected with ancestral NCLDVs. **b** Evolution of phagocytosis then occurred, creating a permanent endomembrane system. **c** The lysis of the symbiont led to HGT from the symbiont to the host, and both symbiont and host genomes mixed. **d** Infection with NCLDV might have triggered ER cisternae fusion because it can instigate the formation of a virion factory surrounded by ER membranes. VF, virion factory; NCLDV, nucleo-cytoplasmic large DNA virus; ER, endoplasmic reticulum; NM, nuclear membrane. Cell walls were omitted in these illustrations

6 Conclusion

Archaea have possessed and interacted with their infectious partners the archaeal viruses, as described in Chap. 3. And eukarya also have possessed, interacted with, and co-evolved with their eukaryal viruses, including giant viruses, as described in Chap. 4. However, in the ancestral era, organisms belonging to these two modern domains should have possessed and interacted with the same viruses. It is still only recently that we have started to recognize the unique, distinctive and astonishingly diverse world of Archaeal viruses. Will the vast world of archaea and their viruses, and the gigantic world of giant viruses reveal to us their mysterious system and ancestral behaviors which led to the emergence of our ancestors?

In 2017, an amazing and interesting paper was published reporting that bacterial viruses (bacteriophages) construct a nucleus-like structure in their infected host bacterium during viral replication (Chaikeeratisak et al. 2017). This nucleus-like structure does not resemble the virion factories constructed by giant viruses, however, this is the first report that prokaryotic viruses can also construct a "nucleus-like" compartment. The biological and evolutionary significance of such findings in the frame of eukaryogenesis must be enormous.

How about archaea and archaeal viruses? We must continue to investigate the immense diversity of these organisms, and try to elucidate how archaea and their viruses may have influenced the evolution of archaeal-eukaryal cell lineage. For instance, hardly anything is known about the viral diversity of recently discovered new archaeal groups (i.e. Asgard superphylum). State of the art metagenomics is a powerful tool to explore the diversity of uncultivable cellular diversity. However, owing to the lack of universal genes in viruses, using metagenomics to understand the natural viral diversity bears critical restrictions. In addition, assigning individual viral reads to their corresponding host organisms, even at the domain level, still remains a difficult task. But by accumulating more knowledge on diverse cellular and viral organisms by both culture dependent and independent analysis, we should be able to draw closer to the ancestral eukaryal lineages, and to the underlying biological mechanism which enabled the rise of eukarya.

References

Abergel C, Legendre M, Claverie JM (2015) The rapidly expanding universe of giant viruses: Mimivirus, Pandoravirus, Pithovirus and Mollivirus. FEMS Microbiol. Rev. fuv037

Andrade KR, Boratto PPVM, Rodrigues FP, Silva LCF, Dornas FP, Pilotto MR, La Scola B, Almeida GMF, Kroon EG, Abrahao JS (2015) Oysters as hot spots for Mimivirus isolation. Arch Virol 160:477–482

Andreani J, Aherfi S, Khalil JYB, Pinto FD, Bitam I, Raoult D, Colson P, La Scola B (2016) Cedratvirus, a double-cork structured giant virus, is a distant relative of Pithoviruses. Viruses 8:300

Archibald J (2014) One plus one equals one. Oxford University Press, Oxford

Arnold HP, Ziese U, Zillig W (2000a) SNDV, a novel virus of the extremely thermophilic and acidophilic archaeon *Sulfolobus*. Virol 272:409–416

Arnold HP, Zillig W, Ziese U, Holz I, Crosby M, Utterback T, Weidmann JF, Kristjanson JK, Klenk HP, Nelson KE (2000b) A novel lipothrixvirus, SIFV, of the extremely thermophilic crenarchaeon *Sulfolobus*. Virol 267:252–266

Arslan D, Legendre M, Seltzer V, Abergel C, Claverie JM (2011) Distant Mimivirus relative with a larger genome highlights the fundamental features of Megaviridae. Proc Natl Acad Sci USA 108:17486–17491

Assis FL, Bajrai L, Abrahao JS, Kroon EG, Dornas FP, Andrade KR, Boratto PVM, Pilotto MR, Robert C, Benamar S, La Scola B, Colson P (2015) Pan-genome analysis of Brazilian lineage A amoebal Mimiviruses. Viruses 7:3483–3499

Bamford DH, Ravantti JJ, Rönnholm G, Laurinavičius S, Kukkaro P, Dyall-Smith M, Somerharju P, Kalkkinen N, Bamford JKH (2005) Constituents of SH1, a novel lipid-containing virus infecting the halophilic euryarchaeon *Haloarcula hispanica*. J Virol 79:9097–9107

Barns SM, Delwiche CF, Palmer JD, Pace NR (1996) Perspectives on archaeal diversity, thermophily and monophyly from environmental rRNA sequences. Proc Natl Acad Sci USA 93:9188–9193

Bath C, Dyall-Smith ML (1998) His1, an Archaeal Virus of the *Fuselloviridae* family that infects *Haloarcula hispanica*. J Virol 72:9392–9395

Bell PJL (2001) Viral eukaryogenesis: was the ancestor of the nucleus a complex DNA virus? J Mol Evol 53:251–256

Bell PJL (2009) The viral eukaryogenesis hypothesis. Ann N Y Acad Sci 1178:91–105

Boratto PVM, Arantes TS, Silva LCF, Assis FL, Kroon EG, La Scola B, Abrahao JS (2015) Niemeyer virus: a new Mimivirus group A isolate harboring a set of duplicated aminoacyl-tRNA synthetase genes. Front Microbiol 6:1256

Boyer M, Yutin N, Pagnier I, Barrassi L, Fournous G, Espinosa L, Robert C, Azza S, Sun S, Rossmann MG, Suzan-Monti M, La Scola B, Koonin EV, Raoult D (2009) Giant Marseillevirus highlights the role of amoebae as a melting pot in emergence of chimeric microorganisms. Proc Natl Acad Sci USA 106:21848–21853

Brochier-Armanet C, Boussau B, Gribaldo S, Forterre P (2008) Mesophilic Crenarchaeota: proposal for a third archaeal phylum, the Thaumarchaeota. Nat Rev Microbiol 6:245–252

Burgers PMJ (2009) Polymerase dynamics at the eukaryotic DNA replication fork. J Biol Chem 284:4041–4045

Campos RK, Boratto PV, Assis FL, Aguiar ERGR, Silva LCF, Albarnaz JD, Dornas FP, Trindade GS, Ferreira PP, Marques JT, Robert C, Raoult D, Kroon EG, La Scola B, Abrahao JS (2014) Samba virus: a novel Mimivirus from a giant rain forest, the Brazilian Amazon. Virol J 11:95

Cavalier-Smith T (1983) A 6-kingdom classification and a unified phylogeny. In: Schermmler W and Schenk HEA (ed) Endocytobiology II, edn. De Gruyter, Berlin, pp. 1027–1034

Cavalier-Smith T (2010) Origin of the cell nucleus, mitosis and sex: roles of intracellular coevolution. Biol Direct 5:7

Chaikeeratisak V, Nguyen K, Khanna K, Brilot AF, Erb ML, Coker JKC, Vavilina A, Newton GL, Buschauer R, Pogliano K, Villa E, Agard DA, Pogliano J (2017) Assembly of a nucleus-like structure during viral replication in bacteria. Science 355:194–197

Chatterjee A, Ali F, Bange D, Kondabagil K (2016) Isolation and complete genome sequencing of Mimivirus bombay, a giant virus in sewage of Mumbai, India. Genomics Data 9:1–3

Claverie JM (2006) Viruses take center stage in cellular evolution. Genome Biol 7:110

Colson P, de Lambellerie X, Fournous G, Raoult D (2012) Reclassification of giant viruses composing a fourth domain of life in the new order Megavirales. Intervirol 55:321–332

Colson P, Pagnier I, Yoosuf N, Fournous G, La Scola B, Raoult D (2013) "Marseilleviridae", a new family of giant viruses infecting amoebae. Arch Virol 158:915–920

Desnues C, La Scola B, Yutin N, Fournous G, Robert C, Azza S, Jardot P, Monteil S, Campocasso A, Koonin EV, Raoult D (2012) Provirophages and transpovirons as the diverse mobilome of giant viruses. Proc Natl Acad Sci USA 109:18078–18083

Doolittle WF, Brown JR (1994) Tempo, mode, the progenote, and the universal root. Proc Natl Acad Sci USA 91:6721–6728

Dos Santos RN, Campos FS, de Albuquerque NRM, Finoketti F, Correa RA, Cano-Ortiz L, Assis FL, Arantes TS, Roehe PM, Franco AC (2016) A new marseillevirus isolated in Southern Brazil from *Limnoperna fortunei*. Sci Rep 6:35237

Doutre G, Philippe N, Abergel C, Claverie JM (2014) Genome analysis of the first Marseilleviridae representative from Australia indicates that most of its genes contribute to virus fitness. J Virol 88:14340–14349

Edgell DR, Malik S-B, Doolittle WF (1998) Evidence of independent gene duplications during the evolution of archaeal and eukaryotic family B DNA polymerases. Mol Biol Evol 15:1207–1217

Elkins JG, Podar M, Graham DE, Makarova KS, Wolf Y, Randau L, Hedlund BP, Brochier-Armanet C, Kunin V, Anderson I, Lapidus A, Goltsman E, Barry K, Koonin EV, Hugenholtz P, Kyrpides N, Wanner G, Richardson P, Keller M, Stetter KO (2008) A korarchaeal genome reveals insights into the evolution of the Archaea. Proc Natl Acad Sci USA 105:8102–8107

Embley TM, Martin W (2006) Eukaryotic evolution, changes and challenges. Nature 440:623–630

Fischer MG, Allen MJ, Wilson WH, Suttle CA (2010) Giant virus with a remarkable complement of genes infects marine zooplankton. Proc Natl Acad Sci USA 107:19508–19513

Forterre P, Prangishvili D (2009) The great billion-year war between ribosome- and capsid-encoding organisms (cells and viruses) as the major source of evolutionary novelties. Ann NY Acad Sci 1178:65–77

Forterre P, Gaïa M (2016) Giant viruses and the origin of modern eukaryotes. Curr Opin Microbiol 31:44–49

Fuerst JA, Sagulenko E (2012) Keys to eukaryality: planctomycetes and ancestral evolution of cellular complexity. Front Microbiol 3:167

Guy L, Ettema TJ (2011) The archaeal 'TACK' superphylum and the origin of eukaryotes. Trends Microbiol 19:580–587

Hage AE, Tollervey D (2004) A surfeit of factors: why is ribosome assembly so much more complicated in eukaryotes than bacteria? RNA Biol 1:10–15

Hallam SJ, Konstantinidis KT, Putnam N, Schleper C, Watanabe Y, Sugahara J, Preston C, de la Torre J, Richardson PM, DeLong EF (2006) Genomic analysis of the uncultivated marine crenarchaeote *Cenarchaeum symbiosum*. Proc Natl Acad Sci USA 103:18296–18301

Häring M, Peng X, Brügger K, Rachel R, Stetter KO, Garrett RA, Prangishvili D (2004) Morphology and genome organization of the virus PSV of the hyperthermophilic archaeal genera *Pyrobaculum* and *Thermoproteus*: a novel virus family, the *Globuloviridae*. Virol 323:233–242

Häring M, Vestergaard G, Rachel R, Chen LM, Garrett RA, Prangishvili D (2005a) Virology: independent virus development outside a host. Nature 436:1101–1102

Häring M, Rachel R, Peng X, Garrett RA, Prangishvili D (2005b) Viral diversity in hot springs of Pozzuoli, Italy, and characterization of a unique archaeal virus, *acidianus* bottle-shaped virus, from a new family, the *Ampullaviridae*. J Virol 79:9904–9911

Horiike T, Hamada K, Kanaya S, Shinozawa T (2001) Origin of eukaryotic cell nuclei by symbiosis of archaea in bacteria is revealed by homology-hit analysis. Nat Cell Biol 3:210–214

Hübscher U, Maga G, Spadari S (2002) Eukaryotic DNA polymerases. Annu Rev Biochem 71:133–163

Iyer LM, Aravind L, Koonin EV (2001) Common origin of four diverse families of large eukaryotic DNA viruses. J Virol 75:11720–11734

Iyer LM, Balaji S, Koonin EV, Aravind L (2006) Evolutionary genomics of nucleo-cytoplasmic large DNA viruses. Virus Res 117:156–184

Janekovic D, Wunderl S, Holz I, Zillig W, Gierl A, Neumann H (1983) TTV1, TTV2 and TTV3, a family of viruses of the extremely thermophilic, anaerobic, sulfur reducing archaebacterium *Thermoproteus tenax*. Mol Gen Genet 192:39–45

Kaniak-Golik A, Skoneczna A (2015) Mitochondria-nucleus network for genome stability. Free Rad Biol Med 82:73–104

King AM, Adams MJ, Carstens EB, Lefkowitz EJ (eds) (2012) Virus taxonomy: classification and nomenclature of viruses: ninth report of the International Committee on taxonomy of viruses. Elsevier, London-Waltham-San Diego

Könneke M, Bernhard AE, de la Torre JR, Walker CB, Waterbury JB, Stahl DA (2005) Isolation of an autotrophic ammonia-oxidizing marine archaeon. Nature 437:543–546

Kuznetsov Y, Klose T, Rossmann M, McPherson A (2013) Morphogenesis of Mimivirus and its viral factories: an atomic force microscopy study of infected cells. J Virol 87:11200–11213

Lake JA, Henderson E, Oakes M, Clark MW (1984) Eocytes: a new ribosome structure indicates a kingdom with a close relationship to eukaryotes. Proc Natl Acad Sci USA 81:3786–3790

Lake JA, Jain R, Rivera MC (1999) Mix and match in the tree of life. Science 283:2027–2028

La Scola B, Audic S, Robert C, Jungang L, de Lamballerie X, Drancourt M, Birtles R, Claverie JM, Raoult D (2003) A giant virus in Amoebae. Science 299:2033

La Scola B, Desnues C, Pagnier I, Robert C, Barrassi L, Fournous G, Merchat M, Suzan-Monti M, Forterre P, Koonin E, Raoult D (2008) The virophage as a unique parasite of the giant Mimivirus. Nature 455:100–104

Legendre M, Bartoli J, Shmakova L, Jeudy S, Labadie K, Adrait A, Lescot M, Poirot O, Bertaux L, Bruley C, Coute Y, Rivkina E, Abergel C, Claverie JM (2014) Thirty-year-old distant relative of giant icosahedral DNA viruses with a pandoravirus morphology. Proc Natl Acad Sci USA 111:4274–4279

Legendre M, Lartigue A, Bertaux L, Jeudy S, Bartoli J, Lescot M, Alempic JM, Ramus C, Bruley C, Labadie K, Shmakova L, Rivkina E, Coute Y, Abergel C, Claverie JM (2015) In-depth study of Mollivirus sibericum, a new 30,000-y-old giant virus infecting Acanthamoeba. Proc Natl Acad Sci USA 112:E5327–E5335

Margulis L (1993) Symbiosis in cell evolution, 2nd edn. Freeman, New York

Margulis L (1996) Archaeal–eubacterial mergers in the origin of eukarya: phylogenetic classification of life. Proc Natl Acad Sci USA 93:1071–1076

Martin A, Yeats S, Janekovic D, Reiter WD, Aicher W, Zillig W (1984) SAV 1, a temperate uv-inducible DNA virus-like particle from the archaebacterium *Sulfolobus acidocaldarius* isolate B12. EMBO J 3:2165

Martin W (2005) Archaebacteria (Archaea) and the origin of the eukaryotic nucleus. Curr Opin Microbiol 8:630–637

Martin W, Müller M (1998) The hydrogen hypothesis for the first eukaryote. Nature 392:37–41

Martin W, Koonin EV (2006) Introns and the origin of nucleus-cytosol compartmentalization. Nature 440:41–45

Mochizuki T, Yoshida T, Tanaka R, Forterre P, Sako Y, Prangishvili D (2010) Diversity of viruses of the hyperthermophilic archaeal genus Aeropyrum, and isolation of the Aeropyrum pernix bacilliform virus 1, APBV1, the first representative of the family Clavaviridae. Virol 402:347–354

Mochizuki T, Krupovic M, Pehau-Arnaudet G, Sako Y, Forterre P, Prangishvili D (2012) Archaeal virus with exceptional virion architecture and the largest single-stranded DNA genome. Proc Natl Acad Sci USA 109:13386–13391

Monier A, Claverie JM, Ogata H (2008) Taxonomic distribution of large DNA viruses in the sea. Genome Biol 9:R106

Moreira D, López-García P (1998) Symbiosis between methanogenic archaea and delta-proteobacteria as the origin of eukaryotes: the syntrophic hypothesis. J Mol Evol 47:517–530

Mutsafi Y, Fridmann-Sirkis Y, Milrot E, Hevroni L, Minsky A (2014) Infection cycles of large DNA viruses: emerging themes and underlying questions. Virol 466–467:3–14

Ngounga T, Pagnier I, Reteno DGI, Raoult D, La Scola B, Colson P (2013) Real-time PCR systems targeting giant viruses of Amoebae and their virophages. Intervirol 56:413–423

Nunoura T, Takaki Y, Kakuta J, Nishi S, Sugahara J, Kazama H, Chee GJ, Hattori M, Kanai A, Atomi H, Takai K, Takami H (2011) Insights into the evolution of Archaea and eukaryotic protein modifier systems revealed by the genome of a novel archaeal group. Nucleic Acids Res 39:3204–3223

Ohyanagi H, Ikeo K, Gojobori T (2008) The origin of nucleus: rebuild from the prokaryotic ancestors of ribosome export factors. Gene 423:149–152

Petitjean C, Deschamps P, Lopez-Garcia P, Moreira D (2015) Rooting the domain Archaea by phylogenomic analysis supports the foundation of the new kingdom proteoarchaeota. Genome Biol Evol 7:191–204

Philippe N, Legendre M, Doutre G, Coute Y, Poirot O, Lescot M, Arslan D, Seltzer V, Bertaux L, Bruley C, Garin J, Claverie JM, Abergel C (2013) Pandoraviruses: amoeba viruses with genomes up to 2.5 Mb reaching that of parasitic eukaryotes. Science 341:281–286

Pietilä MK, Roine E, Paulin L, Kalkkinen N, Bamford DH (2009) An ssDNA virus infecting archaea: a new lineage of viruses with a membrane envelope. Mol Microb 72:307–319

Pina M, Bize A, Forterre P, Prangishvili D (2011) The archeoviruses. FEMS Microbiol Rev 35:1035–1054

Prangishvili D (2013) The wonderful world of archaeal viruses. Ann Rev Microbiol 67:565–585

Prangishvili D, Garrett RA, Koonin EV (2006) Evolutionary genomics of archaeal viruses: unique viral genomes in the third domain of life. Virus Res 117:52–67

Prangishvili D, Basta T, Garrett RA, Krupovic M (2016) Viruses of the Archaea. eLS 1–9

Preston CM, Wu KY, Molinski TF, DeLong EF (1996) A psychrophilic crenarchaeon inhabits a marine sponge: *Cenarchaeum symbiosum* gen. nov., sp. nov. Proc Natl Acad Sci USA 93:6241–6246

Raoult D, Audic S, Robert C, Abergel C, Renesto P, Ogata H, La Scola B, Suzan M, Claverie JM (2004) The 1.2-megabase genome sequence of Mimivirus. Science 306:1344–1350

Rensen EI, Mochizuki T, Quemin E, Schouten S, Krupovic M, Prangishvili D (2016) A virus of hyperthermophilic archaea with a unique architecture among DNA viruses. Proc Natl Acad Sci USA 113:2478–2483

Reteno DG, Benamar S, Khalil JB, Andreani J, Armstrong N, Klose T, Rossmann M, Colson P, Raoult D, La Scola B (2015) Faustovirus, an asfarvirus-related new lineage of giant viruses infecting Amoebae. J Virol 89:6585–6594

Rice G, Stedman K, Snyder J, Wiedenheft B, Willits D, Brumfield S, McDermott T, Young MJ (2001) Viruses from extreme thermal environments. Proc Natl Acad Sci USA 98:13341–13345

Schramm B, Locker JK (2005) Cytoplasmic organization of POXvirus DNA replication. Traffic 6:839–846

Seitz KW, Lazar CS, Hinrichs KU, Teske AP, Baker BJ (2016) Genomic reconstruction of a novel, deeply branched sediment archaeal phylum with pathways for acetogenesis and sulfur reduction. ISME J 10:1696–1705

Snyder JC, Bolduc B, Young MJ (2015) 40 Years of archaeal virology: expanding viral diversity. Virol 479–480:369–378

Spang A, Saw JH, Jørgensen SL, Zaremba-Niedzwiedzka K, Martijn J, Lind AE, van Eijk R, Schleper C, Guy L, Ettema TJ (2015) Complex archaea that bridge the gap between prokaryotes and eukaryotes. Nature 521:173–179

Stanier RY (1970) Some aspects of the biology of cells and their possible evolutionary significance. In: Charles HP, Knight BCJG (eds) Organization and control in prokaryotic and eukaryotic cells. Cambridge University Press, Cambridge, pp 1–38

Staub E, Fiziev P, Rosenthal A, Hinzmann B (2004) Insights into the evolution of the nucleolus by an analysis of its protein domain repertoire. BioEssays 26:567–581

Suzan-Monti M, La Scola B, Barrassi L, Espinosa L, Raoult D (2007) Ultrastructural characterization of the giant volcano-like virus factory of Acanthamoeba polyphaga Mimivirus. PLoS ONE 3:e328

Tahirov TH, Makarova KS, Rogozin IB, Pavlov YI, Koonin EV (2009) Evolution of DNA polymerases: an inactivated polymerase-exonuclease module in Pol ε and a chimeric origin of eukaryotic polymerases from two classes of archaeal ancestors. Biol Direct 4:11

Takemura M (2001) Poxviruses and the origin of the eukaryotic nucleus. J Mol Evol 52:419–425

Takemura M (2011) Function of DNA polymerase α in a replication fork and its putative roles in genomic stability and eukaryotic evolution. In: Kušić-Tišma (ed) Fundamental aspects of DNA replication, InTech-Open Access Publisher, pp 187–204

Takemura M, Yokobori S, Ogata H (2015) Evolution of eukaryotic DNA polymerases via interaction between cells and large DNA viruses. J Mol Evol 81:24–33

Takemura M (2016a) The draft genome sequence of *Tokyovirus*, a member of *Marseilleviridae* isolated from the Arakawa River of Tokyo, Japan. Genome Announc 4:e00429–16

Takemura M (2016b) Morphological and taxonomic properties of Tokyovirus, the first Marseilleviridae member isolated from Japan. Microbes Environ 31:442–448

Takemura M, Mikami T, Murono S (2016) Nearly complete genome sequences of two Mimivirus strains isolated from Japanese fresh water pond and river mouth. Genome Announc 4:e01378–16

Tamura K, Stecher G, Peterson D, Filipski A, Kumar S (2013) MEGA6: molecular evolutionary genetics analysis version 6.0. Mol Biol Evol 30:2725–2729

Thomas V, Bertelli C, Collyn F, Casson N, Telenti A, Goesmann A, Croxatto A, Greub G (2011) Lausannevirus, a giant amoebal virus encoding histone doublets. Environ Microbiol 13:1454–1466

Tolonen N, Doglio L, Schleich S, Locker JK (2001) Vaccinia virus DNA replication occurs in endoplasmic reticulum-enclosed cytoplasmic mini-nuclei. Mol Biol Cell 12:2031–2046

Wang TSF (1991) Eukaryotic DNA polymerases. Annu Rev Biochem 60:513–552

Woese CR, Fox GE (1977) Phylogenetic structure of the prokaryotic domain: the primary kingdoms. Proc Natl Acad Sci USA 74:5088–5090

Woese CR, Kandler O, Wheelis ML (1990) Towards a natural system of organisms: proposal for the domains Archaea, Bacteria, and Eucarya. Proc Natl Acad Sci USA 87:4576–4579

Wood AG, Whitman WB, Konisky J (1989) Isolation and characterization of an archaebacterial viruslike particle from *Methanococcus-voltae* A3. J Bacteriol 171:93–98

Yoosuf N, Yutin N, Colson P, Shabalina SA, Pagnier I, Robert C, Azza S, Klose T, Wong J, Rossmann MG, La Scola B, Raoult D, Koonin EV (2012) Related giant viruses in distant locations and different habitats: Acanthamoeba polyphaga moumouvirus represents a third lineage of the Mimiviridae that is close to the Megavirus lineage. Genome Biol Evol 4:1324–1330

Zaremba-Niedzwiedzka K, Caceres EF, Saw JH, Bäckström D, Juzokaite L, Vancaester E, Seitz KW, Anantharaman K, Starnawski P, Kjeldsen KU, Stott MB, Nunoura T, Banfield JF, Schramm A, Baker BJ, Spang A, Ettema TJ (2017) Asgard archaea illuminate the origin of eukaryotic cellular complexity. Nature 541:353–358

Zillig W, Kletzin A, Schleper C, Holz I, Janekovic D, Hain J, Lanzendörfer M, Kristjansson JK (1993) Screening for Sulfolobales, their plasmids and their viruses in Icelandic solfataras. Syst Appl Microbiol 16:609–628

Inteins as Indicators of Bio-Communication

Shannon M. Soucy and J. Peter Gogarten

1 Introduction to Inteins

Inteins are selfish genetic elements that are found in all three domains of life and also in viruses (Pietrokovski 2001; Gogarten et al. 2002; Perler 2002; Swithers et al. 2009; Soucy et al. 2014). Inteins often are found in multiple sites within a single gene. The invaded gene that encodes the host protein is called an extein (Fig. 1a). Inteins are named after the gene they invade, followed by a letter indicating the insertion site. For example, *Hvo polB*-c denotes the intein located in insertion site c of the gene in *Haloferax volcanii* encoding DNA polymerase B. Each intein insertion site is referred to as an intein allele, though this term is misleading, as phylogenetic analysis of all inteins in the Halobacteria (colloquially referred to as the Haloarchaea) shows strong support for an independent origin of each intein allele (Soucy et al. 2014); *i.e.*, inteins inserted into a same site of the gene are much more similar to one another than to inteins occupying different insertion sites in the same gene. The letters denoting the intein allele are given in order of the discovery of the intein insertion site, and not necessarily in the order they are found in the gene. For example, the ribonucleotide reductase gene has 5 insertion sites in the Haloarchaea (Soucy et al. 2014); the order they appear in the gene is: *rir*-1 l, *rir*-1 k, *rir*-1b, *rir*-1 g, and *rir*-1 m. Inteins inserted into the other sites (e.g., *rir*-1a) so far have not been described as occupied in Haloarchaea.

S.M. Soucy (✉)
Dartmouth College, 78 College St, Hanover NH 03755, USA
e-mail: Shannon.Soucy@dartmouth.edu

J.P. Gogarten
University of Connecticut, 91 North Eagleville Rd, Storrs CT 06250, USA
e-mail: Gogarten@uconn.edu

G. Witzany (ed.), *Biocommunication of Archaea*,
DOI 10.1007/978-3-319-65536-9_16

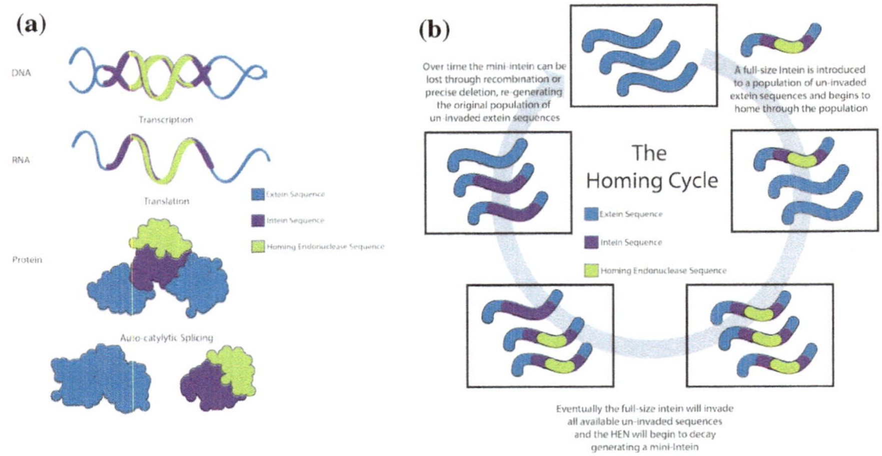

Fig. 1 Overview of intein structure and function (Panel A) and the homing cycle (A)

1.1 Intein Distribution and Patterns in Extein Functions

Inteins are found in the most conserved parts of the most conserved proteins (Swithers et al. 2009), and most often in genes related to replication, recombination, and repair (COG category L) and to a lesser extent in other genes involved in nucleotide metabolism (COG category F) (Novikova et al. 2015). Many of the invaded genes encode proteins that possess nucleotide binding sites (Gogarten et al. 2002). In Archaea, inteins were mainly found in the Euryarchaeota (68% of archaeal inteins) and Crenarchaeota (34% of archaeal inteins) (Novikova et al. 2015). At present debate continues regarding the question, if the observed intein distribution directly reflects the function of the genes, or only the between species conservation of the invaded genes. Different processes have been described to explain the observed distribution of inteins:

1. All sites in all genes are targeted by inteins; however, only genes and insertions that do not tolerate substitutions or small deletions without affecting important cellular functions (replication, recombination, repair, ATP synthesis and nucleotide metabolism) allow the inteins to persist over long periods of time (Gogarten et al. 2002; Swithers et al. 2009). In less important sites, small deletions may inactivate and remove the intein.
2. Inteins have evolved to target conserved sites in conserved genes because these sites are likely to have a similar sequence in an orthologous gene even in other, distantly related species (Swithers et al. 2009). This allows inteins to invade homologous genes present in other species, even across phyla and domain boundaries (Swithers et al. 2013).
3. The splicing activity of some inteins was found to be sensitive to redox and other conditions in the cell, and it was recently proposed that inteins could

respond to the altered redox state during cellular stress, effectively pausing important cellular functions until conditions return to normal (Topilina et al. 2015; Novikova et al. 2015); however, the often limited within species distribution of inteins (Soucy et al. 2014; Naor et al. 2016) argues against inteins providing a strong selective advantage.

More work needs to be done to investigate the extent to which these three possible processes gave rise to the observed distribution of inteins.

1.2 Intein Symbiotic State and the Homing Cycle

Most inteins are composed of two domains, a homing endonuclease domain, and a splicing domain which flanks the homing endonuclease domain on either side (Figs. 1a and 2). The splicing domain is the main component of the intein, this domain removes the intein sequence from the host protein sequence, also called extein, after translation. The homing endonuclease domain is considered an accessory domain, and is sometimes absent from intein sequences, these inteins are called mini-inteins (Derbyshire et al. 1997); inteins with homing endonuclease domains are called full-sized inteins (Soucy et al. 2014). The homing endonuclease domain initiates the conversion of uninvaded alleles into intein containing alleles.

The canonical model of intein invasion and loss, called the homing cycle (Goddard and Burt 1999), starts with an intein containing allele encountering an uninvaded allele. The homing endonuclease makes a double-strand break at the

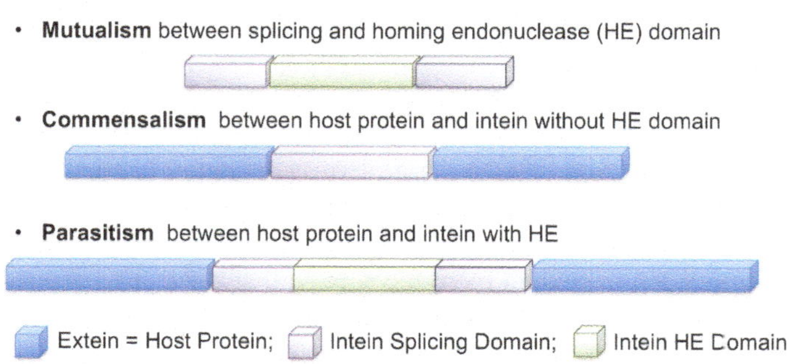

Fig. 2 Types of symbiotic relationships between intein domains and the host protein. Most inteins contain two protein domains with distinct functions: the splicing domain, which removes the intein from the host protein after translation, and the homing endonuclease (HE) domain. In the linear sequence of the gene the HE domain is inserted inside the sequence encoding the splicing domain; however, in the folded protein the splicing domain from a compact protein structure, distinct from the HE domain. The splicing domain prevents a high fitness cost due to interruption of the host protein, the HE domain provides mobility to the intein, because it allows invasion of uninvaded alleles. Inteins that have lost the HE domain (*middle*) have a commensal relationship to the host protein, whereas inteins with HE activity can be considered parasites of the host protein and the host organism (Gogarten and Hilario 2006)

invasion site specific for the homing endonuclease. Using the invaded allele as a template the host DNA repair machinery copies the intein into the invasion site, thereby interrupting the invasion site and preventing future homing endonuclease activity (Fig. 1b). The homing cycle asserts that the invasion of a population goes to completion, and that subsequently homing endonuclease activity decays due to a lack of selection for function, because the opportunity to invade a new allele no longer exists. Recent findings revealed limitations of the homing cycle model. The different versions of the intein insertion site are in an intransitive fitness relationship (Barzel et al. 2011): The HE converts empty target sites, mini inteins are smaller than inteins with HE and therefore are expected to place less of a fitness burden on the host, and organisms with empty target sites may outcompete carriers of mini-inteins for the same reason. Theoretical studies showed that a wide range of fitness relationships between the three states of the intein insertion site allow for their coexistence over long periods of time even in well mixed populations (Yahara et al. 2009; Barzel et al. 2011). These theoretical considerations suggest that invasion might never go to completion and that some empty insertion sites might remain in the population. The persistence of empty target sites at low frequencies also solves the enigma of how the intein encoding part of a gene can be lost by a precise deletion—the last step of the homing cycle. In case of introns, a precise deletion may occur via recombination with a processed mRNA (Jeffares et al. 2006); however, in case of inteins the splicing reaction only occurs after translation, and how this information could flow back from the protein to DNA level remains enigmatic. The precise deletion through recombination with an intein free homolog, either persisting in the population at low copy numbers, or acquired through horizontal gene transfer provides a solution to this puzzle. The theoretical considerations also are in agreement with finding empty target sites and inteins with functioning homing endonuclease in the same population (Soucy et al. 2014; Naor et al. 2016) and with the persistence of functioning homing endonucleases in lineages over long periods of time (Gogarten and Hilario 2006; Butler et al. 2006).

In either of the above scenarios of intein invasion an intein without homing endonuclease, a mini-intein, reflects a long-term association of the intein with the host protein via vertical transmission from mother to daughter cells. Furthermore, since intein invasion relies on the presence of a functioning homing endonuclease, a full-size intein invaded the extein more recently so that a selective pressure existed to maintain the homing endonuclease domain. Therefore, the presence of an intact homing endonuclease is a sign of horizontal gene transfer (HGT) in the recent (evolutionary time) history of the organism. The presence of a homing endonuclease reflects the symbiotic state of the intein (commensal or parasite, Fig. 2), and this symbiotic state provides a marker for the evolutionary history of that intein (Soucy et al. 2014). Curiously, the substitution rate of inteins are much higher than the adjacent extein sequence (Swithers et al. 2009). Except for the terminal residues required for splicing activity, the intein sequence accumulates substitutions at a high rate, leading to decay of the homing endonuclease sequence. This elevated substitution rate also creates informative characters, providing resolution to phylogenies, even between closely related species. The high substitution rate combined with the symbiotic state of inteins makes them valuable as markers of horizontal gene transfer, as this information provides information about which organisms were sharing genes (Swithers et al. 2013; Soucy et al. 2014).

1.3 Inteins as Tools to Trace Gene Flow

Genes acquired through horizontal gene transfer can have several selective outcomes, they can be beneficial to the recipient, they may be nearly neutral, or they may be detrimental. Undoubtedly the effect on the fitness of the host has an impact on the survival of the transferred gene in the recipient lineage. Therefore, an ideal tracer for naturally occurring gene transfer frequencies and patterns should not have an impact on the fitness of the recipient. The intein's self-splicing activity insures that the host protein remains functional, thereby avoiding a strong negative impact on the impact on the fitness of the host organism.

Transferred genes also need to integrate into the recipient genome, which can occur either through homologous recombination, or as an additive transfer. A good tracer should integrate into the recipient genome with a constant probability. Inteins force integration into the recipient genome through cutting the target site in the recipient genome. Another intein feature useful to reveal patterns of naturally occurring gene transfer is that inteins do not possess their own machinery to catalyze the transfer between cells. Therefore, their transfer provides a measure for the genetic exchange occurring in the host cells.

A potential problem is that inteins can only invade if an empty target site is present. A target site that is already occupied cannot be invaded again. However, observation shows that even within populations of the same species, empty target sites frequently persist (Soucy et al. 2014; Naor et al. 2016). Another concern is that full-size inteins can increase mating and recombination frequency of organisms that carry them (Giraldo-Perez and Goddard 2013; Naor et al. 2016); however, this does not detract for the inteins tracing transfer, even though the transfer rate maybe higher in the presence of the intein than in its absence.

2 Inteins as Markers of Bio-Communication

2.1 Inteins as Markers of Bio-Communication in Closely Related Organisms

Part of the intein's life cycle (see Sect. 1.2 and Fig. 1) involves transmission to new hosts via horizontal gene transfer. Therefore, we can use the distribution of inteins to identify HGT events between both closely related and more distantly related organisms (Swithers et al. 2013; Soucy et al. 2014; Fullmer et al. 2014;

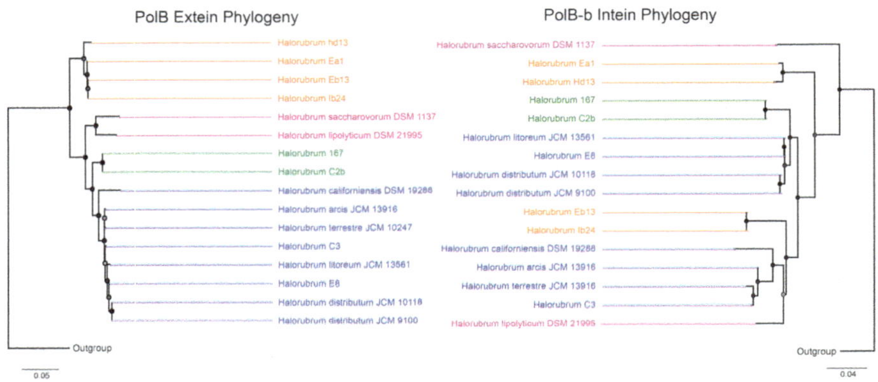

Fig. 3 Comparison of the extein and intein evolutionary histories. The tree on the left gives the reconstructed phylogeny of the DNA polymerase B extein in the *Halorubrum* genus, the tree on the right gives the reconstructed phylogeny of the intein. Sequences were aligned using Muscle v.3.8.31(Edgar 2004), and phylogenies were constructed using FastTree (Price et al. 2010). Strains and species names are given in the same color/grayscale value for both trees. The strain order in the extein phylogeny is strikingly different than in the intein phylogeny, indicating a plethora of horizontal gene transfer occurring within the Halorubrum

Naor et al. 2016). Especially between closely related organisms with high sequence similarity, gene transfer is difficult to detect using phylogenetic approaches. In these cases, the fast evolving intein sequences can serve as markers of bio-communication between organisms (inter organismic communication; Witzany 2016). Most intein alleles are shared by closely related species. In a 2014 survey of inteins in Haloarchaea, six out of 24 inteins were found exclusively in the Haloarchaea (i.e., they had not been found outside the Haloarchaea), and 16 out of 24 had the majority of sequences in the Haloarchaea (Soucy et al. 2014). Seven out of the ten inteins that had a majority in the Haloarchaea, but were shared with taxa outside the group were shared predominately with other Euryarchaeota (Soucy et al. 2014). This indicates that most inteins, at least in the Haloarchaea, are shared within phyla. In a similar analysis comparing intein and extein phylogenies for the *vma1*-b Swithers et al. showed that the intein in *Thermococcus litoralis* had been acquired from within the *Pyrococcus,* another genus within the Thermococcaceae (Swithers et al. 2013). Even more striking is the phylogenetic conflict between *polB* extein and *polB*-b intein in different *Halorubrum* strains, revealing transfer within the *Halorubrum* genus (Fig. 3), and the frequent transfer of the *pol II*-a between strains belonging to the *Halorubrum* genus and other Haloarchaea (Fig. 4). It is noteworthy that even though the *polB*-b and *pol II*-a inteins were frequently transferred within Halorubrum and between *Halorubrum* and other Haloarchaea, these inteins were present in only 23 and 14 out of 37 *Halorubrum* genomes, respectively.

Fig. 4 Maximum likelihood phylogeny of the intein in the large subunit of the DNA polymerase II, insertion site a (*pol II*-a). The phylogeny was calculated with PhyML (Guindon et al. 2010) using the LG, Gamma + I model and sequences as described in (Soucy et al. 2014). Branches with a support values (Anisimova and Gascuel 2006) below 0.85 are shown in *gray*. Sequences are as described in (Soucy et al. 2014). Font colors/grayscale denote different groups of archaea. Intein sequences from the genus *Halorubrum* group in several distinct groups within the Halobacteria and these groups, indicated by lines next to the strain names, are separated by well supported branches. One intein from a Methanomicrobion (*Methanoregula formicica* SMSP) groups within one of these *Halorubrum* clusters, whereas the other *pol II*-a inteins from Methanomicrobia group as a sistergroup to the halobacterial homologs, as is expected from the relationship between Haloarchaea and Methanomicrobia based on the phylogeny of the translation machinery (Woese 1987; Lasek-Nesselquist and Gogarten 2013)

2.2 Detecting the Direction of Bio-Communication

Phylogenetic comparison between intein and extein phylogenies can provide clear indications that the intein has been transferred (Figs. 3 and 4); however, conflict between intein phylogeny and that of the host or host protein often is insufficient to determine the direction of transfer. For example, from the phylogenies depicted in Fig. 4 it is not always clear when a Halorubrum species was the donor and when the recipient of an intein transfer; however, the fact that the inteins from Halorubrum group in 5 well separated locations in this phylogeny reveals repeated transfer of this intein between Haloarchaeal genera. Sometimes the tree topology does provide a clue: the single intein from a methanomicrobia grouping within a group of inteins from *Halorubrum* strains, strongly suggest that this methanomicrobion, *Methanoregula formicica*, acquired the intein through genetic exchange with a Haloarchaeon, likely a member of the genus *Halorubrum* (Fig. 4). The comparison between *polB*-b intein and *polB* extein phylogenies (Fig. 3) suggests that the ancestor of *Halorubrum* Eb13 and Ib24 may have acquired the intein through transfer from the ancestor of *Hrr. californiensis*, *Hrr. arcis*, *Hrr. terrestre* and *Hrr. sp* C3. Similarly, the grouping of the *Thermococcus litoralis vma1*-b intein within a group of homologous inteins from several *Pyrococcus* species reveals *T. litoralis* as the recipient of the transfer (Swithers et al. 2013). A more sophisticated analysis involving the ratio of the pairwise phylogenetic distance between intein sequences and a reference sequence (the associated extein can sometimes serve as a reference) can help to infer additional transfers and to determine the direction of transfer (Swithers et al. 2013). The underlying rational is that if two intein sequence are more similar to each other than expected from the similarity between the two corresponding extein sequences, then the intein was likely acquired through horizontal transfer and not only through vertical inheritance. However, this approach requires calibration of substitution rates for inteins that have been vertically transferred. Swithers et al. used a mini-intein to calibrate their dataset, but mini-inteins are not available for many intein datasets. More needs to be done to develop this method to determine and study gene transfer events between close relatives.

2.3 Inteins as Markers of Bio-Communication in Distantly Related Organisms

Perhaps more interesting than gene transfers between closely related organisms are transfer events that involve distantly related organisms. Many questions about the boundaries of gene transfer remain to be answered. For transfers between distantly related lineages the direction of transfer can often be inferred from the distribution of the intein. If there is a clear majority of close relatives of one of the partners, it is likely that organism is the donor. In the Haloarchaea 14 of 24 inteins are shared with at least one bacterial species. Surprisingly the shared inteins are not limited to other salt tolerant microbes but are found in bacteria from six different phyla (Soucy et al. 2014). Six of the 24 inteins found in the Haloarchaea are found in more bacterial species than archaeal species, indicating the intein may have been transferred from the bacteria and into the Haloarchaea. The other eight intein alleles that are shared between Bacteria and Haloarchaea have far more representatives in Haloarchaea than Bacterial, indicating that bio-communication is occurring in both directions. Thirteen out of 24 inteins in the Haloarchaea are shared with other Euryarchaeota (Soucy et al. 2014). The two inteins with the highest number of other Euryarchaeota, cdc21-a and pol-IIa, also have the highest proportion of mini-inteins, implying these inteins have an ancient association with the euryarchaeal lineage (Soucy et al. 2014). The intein with the highest proportion of euryarchaeal sequences outside the Haloarchaea, *rfc*-a, also has several mini-inteins (Soucy et al. 2014). In contrast, 10 of the 14 inteins shared between Haloarchaea and Bacteria have no mini-inteins in the Haloarchaea, and few mini-inteins in the Bacteria (Soucy et al. 2014).

These findings support ancient and ongoing bio-communication through gene transfer with other Euryarchaeota and more recent bio-communication with bacterial species, supporting earlier evidence that haloarchaeal evolution included extensive gene transfers with the bacteria (Williams et al. 2012; Nelson-Sathi et al. 2012, 2014; Becker et al. 2014). No clear pattern has emerged regarding the types of genes that are invaded by inteins shared by Bacteria and Archaea, and those that are found only in Archaea. As with most inteins, most of the shared inteins fall into the replication, recombination, and repair COG category.

3 Conclusion

Inteins are valuable indicators of bio-communication through gene transfer within and between species. The phylogenetic information provided by inteins due to their high substitution rate has been used to identify gene transfer between organisms at all level of relationship: within populations, between related species belonging to the same order, between orders, phyla, and even between organisms from different domains. More needs to be done to integrate information of symbiotic state and substitution rates with the comparative analysis of topologies of gene and species phylogenies. To date, conclusions from using inteins as indicators of gene flow confirm results obtained from using whole genome data and individual gene phylogenies: Genes are transferred much more frequently between closely related organism than between divergent organisms (Andam and Gogarten 2011; Williams et al. 2012), but even transfers across domain boundaries have occurred, and some of these greatly impacted the physiology and ecology of the recipient organisms (Soucy et al. 2015).

References

Andam CP, Gogarten JP (2011) Biased gene transfer in microbial evolution. Nat Rev Microbiol 9:543–55. Research Support, U.S. Gov't, Non-P.H.S

Anisimova M, Gascuel O (2006) Approximate likelihood-ratio test for branches: a fast, accurate, and powerful alternative. Syst Biol 55:539–552. doi:10.1080/10635150600755453

Barzel A, Obolski U, Gogarten JP, Kupiec M, Hadany L (2011) Home and away-the evolutionary dynamics of homing endonucleases. BMC Evol Biol 11:324. doi:10.1186/1471-2148-11-324

Becker EA, Seitzer PM, Tritt A, Larsen D, Krusor M, Yao AI, Wu D, Madern D, Eisen JA, Darling AE, Facciotti MT (2014) Phylogenetically driven sequencing of extremely halophilic archaea reveals strategies for static and dynamic osmo-response. PLoS Genet 10:e1004784. doi:10.1371/journal.pgen.1004784

Butler MI, Gray J, Goodwin TJD, Poulter RTM (2006) The distribution and evolutionary history of the PRP8 intein. BMC Evol Biol 6:42. doi:10.1186/1471-2148-6-42

Derbyshire V, Wood DW, Wu W, Dansereau JT, Dalgaard JZ, Belfort M (1997) Genetic definition of a protein-splicing domain: functional mini-inteins support structure predictions and a model for intein evolution. Genetics 94:11466–11471

Edgar RC (2004) MUSCLE: multiple sequence alignment with high accuracy and high throughput. Nucleic Acids Res 32:1792–1797. doi:10.1093/nar/gkh340

Fullmer MS, Soucy SM, Swithers KS, Makkay AM, Wheeler R, Ventosa A, Gogarten JP, Papke RT (2014) Population and genomic analysis of the genus Halorubrum. Extrem Microbiol. doi:10.3389/fmicb.2014.00140

Giraldo-Perez P, Goddard MR (2013) A parasitic selfish gene that affects host promiscuity. Proc Biol Sci 280:20131875. doi:10.1098/rspb.2013.1875

Goddard MR, Burt A (1999) Recurrent invasion and extinction of a selfish gene. Proc Natl Acad Sci U S A 96:13880–13885

Gogarten JP, Hilario E (2006) Inteins, introns, and homing endonucleases: recent revelations about the life cycle of parasitic genetic elements. BMC Evol Biol 6:94. doi:10.1186/1471-2148-6-94

Gogarten JP, Senejani AG, Zhaxybayeva O, Olendzenski L, Hilario E (2002) Inteins: structure, function, and evolution. Annu Rev Microbiol 56:263–287. doi:10.1146/annurev.micro.56.012302.160741

Guindon S, Dufayard J-F, Lefort V, Anisimova M, Hordijk W, Gascuel O (2010) New algorithms and methods to estimate maximum-likelihood phylogenies: assessing the performance of PhyML 3.0. Syst Biol 59:307–321. doi:10.1093/sysbio/syq010

Jeffares DC, Mourier T, Penny D (2006) The biology of intron gain and loss. Trends Genet 22:16–22. doi:10.1016/j.tig.2005.10.006

Lasek-Nesselquist E, Gogarten JP (2013) The effects of model choice and mitigating bias on the ribosomal tree of life. Mol Phylogenet Evol 69:17–38. doi:10.1016/j.ympev.2013.05.006

Naor A, Altman-Price N, Soucy SM, Green AG, Mitiagin Y, Turgeman-Grott I, Davidovich N, Gogarten JP, Gophna U (2016) Impact of a homing intein on recombination frequency and organismal fitness. Proc Natl Acad Sci U S A 113:E4654–E4661. doi:10.1073/pnas.1606416113

Naor A, Lapierre P, Mevarech M, Papke RT, Gophna U (2012) Low species barriers in halophilic archaea and the formation of recombinant hybrids. Curr Biol 22:1444–1448. doi:10.1016/j.cub.2012.05.056

Nelson-Sathi S, Dagan T, Landan G, Janssen A, Steel M, McInerney JO, Deppenmeier U, Martin WF (2012) Acquisition of 1000 eubacterial genes physiologically transformed a methanogen at the origin of Haloarchaea. Proc Natl Acad Sci U S A 109:20537–20542. doi:10.1073/pnas.1209119109

Nelson-Sathi S, Sousa FL, Roettger M, Lozada-Chávez N, Thiergart T, Janssen A, Bryant D, Landan G, Schönheit P, Siebers B, McInerney JO, Martin WF (2014) Origins of major archaeal clades correspond to gene acquisitions from bacteria. Nature. doi:10.1038/nature13805

Novikova O, Jayachandran P, Kelley DS, Morton Z, Merwin S, Topilina NI, Belfort M (2015) Intein clustering suggests functional importance in different domains of life. Mol Biol Evol. doi:10.1093/molbev/msv271

Perler FB (2002) InBase: the intein database. Nucleic Acids Res 30:383–384

Pietrokovski S (2001) Intein spread and extinction in evolution. Trends Genet 17:465–472

Price MN, Dehal PS, Arkin AP (2010) FastTree 2–approximately maximum-likelihood trees for large alignments. PLoS ONE 5:e9490. doi:10.1371/journal.pone.0009490

Soucy SM, Fullmer MS, Papke RT, Gogarten JP (2014) Inteins as indicators of gene flow in the halobacteria. Front Microbiol 5:299. doi:10.3389/fmicb.2014.00299

Soucy SM, Jinling H, Gogarten JP (2015) Horizontal gene transfer: building the web of life

Swithers KS, Senejani AG, Fournier GP, Gogarten JP (2009) Conservation of intron and intein insertion sites: implications for life histories of parasitic genetic elements. BMC Evol Biol 9:303. doi:10.1186/1471-2148-9-303

Swithers KS, Soucy SM, Lasek-Nesselquist E, Lapierre P, Gogarten JP (2013) Distribution and evolution of the mobile vma-1b intein. Mol Biol Evol. doi:10.1093/molbev/mst164

Topilina NI, Novikova O, Stanger M, Banavali NK, Belfort M (2015) Post-translational environmental switch of RadA activity by extein–intein interactions in protein splicing. Nucleic Acids Res 43:6631–6648. doi:10.1093/nar/gkv612

Williams D, Gogarten JP, Papke RT (2012) Quantifying homologous replacement of loci between haloarchaeal species. Genome Biol Evol 4:1223–1244. doi:10.1093/gbe/evs098

Witzany G (2016) The biocommunication method: on the road to an integrative biology. Commun Integr Biol 9:e1164374. doi:10.1080/19420889.2016.1164374

Woese CR (1987) Bacterial evolution. Microbiol Rev 51:221–271

Yahara K, Fukuyo M, Sasaki A, Kobayashi I (2009) Evolutionary maintenance of selfish homing endonuclease genes in the absence of horizontal transfer. Proc Natl Acad Sci USA 106: 18861–18866. doi:10.1073/pnas.0908404106

Riboswitches: Regulatory ncRNAs in Archaea

D. Swati

1 Introduction

Non-coding RNAs have caught the attention of Molecular Biologists at last. Initially dismissed as mere messengers of information transfer from the DNA molecule to amino acids, in order to build proteins, with the advent of sophisticated next generation sequencing methods, microarrays and other experimental innovations and sophisticated algorithms for finding ncRNA in silico, ncRNAs finally become the focus of concerted study.

The Central Dogma of Crick has been turned on its head, as new types of RNA are discovered with varied structure and function. The most ancient of these ncRNA is thought to be the ribozyme, which is a catalytic RNA that cleaves and ligates the DNA molecule for the replication and transcription processes. These important processes related to information transfer, could not take place without the existence of Hammerhead (HHR) ribozymes, or Ribonuclease P (RnaseP) and RnaseA. A ribozyme performs its catalyzing process without the aid of protein factors (Seehafer et al. 2011).

Riboswitches, regulatory ncRNA elements are also thought to be relics of an RNA world, before the evolution of protein factors involved in regulation of gene expression. These ncRNAs are termed as "riboswitch" because they possess the ability to function as switches that regulate gene expression, by altering their conformation in response to a ligand or a small molecule. They are located in the untranslated regions of mRNAs. They contain a specific conserved ligand-binding domain labelled aptamer, along with a variable sequence, termed the expression platform which enables regulation of the downstream coding sequences. When the cellular concentration of a metabolite exceeds an optimum level, its binding to the riboswitch aptamer domain

D. Swati (✉)
Department of Physics and Bioinformatics, MMV, Banaras
Hindu University, Varanasi 221005, India
e-mail: swatid@gmail.com

G. Witzany (ed.), *Biocommunication of Archaea*,
DOI 10.1007/978-3-319-65536-9_17

induces a conformational change in the expression platform, leading to the modulation of downstream events. The type of conformational change in the expression platform is decided by the switching sequence, which usually lies between the aptamer and the expression platform. Riboswitches have become recognized as important and widespread elements in the control of gene expression in numerous Bacteria, Archaea, Plants, Fungi and Algae (Sudarsan et al. 2003). The first identifications of vitamin-specific riboswitches (Mironov et al. 2002; Winkler et al. 2002; Mandal et al. 2003) were followed by discoveries of many other riboswitch types (Rodionov et al. 2003; Nahvi et al. 2004; Nudler and Mironov 2004). Currently, riboswitches are known to sense purines and their derivatives, protein coenzymes, amino acids, and a phosphorylated sugar. Some riboswitches respond specifically to inorganic ligands, including metals (Mg^{2+} cations) (Cromie et al. 2006; Dann et al. 2007) that shield the negative charge of the sugar-phosphate backbone in RNA. There are specific ncRNA elements, earlier known as GEMM and ydaO, which have been shown fairly recently to control the concentration level of second messenger molecules, cyclic-di-guanosine monophosphate (c-di-GMP) (Sudarsan et al. 2008; Smith and Strobel 2011), and cyclic-di-adenosine monophosphate (c-di-AMP) (Romling 2008; Nelson et al. 2013; Ren and Patel 2014) respectively, whose presence and concentrations are critical for a bacteria's survival. The importance of these and other second messenger signaling molecules is being recognised to merit a conference dedicated to their research (Hengge et al. 2016).

1.1 Regulating ncRNAs: Riboswitches

The selectivity of a riboswitch depends upon the conserved aptameric domain in which the ligands bind. Riboswitches recognize chemically diverse ligands and do not possess a uniform metabolite recognition mechanism. The majority of riboswitches form tight binding pockets that are complementary to their cognate ligands, with small ligands completely encapsulated in such pockets. The pockets contain conserved nucleotides and non-canonical base pairs that belong to a widened irregular helix or to converging stems. The individual bases in these helices may vary due to mutations, but the base pairing is conserved over evolutionary distant species, so that the tertiary structure of the aptamer or binding pocket is conserved (Barrick and Breaker 2007). Switching sequence is present between the aptamer domain and the expression platform and sometimes is embedded in the aptamer domain. It controls which secondary structure the expression platform will adopt after the binding of the ligand to the aptamer domain and thus decides the level of expression of mRNA. Binding of a ligand in an aptamer region induces conformational change in the expression platform. This expression platform is not conserved and sometimes shows variability among the same riboswitch class (Mandal and Breaker 2004a) (Figs. 1 and 2).

Riboswitch mediated gene regulation may involve either of the following mechanisms:

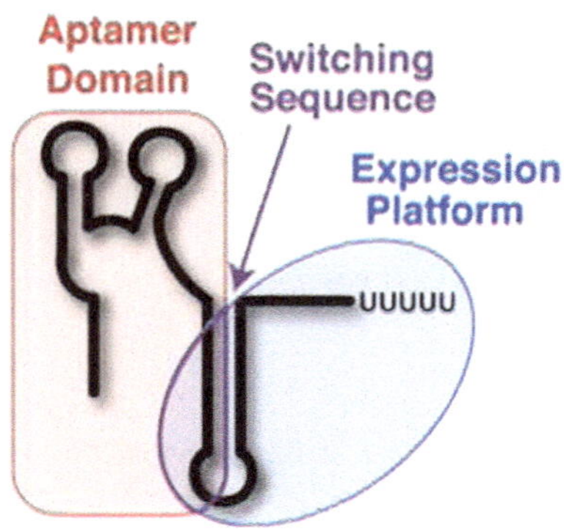

Fig. 1 Architecture of riboswitch

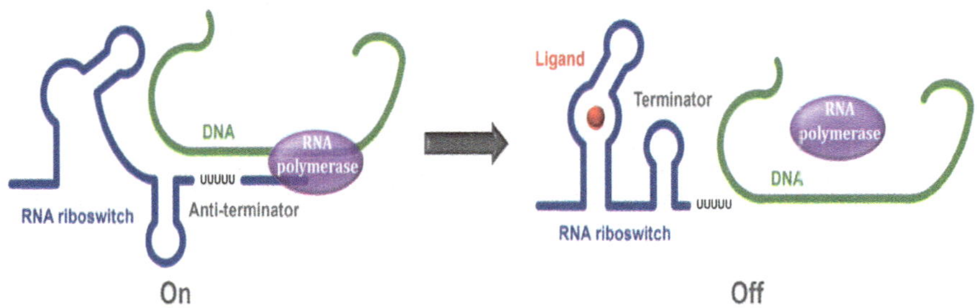

Fig. 2 Transcriptional inhibition by binding of ligand (http://2015.igem.org/Team:Exeter/RNA_Riboswitches)

1. **Transcription termination**: It involves formation of a transcription terminator loop. For transcriptional riboswitches, the ligand binding promotes the activity of an intrinsic transcriptional terminator, which consists of a G + C-rich helix, followed by a run of uridines in the nascent RNA that form a U-A RNA-DNA hybrid within the transcription elongation complex. Ligand binding alters the conformation of the expression platform and causes the dissociation of RNA polymerase (RNAP) from the DNA template, which leads to transcription termination. In the absence of the ligand, the RNA folds into an alternate anti-terminator structure that allows the transcription of the downstream gene (Mironov et al. 2002).
2. **Translational initiation**: This process involves the exposing of the ribosome binding site (RBS) or Shine Dalgarno (SD) sequence in bacteria by forming base paired structures. The ligand binding induces the conformational change in the terminator helix of typical transcription attenuator riboswitch and forms base pairs with the Shine-Dalgarno sequence to form a sequestered helix and prevents ribosomal binding. In the absence of the ribosome, the translation process is

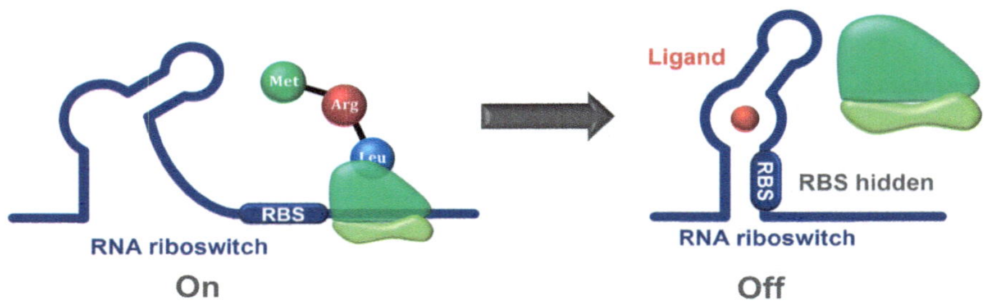

Fig. 3 Translational inhibition by binding of ligand (http://2015.igem.org/Team:Exeter/RNA_Riboswitches)

stopped. There are different mechanisms by which RBS is exposed or sequestered. One of these ways is through cleavage of the riboswitch. In the absence or presence of a certain ligand, the riboswitch can adopt a conformation in which a cleavage site is revealed. If the riboswitch becomes cleaved, then the RBS can be released and accessed by a ribosome, which can then read the protein coding region of the mRNA. The cleavage of this riboswitch can be carried out by a protein or ribozyme, or in some cases by the riboswitch itself (Fedor 2009). Another way in which an RBS can be sequestered is by being placed within a loop structure. When in a loop structure, the ribosome is unable to bind to the RBS sequence, and hence the gene cannot be expressed. If the concentration of the ligand changes, then the conformation of the riboswitch changes to remove the loop and then the ribosome is able to bind to the RBS and read the rest of the mRNA (Fig. 3).

1.2 Types of Riboswitches

There are about 40 Riboswitches reported till date in the Rfam database (Nawarocki et al. 2014), and few more that have not been included in the database, like the Glutamine (glnA) binding riboswitch (Ames and Breaker 2011). These Riboswitches are classified on the basis of the metabolites that they recognize, and are tabulated as given (Table 1).

The results for a restricted number of representative riboswitches from our study are given and discussed below. These are:

1. The TPP (enzyme cofactor) Riboswitch
2. The Purine (nucleotide) Riboswitch
3. The Lysine (aminoacid) Riboswitch
4. The GlmS Riboswitch, which also functions as a ribozyme
5. The cyclic-guanosine-monophosphate Riboswitch
6. The cyclic-adenosine-monophosphate Riboswitch.

Table 1 Classification of regulatory RNA elements in archaea

S.No.	Type of riboswitch	Rfam id	Classification	Description
1.	AdoCbl-variant	RF01689	Cis-reg; riboswitch	AdoCbl variant RNA
2.	AdoCbl_riboswitch	RF01482	Cis-reg; riboswitch	AdoCbl riboswitch
3.	Cobalamin	RF00174	Cis-reg; riboswitch	Cobalamin riboswitch
4.	drz-agam-1	RF01787	Cis-reg; riboswitch	drz-agam-1 riboswitch
5.	drz-agam-2-2	RF01788	Cis-reg; riboswitch	drz-agam-2-2 riboswitch
6.	FMN	RF00050	Cis-reg; riboswitch	FMN riboswitch (RFN element)
7.	glmS	RF00234	Cis-reg; riboswitch	glmS glucosamine-6-phosphate activated ribozyme
8.	Glycine	RF00504	Cis-reg; riboswitch	Glycine riboswitch
9.	Lysine	RF00168	Cis-reg; riboswitch	Lysine riboswitch
10.	MFR	RF01510	Cis-reg; riboswitch	M. florum riboswitch (hypoxanthine)
11.	Mg_sensor	RF01056	Cis-reg; riboswitch	Magnesium Sensor
12.	mini-ykkC	RF01068	Cis-reg	mini-ykkC RNA motif
13.	MOCO_RNA_motif	RF01055	Cis-reg; riboswitch	Moco (molybdenum cofactor) riboswitch
14.	PreQ1	RF00522	Cis-reg; riboswitch	PreQ1 riboswitch
15.	preQ1-II	RF01054	Cis-reg; riboswitch	preQ1-II (pre queuosine) riboswitch
16.	Purine	RF00167	Cis-reg; riboswitch	Purine riboswitch (adenine, guanine, 2′-deoxyguanosine)
17.	ROSE	RF00435	Cis-reg; thermoregulator	Repression of heat shock gene expression (ROSE) element
18.	ROSE_2	RF01832	Cis-reg; thermoregulator	Repression of heat shock gene expression (ROSE) element
19.	ROSE_3	RF02523	Cis-reg; thermoregulator	Repression of heat shock gene expression (ROSE) element
20.	S15	RF00114	Cis-reg; leader	Ribosomal S15 leader
21.	SAH_riboswitch	RF01057	Cis-reg; riboswitch	S-adenosyl-L-homocysteine riboswitch
22.	SAM	RF00162	Cis-reg; riboswitch	SAM riboswitch (S box leader)

(continued)

Table 1 (continued)

S.No.	Type of riboswitch	Rfam id	Classification	Description
23.	SAM-Chlorobi	RF01724	Cis-reg	SAM-Chlorobi RNA
24.	SAM-I-IV-variant	RF01725	Cis-reg; riboswitch	SAM-I/IV variant riboswitch
25.	SAM-II_long_loops	RF01726	Cis-reg element	SAM-II long loop element
26.	SAM-IV	RF00634	Cis-reg; riboswitch	S-adenosyl methionine (SAM) riboswitch,
27.	SAM-SAH	RF01727	Cis-reg; riboswitch	SAM/SAH riboswitch
28.	SAM_alpha	RF00521	Cis-reg; riboswitch	SAM riboswitch (alpha-proteobacteria)
29.	SAM-V	RF01826	Cis-reg; riboswitch	SAM-V riboswitch
30.	SMK_box_riboswitch	RF01767	Cis-reg; riboswitch	SMK box translational riboswitch
31.	THF	RF01831	Cis-reg; riboswitch	THF riboswitch
32.	Thr_leader	RF00506	Cis-reg; leader	Threonine operon leader
33.	TPP	RF00059	Cis-reg; riboswitch	TPP riboswitch (THI element)
34.	yjdF	RF01764	Cis-reg	yjdF RNA
35.	ykkC-III	RF01763	Cis-reg	ykkC-III RNA
36.	ykkC-yxkD	RF00442	Cis-reg	ykkC-yxkD leader
37.	ykoK	RF00380	Cis-reg	ykoK leader
38.	yybP-ykoY	RF00080	Cis-reg	yybP-ykoY leader
39.	ybhL	RF00520	Cis-reg	ybhL leader
40.	ydaO/c-di-AMP	RF00379	Cis-reg	Binds the cyclic-di-adenosine monophosphate
41.	c-di-GMP-I	RF01051	Cis-reg	Cyclic di-GMP-I riboswitch
42.	c-di-GMP-II	RF01786	Cis-reg; riboswitch	Cyclic di-GMP-II riboswitch

The rationale for choosing this representative set of riboswitch will be given when the results are shown and discussed.

1.3 The Domain Archaea

Archaea was introduced as the third domain by Woese and Fox on the basis of 16S rRNAs (Woese and Fox 1977). Sustained studies and investigations have revealed many remarkable things about Archaea-their unique cellular components and pathways, their abundance and their surviving in diverse natural environments, and their essential role in shaping the evolutionary path of life on Earth.

The information processing system (DNA replication, transcription and translation) of the Archaea are similar to that of the Eukaryotes, but their metabolic and other operational processes are similar to that of the Bacteria (Woese et al. 1990). This similarity does not necessarily reflect closely shared common ancestor between Bacteria and Archaea or Eukaryotes and Archaea but the similarity between these species could have arisen in many ways. These include loss of genes, extensive sequence divergence or Horizontal Gene Transfer (HGT) which makes it difficult to make a simple phylogenetic analysis.

Archaea are a diverse group of organisms that show many idiosyncratic properties, ranging from their survival at extreme habitats to their unique ability of methanogenesis to produce biological methane and to their unconventional morphology like the square shaped haloarchaeon *Haloquadratum walsbyi* (Burns et al. 2007). Their habitats are sometimes similar and sometimes different from those of the Bacteria. Regions of extremely high temperature, volcanic eruptions, deep sea thermal vents, salt lakes and highly acidic environments as habitats clearly place them as extremophiles, and suggest that they could be a life form that survived three billion years ago.

Environmental surveys indicate that the Archaea are diverse and abundant not only in extreme environments, but also in soil, oceans (floor as well as surface) and freshwater, where they may fulfill a key role in the biogeochemical cycles of the planet (Offre et al. 2013). Moreover, Archaea are also found to inhabit human gut (Eckburg et al. 2005; Gill et al. 2006; Gaci et al. 2014). Molecular biology, genomics and phylogenetic data strengthen Woese's definition of Archaea as a third domain of life in addition to Bacteria and Eukarya. Phylogenomics analyses of the constituents of different molecular systems highlight a core of mainly vertically inherited genes in Archaea, unlike Bacteria.

1.4 Diversity and Habitat of Archaea

The Archaea have traditionally been grouped into methanogens, extreme halophiles and thermoacidophiles based on their metabolism. The methanogens are found in anaerobic marine and freshwater environments and in the gastrointestinal tracts of animals, where they participate in the conversion of organic matter by utilizing the metabolic products of bacteria (for example, CO_2, H_2, acetate and formate) and other simple methyl compounds that are available in the environment (for example, methylamines and methanol) and converting them into methane (CH_4). Haloarchaea reside in hypersaline environments (such as salterns, lakes and the Dead Sea), where they grow as heterotrophs, and often in association with phototrophic algae. Thermoacidophiles (including hyperthermophiles, which grow fastest at temperatures above 80 °C) colonize volcanic terrestrial environments and deep-sea hydrothermal vents, growing aerobically or anaerobically as heterotrophs or autotrophs, and often deriving energy by sulphur oxidation or reduction. One of the remarkable features of Archaeal domain is the varying kind of habitats in which the archaeal organisms survive ranging from normal to inhospitable conditions. They are categorized as below:

- Thermophiles and Hyperthermophiles: Thermophiles, are organisms with an optimum growth temperature of 50 °C or more, a maximum of up to 70 °C or more, and a minimum of about 40 °C. Hyperthermophiles, require a very high temperature (80–105 °C) for growth. Their membranes and proteins are unusually stable at these extremely high temperatures. e.g. *Sulfolobus solfataricus* and *Aeropyrum pernix.*
- Halophiles: They require high salt concentrations to grow. The extremely halophilic, aerobic members of Archaea are classified within the family Halobacteriaceae, order Halobacteriales in Class III. Halobacteria belong to the phylum Euryarchaeota. e.g. *Halalkalicoccus jeotgali, Haloarcula hispanica.*
- Psychrophiles: They are capable of growth and reproduction in cold temperatures, ranging from −20 °C to +10 °C, and often thrive at temperatures near the freezing point of water e.g. *Methanococcoides burtonii* and *Methanogenium frigidum.*
- Acidophiles: Thrive under highly acidic conditions (usually at pH 2.0 or below). Examples are *Acidianus hospitalis* and *Ferroplasma acidarnamus* Most acidophiles have evolved extremely efficient mechanisms to pump protons out of intracellular space in order to keep the cytoplasm at or near neutral pH. Therefore, intracellular proteins do not need to develop acid stability through evolution.
- Ammonia Oxidizers: This group includes organisms that are identified as chemo-lithoautotrophic ammonia oxidizers. *Nitrosopumilus maritimus, Nitrososphaera viennensis* and *Nitrososphaera gargensis*, of phylum Thaumarchaeota, are also ammonia oxidizers.

1.5 Classification of Archaeal Genomes

There are 649 Sequenced Archeal Genomes are reported to date in the NCBI Genome database (Fig. 4).

Recently Archaea have been classified into four main groups:

- **Asgard Group**: Asgard superphylum is a recently classified group of uncultivated archaea. Asgard archaea affiliate with eukaryotes in phylogenomic analyses, and their genomes are enriched for proteins formerly considered specific to eukaryotes. Zaremba-Niedzwiedzka et al. (2017). This group contains 10 members.
- **Thorarchaeota**: Thorarchaeal genomes encode several homologues of eukaryotic membrane-trafficking machinery components, including Sec23/24 and TRAPP domains. Furthermore, Thorarchaeal proteins show similar features

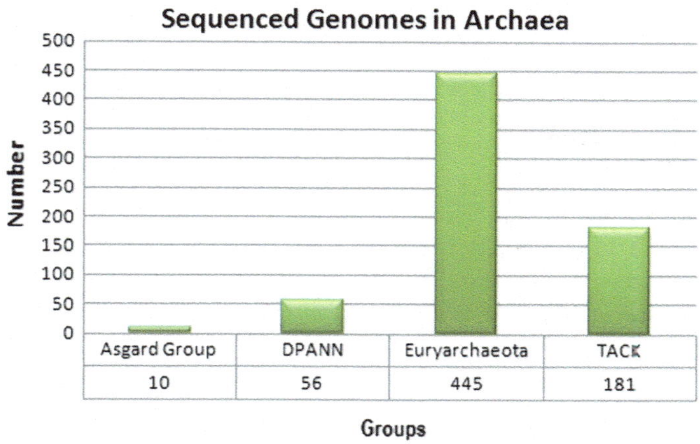

Fig. 4 Total number of sequenced genomes in different archaeal groups

to eukaryotic coat proteins involved in vesicle biogenesis (Zaremba-Niedzwiedzka et al. 2017).

- **Lokiarchaeota**: Lokiarchaeum genome revealed that it encodes a multitude of genes that were previously regarded specific to eukaryotes. These so-called eukaryotic signature proteins (ESP) include several cytoskeletal components (actin homologues and gelsolin-domain proteins), ESCRT complex proteins (including ESCRT-I, -II and -III components), and a wide variety of small GTPases including Gtr/Rag family GTPase orthologues—proteins that in eukaryotes are involved in various regulatory processes including cytoskeleton remodelling, signal transduction, nucleo-cytoplasmic transport and vesicular trafficking (Spang et al. 2015).
- **Odinarchaeota**: These belong to the Asgard group, and await further analysis.
- **Heimdallarchaeota**: These also belong to the Asgard group, and await further analysis.
- **TACK GROUP**: This is also called as TACK Superphylum that comprises the Thaumarchaeota, Crenarchaeota and Korarchaeota as well as the recently proposed phylum Aigarchaeota. Studies suggest this TACK superphylum to be the parent of Eukaryotic lineage (Guy and Ettema 2011).

  - **Bathyarchaeota**: Genes involved in nitrite reduction were detected in all Bathyarchaeota subgroups and indicate a potential for dissimilatory nitrite reduction to ammonium (Lazar et al. 2016).
  - **Verstraetarchaeota**: These archaea encode the genes required for methylotrophic methanogenesis, and may conserve energy using a mechanism similar to that proposed for the obligate H_2-dependent methylotrophic Methanomassiliicoccales and recently described Candidatus 'Methanofastidiosa' (Vanwonterghem et al. 2016).

- **Thaumarchaeota**: Members are chemolithoautotrophic ammonia-oxidizers that play an important role in biogeochemical cycles in both aquatic and terrestrial environments, such as the nitrogen cycle and the carbon cycle. Thaumarchaeota were intially referred to (mistakenly) as 'mesophilic or low-temperature Crenarchaeota' and were first discovered in marine environments, but now also known to reside in terrestrial habitats (Spang et al. 2010). Originally, the Thaumarchaeota was envisioned to be the deepest rooting archaeal phylum based on a phylogenetic analysis of concatenated protein datasets (Brochier-Armanet et al. 2008), for example *Crenarchaeum symbiosum* and *Nitrosopumilus maritimus* but T. Ettema's group recently have corrected that by identifying the Lokiarchaeota as the deepest rooting Archaeal phylum to date (Spang et al. 2015; Zaremba-Niedzwiedzka et al. 2017).

- **Crenarchaeota**: Well-characterized archaeal phylum first described by Woese et al. (1990), comprising mostly of acido-thermophilic anaerobes, although some aerobic (Sulfolobales) and micro-aerophilic lineages exist. Crenarchaea of the orders Sulfolobales and Desulfurococcales utilize a unique Cdv cell division machinery that is related to eukaryotic membrane remodeling systems. Phylogenetically, Crenarchaeota represents a sister group to the clade comprising Thaumarchaeota (once known as 'low-temperature Crenarchaeota') and the candidate phylum Aigarchaeota. e.g. *Sulfolobus solfataricus* and *Aeropyrum pernix.*, *Pyrobaculum aerophilum.*
- **Korarchaeota**: Candidate phylum comprising a group of deep-branching Archaea with ultra-thin, needle-shaped cells measuring up to 100 μm in length, which are geographically restricted to terrestrial and marine thermal environments. Based on the genome sequences that are currently available, this has been categorized as a separate phylum (Lazar et al. 2008). Candidatus *Korarchaeum cryptofilum* (Elkins 2008) relies on a simple mode of peptide fermentation for generation of energy and biomass. Members of this group are most abundant in hot springs with temperatures greater than 55 °C and a pH range of 5.7–7.0.
- **DPANN GROUP**: The group has been named after the initials of candidate organisms found in this group. This group contains 54 members.
- Aenigmarchaeota
- Diapherotrites
- Nanohaloarchaeota
- Micrarchaeota
- Pacearchaeota
- Woesearchaeota
- Parvarchaeota
- Nanoarchaeota
- **Euryarchaeota**: This group is the largest group of Archaeal organisms containing 432 members. It is considered to be more closely related to Eukaryotic domain, because of the occurrence of homologues of Histone proteins in these organisms (Fig. 5).

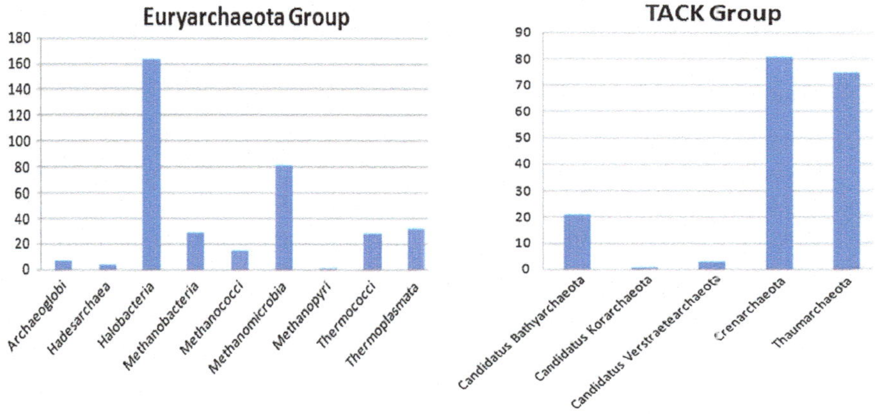

Fig. 5 Distribution of different archaeal groups

- Archaeoglobi
- Hadesarchaea
- Halobacteria
- Methanogens
- Thermococci.
- Thermoplasmata

2 Methodology

Completely sequenced Archaeal Genomes present in the NCBI repository were used for studying Riboswitches and other ncRNAs. The sequence files in FASTA format were downloaded from NCBI Genome database and information related to these genomes were retrieved from Genbank. To identify riboswitches and other ncRNAs three bioinformatic approaches were used namely, ab initio, consensus structure prediction and statistical model based prediction.

1. **Ab initio structure prediction**

This method predicts the structure of a single sequence. The input in this approach is a known sequence in the FASTA format which is used in BLAST (Altschul et al. 1990) algorithm for similarity check and homology search. Homologous sequences with maximum percentage identity are selected and they are further aligned with the known reference sequence. Sequences that are aligned with the reference sequence are selected and folded into the secondary structure using Mfold or RNAFold. Sequences that fold correctly into the corresponding secondary structure are selected and their free energy is determined. If the free energy is considerable then their 3-D structure is visualized and compared to known structure and confirmed as a putative new hit.

2. **Consensus structure prediction**

This method predicts the structure shared by more than one sequences. Input in this approach is a set of known sequences which are aligned with the putative sequences of new hits. Multiple sequence alignment is done showing the stems and loops. Then, a common secondary structure is deduced from the MSA whose free energy is determined and compared to the known structure.

3. **Statistical model-based prediction and alignment**

This approach is used to search for desirable structures in whole genomes or data bases. Stochastic grammars can describe models defined by HMMs. In this approach, alignment in Stockholm format taken from Rfam is used as an input. This alignment is used as query in computational programs like INFERNAL. The program is run against individual genome files in FASTA format. On the basis of alignment score the putative hits are selected and then folded into secondary structure. Their free energy is calculated and the 3-D structure thus generated is compared to previously known structure.

3 Results and Discussion

This chapter is based on a thesis—"A study of riboswitches and other non-coding RNAs in Archaeal genomes" by my student, Angela Gupta (2016, partly published). Twenty different types of riboswitches were found in that investigation of Archaeal genomes and metagenomes that have been sequenced till December 2016.

Of these, only the findings of representative riboswitches are discussed. These include the enzyme cofactor, nucleotide, aminoacid, that is, the Thymine pyrophosphate, Purine (Adenine) and Lysine riboswitches. There is only one riboswitch that functions as a ribozyme as well, the glmS riboswitch. Its function is to regulate the level of glucosamine phosphate (GlcN6P) in the cell. That has been included, as well as the two riboswitches, that were earlier labeled as GEMM and ydao RNA elements, that control the level of second messenger molecules-cyclic di-guanosine monophosphate(c-di-GMP) and cyclic di-adenosine monophosphate (c-di-AMP) in the organism's cell, and are representatives of a direct response through a ncRNA element to external environmental stress.

3.1 TPP Riboswitch

We have chosen to discuss the TPP riboswitch, because it occurs most frequently, in the bacterial as well as eukaryotic domain (Yadav et al. 2015). Thiamine pyrophosphate (TPP) is a coenzyme derived from vitamin B1 (thiamine), synthesized by bacteria, archaea, fungi, algae and plants (Wachter 2007). TPP-sensitive mRNA domains function as the most prevalent riboswitches known. Previous studies have also reported TPP riboswitches in a few Archaea (Serganov and Patel 2007).

The aptamer domain of the TPP riboswitch is highly conserved across all three domains of life. It is present in the 5′ UTR (untranslated region) of the gene to be regulated in prokaryotes and in the 3′ UTR, downstream of the gene, at 70 bp (base pairs) before the polyadynylation signal in eukaryotes (Bocobza et al. 2007). The ligand-free state of the TPP riboswitch is a partially-folded RNA structure but undergoes conformation on binding of the ligand, and thus switches off the transcription of the THI C and THI I genes, which is expressed as thiamine pyrophosphate (Montange and Batey 2008) (Fig. 6).

To identify new hits of TPP Riboswitch in Archaea, all the Thiamine biosynthesis genes in Archaea were extracted from NCBI Gene database. 216 THI I and 55 THI C genes were present in Archaeal organisms. In a comparative proteomics study, Ananthraman et al. (2002) have found that THI I genes are found frequently in Archaea, and rarely in bacteria. Bacteria have the THI C gene. Generally, TPP riboswitch is found in the UTR of THI I gene. As a first step, a general similarity search was done using BLAST. The hits obtained were already identified previously, no new hits were found. Then context based search was done to identify new hits of the TPP riboswitch. The results also include the previously known instances. The results are shown in Table 2.

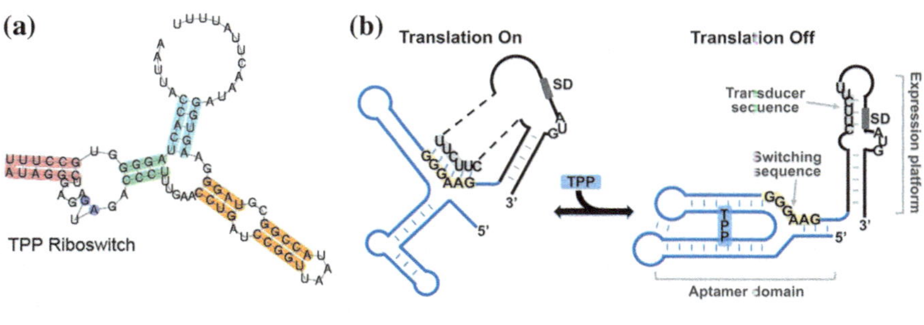

Fig. 6 **a** Secondary structure of TPP riboswitch, **b** mechanism of TPP riboswitch function (http://www.the-scientist.com/?articles.view/articleNo/15675/title/RNAs-Running-the-Show/)

Table 2 Representative riboswitches in archaeal organisms

S.No	Riboswitch	Name	Phylum	Minimum free energy (Kc/mol)
1	TPP	*Methanocorpusculum labreanum z*	Euryarchaeota	−36.00
		Picrophillus torridus	Euryarchaeota	−26.96
		Thermoplasma acidophilum	Euryarchaeota	−32.61
		Thermoplasma volcanium	Euryarchaeota	−33.68
		Ferroplasma acidalmanous	Euryarchaeota	−28.65
		C. Korarchaeum cryptofilum OPF8	Korarchaeota	−41.82
2	Lysine	*C. Korarchaeum cryptofilum OPF8*	Korarchaeota	−38.22
		Halogeometricum borinquense DSM 11551	Euryarchaeota	−46.89
		Halophilic archaeon DL31	Euryarchaeota	−49.91
		Haloquadratum walsbyi DSM 16790	Euryarchaeota	−34.19
3	Adenine	*Sulfolobus acidocaldarius*	Crenarchaeota	−20.82
4	glmS	*Haloarcula hispanica N601*	Euryarchaeota	−55.67
		Haloarcula marismortui ATCC	Euryarchaeota	−57.00
		Thermococcus barophilus MP	Euryarchaeota	−31.53
5	c-di-GMP	*Sulfolobus acidocaldarious*	Crenarchaeota	−18.10
		Pyrococcus furiousus	Euryarchaeota	−15.40
		Methanopyrus kandleri	Euryarchaeota	−16.10
		Ferroplasma acidarmanus	Euryarchaeota	−16.60
		Thermofilum sp. *1910B*	Euryarchaeota	−16.20
6	ydaO/c-di-AMP	*Halobacterium salinarum*	Euryarchaeota	−34.8

3.2 *Purine (Adenine) Riboswitch*

The collection of purine-sensing riboswitches (Kim and Breaker 2008) together rank as the seventh most common family of validated riboswitches in Bacteria (Breaker, 2011). Riboswitches recognizing guanine, purine, 2′-deoxyguanosine and preQuenosine(preQ1) are members of this family. They are recognized by their architecture, consisting of a three-stem junction wherein the loops of two stems associate through tertiary interactions (Mandal and Breaker 2004b; Serganov et al. 2004). Representatives can selectively bind guanine or adenine (Mandal and Breaker 2004b), or 2′-deoxyguanosine (Kim et al. 2007), depending on the identities of nucleotides within the regions joining the stems (J1-2, J2-3, and J3-1). Riboswitches for Adenine and Guanine are very ligand-specific, to the extent that a change of one key nucleotide at position 74 of junction J3-1 decides which ligand will be bound. If it is C, then Guanine is bound, and if it is U, then Adenine is

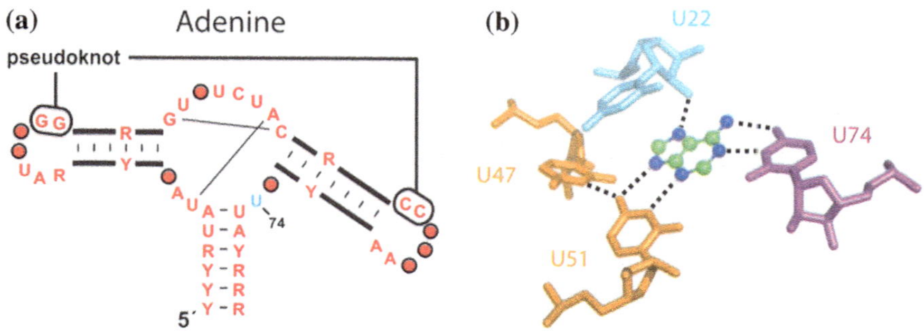

Fig. 7 **a** Secondary structure of adenine riboswitch, **b** binding pocket of adenine riboswitch (Roth and Breaker 2009)

bound. The riboswitch binding 2′-deoxyguanosine has been detected in only one bacterial species, Mesoplasma florum at present (Batey et al. 2004). The crysral structure of the adenine riboswitch, controlling the expression of *addA* gene in *Vibrio vulnificus* has been known for some time (Serganov et al. 2004), and the mechanism of binding of adenine or guanine to the respective riboswitch have been studied by Molecular Dynamics simulation studies (Sharma et al. 2009).

Since the biosynthesis of purines is necessary for the survival of an organism, it is surprising to find only one instance of Adenine binding riboswitch in Archaea. Brown et al. (2011) have studied the distribution of genes involved in the purine biosynthesis pathway, and have found that these genes are absent in seven archaea from different phyla, like *Aeropyrum pernix, Nanoarchaem equitans* and *Staphylothermus marinus*, for example. It is found that *A. pernix* needs adenine in its medium to grow. Such organisms must depend on their environment, or evolve salvage pathways for the biosynthesis of purines. This scattered distribution of lack of genes essential for purine biosynthesis suggests that this particular pathway in Arharea arose from horizontal gene transfer (HGT), gene duplication and gene loss (Brown et al. 2011) (Fig. 7).

3.3 *Lysine Riboswitch*

The Lysine riboswitch is a metabolite binding RNA element found within certain messenger RNAs that serve as a precision sensor for the amino acid lysine. This riboswitch is found in a number of genes involved in lysine metabolism, including LysC (Rodionov et al. 2003). In Archaea three types of Lysine biosynthesis genes are present namely—LysC, LysX and LysW. There were a large number of sequence hits amongst Archaea, which were aligned with the corresponding conserved sequence regions of bacteria. However, not all the sequences corresponded to the Lysine riboswitch, since they did not fold into the secondary structure required to bind Lysine in the aptamer domain. A few sequences were confirmed as

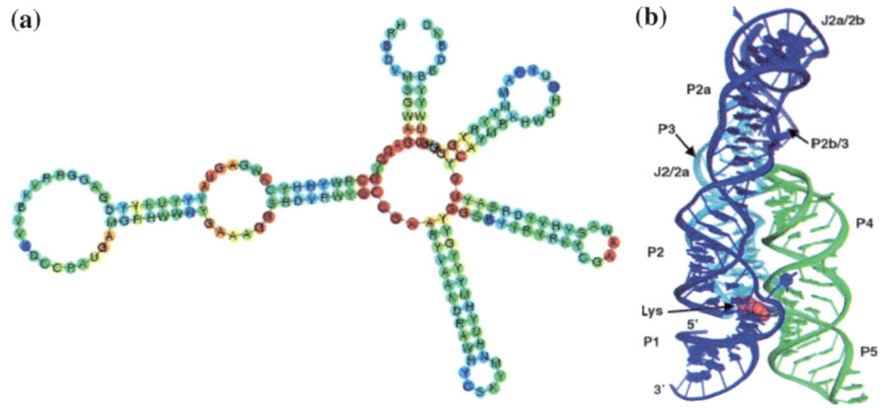

Fig. 8 **a** Secondary structure of lysine riboswitch, **b** tertiary structure of lysine riboswitch, showing the ligand in pink (http://www.colorado.edu/lab/bateygroup/structures/lysine)

the Lysine riboswitch after analyzing the structural similarity, and calculating the mean free energy. The minimum threshold for Gibbs free energy is taken as −10 kcal/mole, since any structure with higher values of free energy, corresponds to unstable helical structures which are unable to fold into the scaffold required to bind the ligand (Seehaefer et al. 2010). The putative Lysine riboswitches are listed in Table 2 and Fig. 8.

3.4 glmS Riboswitch

The glmS riboswitch is located in the 5′-UTR region of the gene encoding glucosamine-6-phosphate (GlcN6P) synthetase (Winkler et al. 2004), which converts glutamine and fructose-6-phosphate to glucosamine-6-phosphate (GlcN6P). The glmS riboswitch is a ribozyme with activity triggered by binding of the metabolite GlcN6P. Like many other riboswitches that control downstream gene expression, the glmS riboswitch negatively regulates GlcN6P synthetase production (Winkler et al. 2004). High GlcN6P concentrations activate the riboswitch, resulting in reduced synthetase expression. However, unlike every other riboswitch identified to date, the glmS riboswitch does not appear to undergo any structural rearrangement upon ligand binding, suggesting that the binding pocket is pre-formed in the absence of the ligand (Hampel and Tinsley 2006). Instead, GlcN6P binding results in the specific cleavage of the glmS mRNA at a single site 5′ of the riboswitch sequence. The glmS riboswitch is a ribozyme and like other autolytic ribozymes, the cleavage reaction involves nucleophilic attack of a vicinal 2′-OH on the scissile phosphate to produce scission products with a 2′–3′ cyclic phosphate and a 5′-OH (Collins et al. 2007; Xi and Hamelberg 2010). The ligand is

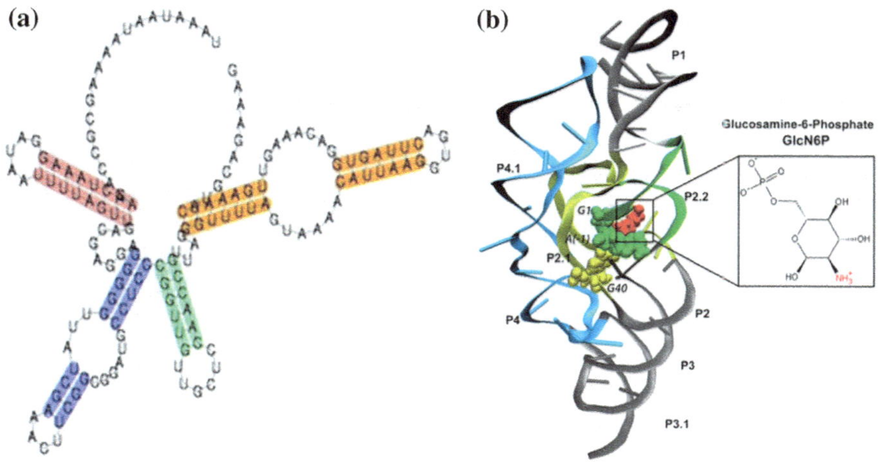

Fig. 9 **a** Secondary structure of glmS riboswitch, **b** tertiary structure of glmS riboswitch showing the binding of glcN6P in the aptamer domain (from Xin and Hamelberg 2010)

absolutely required for glmS ribozyme activation. GlcN6P addition increases the cleavage rate 100,000-fold over the background hydrolysis (D'Amaré and Scott 2010).

The protocol for finding riboswitches was applied, and five instances of the glmS riboswitch/ribozyme were found. The results are shown in Table 2 and Fig. 9.

3.5 *c-di-GMP Riboswitch*

Cyclic di-guanosine monophosphate (c-di-GMP) is a circular RNA dinucleotide that functions as a second messenger in diverse species of bacteria to trigger wide-ranging physiological changes, including cell differentiation, conversion between motile and biofilm lifestyles, and expression of virulences. Cyclic di-GMP-I riboswitches were the first kind of riboswitch to be discovered whose role is not primarily in regulating metabolism, but is instead a part of a signaling pathway. Even though the importance of c-di-GMP as a second messenger was recognized few decades back, only protein facors controlling its expression were studied. Sudarsan et al. (2008) showed that the GEMM ncRNA could act as a riboswitch binding this macromolecule, controlling its levels within the cell. Sudarsan et al. (2008) have shown that the survival capability of *Vibrio cholera* within the human host is fine-tuned to an optimum level of c-di-GMP within its cells, exceeding that level, may be conducive to the death of the pathogen. This has researchers investigating this riboswitch as a putative drug target.

Besides regulating virulence genes, this second messenger also controls the genes for synthesis and degradation of enzymes that promote growth of flagalle and

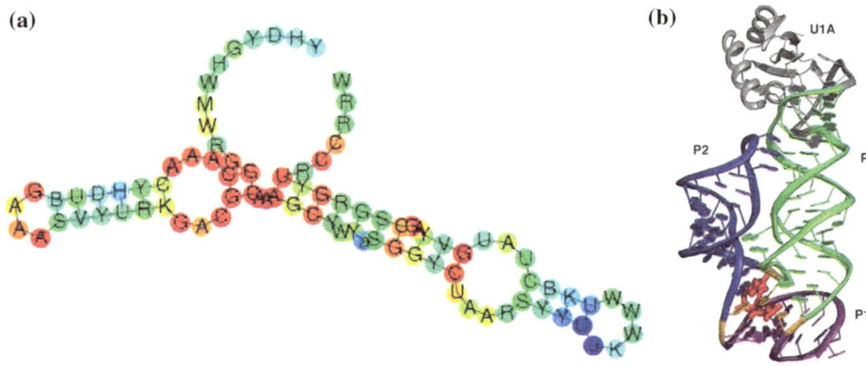

Fig. 10 **a** Secondary structure of c-di-GMP riboswitch, **b** tertiary structure of c-di-GMP Riboswitch (www.nature.com/nsmb/journal/v16/n12fig_tab/nsmb.1702_F1.html)

pili and sporulation. Besides these, they also control the genes that are required for an organism to grow biofilms (Smith and Strobel 2011) and those involved in ion transport and aminoacid transport. Since recently biofilms of Archaea have been reported (Orell et al. 2013), it was surmised that this riboswitch could be also found in Archaea. A context-specific search has yielded five instances of c-di-GMP riboswitch among the Archaea in the present study (see Table 2 and Fig. 10).

3.6 c-di-AMP Riboswitch

The ydaO/yuaA leader (now called the cyclic di-AMP ribsowitch) is a conserved RNA structure found upstream of the ydaO and yuaA genes in *Bacillus subtilis,* and related genes in other bacteria (Block et al. 2010). The ydaO ncRNA element was first erroneously identified as an ATP-binding riboswitch, but Nelson et al. (2013) showed that it was actually the c-di-AMP, second messenger molecule that was the ligand. The level of di-adenosine monophosphate controls genes for DNA repair, cell-wall metabolism, osmotic stress, potassium homeostasis and sporulation. It was found by several researchers that c-di-AMP is an essential molecule for Firmicutes, and if the concentration level rose over the optimum value, it became toxic for the bacterium (Commichau et al. 2015). It appears to be essential for the Mycoplasmas also, even though they lack a cell wall.

It was confirmed by simultaneous structural studies(Ren and Patel 2014; Gao and Serganov 2014), published together that it was a riboswitch that bound two ligands of cyclic-di-adenosine monophosphate (c-di-AMP) together in a 'pseudo-symmetric' manner, as shown in Fig. 11b. In an ongoing in silico experiment, we have identified one occurrence of this riboswitch in an Archaea.

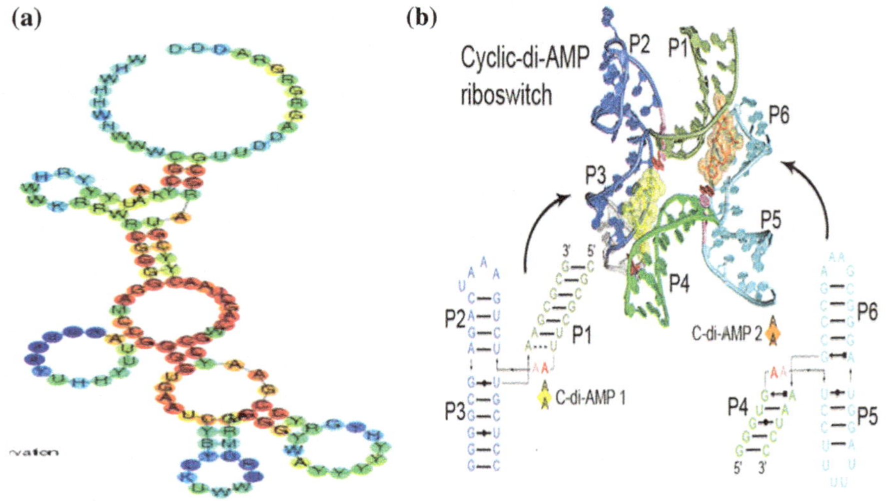

Fig. 11 **a** Secondary structure of c-di-AMP riboswitch, **b** binding mechanism of c-di-AMP riboswitch (Gao and Serganov 2014)

4 Conclusions

The capacity to process signals is necessary for an organism's survival. Whether the signal is internal, like the fluctuation of the concentration level of a metabolite, or the ion balance and osmotic pressure inside the cell, or whether it is a communication through a second messenger, of environmental stress, the response to the change has to be fast. The regulation of genes, in order to suppress(negative feedback) or enhance (positive feedback) the level of the corresponding protein needed in a metabolic pathway or signal transduction cascade may be achieved by the use of protein factors (as is prevalent in the eukaryotic domain) or by the use of riboswitches.

The moot point is, are riboswitches relics from a primordial RNA world before proteins could be synthesized in nature and have survived as such, as some authors would like to state (Vitrescak et al. 2004; Breaker 2012), or did they evolve several times, as and when needed. Breaker (2012) opines that the riboswitches found most frequently are TPP, FMN (flavin mononucleotide) and Adocbl(adenosyl cobalamine), all types of vitamine B, have complex, elaborate binding scaffolds or aptamers, and so probably existed in the same form in the RNA world. On the other hand, the riboswitches with smaller and simpler structure, like the purine riboswitch, is probably the type to have evolved according to the need of the organism, several times.

Riboswitches are found to be more frequent in the Euryarchaeota, than in the other well-studied Archaeal phylum Crenarchaeota. This maybe seen from Tables 2 and A.1 of the Appendix. But the pattern of occurrence defies any attempt to systematize the distribution of riboswitches.

They occur more frequently in bacteria and less in archaea, and even lesser in eukaryotes. This fact alone, definitely does not give us a vertical evolutionary tree picture involving the three domains, based on the occurrences (or non-occurences) of riboswitches.

The discovery and classification of Archaeal genomes is a rapidly expanding field, and further, more efficient algorithms based on stochastic context free grammar (SCFG) may soon be in the public domain. To date, INFERNAL (Nawrocki and Eddy 2013) based on SCFG, is the best riboswitch finder and validator. Primary sequence level conservation is just a mere indicator of similarity, as we found while searching for Lysine riboswitches in Archaea. For the sequence to fold into the appropriate secondary structure involving the right number of helices, hairpins and sometimes pseudoknots, cannot be predicted by one software, and sometimes scripts have to be written, as in RNABOB (Eddy 2015), to manually correct the 'overhang' in aligned RNA sequences. Earlier algorithms based on profile Hidden Markov Models(pHMM) like Mfold (Zuker 2003) have been less successful than INFERNAL, but still somewhat useful in a preliminary study of riboswitches.

The results may be modified, as more genes are identified, and deposited in gene banks of repositories like NCBI, to perform a context-based search of the relevant riboswitch using a more efficient software, as also whenever new Archaeal genomes are classified and categorized from environmental samples. Apart from the boom in the numbers and sequences of Archaeal genomes, the number of riboswitches and ribozymes are also increasing, with a new field of synthetic riboswitches already established. In the first decade (2002–2011) of discovery and analysis of riboswitches in Bacteria, it was stated that the large interesting clade of Cyanobacteria did not have any riboswitches. Then Nakahira and coworkers (Nakahira et al. 2013) synthesized a riboswitch for the base Theophylline (1-3 di-methyl xanthine), and showed that it controlled the expression of genes(caiK) for circadian clock behavior of Cyanobacteria.

Ames and Breaker (2011) and coworkers have found a riboswitch for glutamine by purely in silico methods. Glutamine is an important amino acid found in intracellular pools and a precursor to nucleotide, aminoacids and protein biosynthesis in the human body, as well as other organisms. Similar research on synthetic aptamers is flourishing, and is bound to uncover new classes of riboswitches, some of which might be found in Archaeal genomes also. The intersection of two burgeoning fields, Archaeal genomes and Riboswitches, make this present study, by definition, a work in progress.

Acknowledgements The results of this study are from the thesis of Angela Gupta, Ph.D. scholar of Bioinformatics Department, MMV, Banaras Hindu University. The figures and tables were provided be her.

Appendix

See Fig. A.1 and Table A.1.

Table A.1 Riboswitches in archaeal genomes and metagenomes

S.No	Riboswitch	Name	Phylum	Minimum free energy (Kc/mol)
1.	TPP	*Methanocorpusculum labreanum*	Euryarchaeota	−36.00
		Picrophillus torridus	Euryarchaeota	−26.96
		Thermoplasma acidophilum	Euryarchaeota	−32.61
		Thermoplasma volcanium	Euryarchaeota	−33.68
		Ferroplasma acidalmanous	Euryarchaeota	−28.65
		C. Korarchaeum cryptophylum	Korarchaeota	−41.82
		Uncultured_archaeon_clone_ASS_A1	Unclassified	−28.63
2.	Lysine	*C. Korarchaeum cryptofilum*	Thaumarchaeota	−38.22
		Halogeometricum borinquense	Euryarchaeota	−46.89
		DSM 11551	Euryarchaeota	−49.91
		Halophilic archaeon DL31	Euryarchaeota	−34.19
		Haloquadratum walsbyi DSM 16790		
3.	Adenine	*Sulfolobus acidocaldarius*	Crenarchaeota	−20.82
4.	glmS	*Haloarcula hispanica N601*	Euryarchaeota	−55.67
		Haloarcula marismortui ATCC	Euryarchaeota	−57.00
		Thermococcus barophilus	Euryarchaeota	−31.53
5.	YybP	*Natronobacterium gregoryi SP2*	Crenarchaeota	−44.30
		C. Korarchaeum cryptofilum OPF8	Korarchaeota	−23.28
6.	YkkC-YxkD	*Archaeoglobus profundus*	Euryarchaeota	−28.90
		Caldisphaera lagunensis	Crenarchaeota	−21.20
		Caldivirga maquilingensis	Crenarchaeota	−24.49
		Candidatus nitrosoarchaeum limnia	Thaumarchaeota	−25.80
		Halopiger xanaduensis	Euryarchaeota	−31.00
		Thermoproteus uzoniensis	Euryarchaeota	−30.93
7.	PreQ1	*Sulfolobus acidocaldarius*	Crenarchaeota	−11.97
		Pyrobaculum oguniense TE7	Euryarchaeota	−16.90
8.	Cobalamin	*Nanoarchaeum equitans*	Nanoarchaeota	−40.00
9.	FMN	*Methanobacterium* sp. *SWAN-1*	Euryarchaeota	−23.5
		Methanobrevibacter ruminantium	Euryarchaeota	−35.2
		Methanobrevibacter smithii	Euryarchaeota	−32.4
10.	Purine	*Sulfolobus acidocaldarius*	Crenarchaeota	−20.82
11.	SAH	*Halorhabdus tiamaetia*	Euryarchaeota	−24.50
12.	Four-U thermometer	*Ignesphaera aggregans*	Euryarchaeota	−16.90
		Natrinema pellibrum	Euryarchaeota	−21.40
		Tthermoproteus tenax	Euryarchaeota	−14.40

(continued)

Table A.1 (continued)

S.No	Riboswitch	Name	Phylum	Minimum free energy (Kc/mol)
13.	ybhL	*C. Korarchaeum cryptofilum OPF8*	Korarchaeota	−19.40
14.	Mini-yKkC	*Halomicrobium mukohataei DSM 12286* *Natronobacterium gregoryi SP2* *Natrinema* sp. *J7-2* *Haloterrigena turkmenica*	Euryarchaeota Euryarchaeota Euryarchaeota Euryarchaeota	−20.06 −19.61. −25.15 −27.29
15.	Mg sensor	*Halobacterium* sp. *Haloterrigena turkmenica*	Euryarchaeota Euryarchaeota	−35.10 −29.00
16.	ROSE elements	*Halobacterium salinarum* *Pyrobaculum oguniense TE7* *Sulfolobus tokodii* *Halorhabdus utahensis* *Haloterrigena turkmenika*	Euryarchaeota Euryarchaeota Crenarchaeota Euryarchaeota Euryarchaeota	−53.10 −45.60 −36.10 −51.20 −43.72
17.	glmS	*Haloarcula hispanica N601* *Haloarcula marismortui ATCC* *Thermococcus barophilus MP*	Euryarchaeota Euryarchaeota Euryarchaeota	−55.67 −57.00 −31.53
18.	ydaO/c-di-AMP	*Halobacterium salinarum*	Euryarchaeota	−34.8
19.	c-di-GMP	*Sulfolobus acidocaldarious* *Pyrococcus furiousus* *Methanopyrus kandleri* *Ferroplasma acidarmanus* *Thermofilum* sp. *1910B*	Crenarchaeota Euryarchaeota Euryarchaeota Euryarchaeota Euryarchaeota	−18.10 −15.40 −16.10 −16.60 −16.20
20.	Ado-cbl	*Uncultured archaeon clone gwa2_scaffold_75* *Uncultured archaeon clone GWB1_scaffold_43* *Archaeon GW2011_AR15* *Uncultured marine thaumarchaeote KM3_88_D08* *Alpha proteobacterium HIMB114*	Environmental sample Environmental sample Environmental sample Environmental sample	−28.20 −28.20 −28.20 −27.20 −26.60

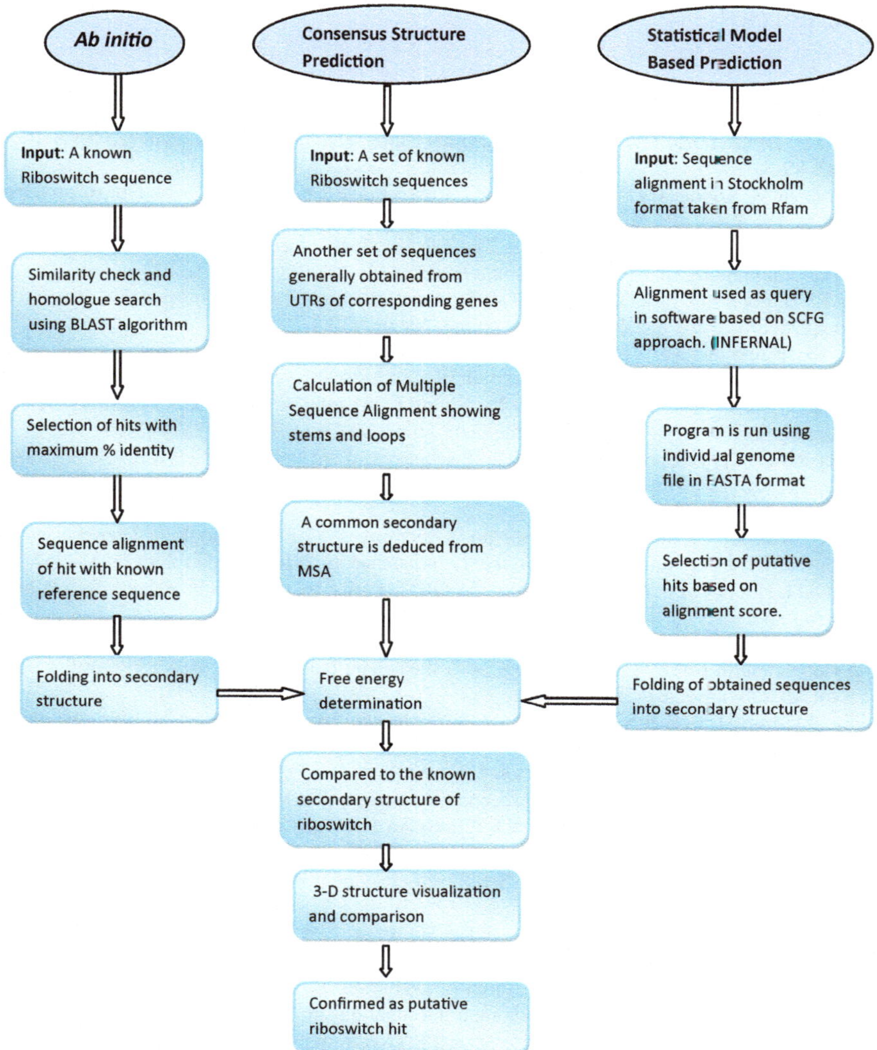

Fig. A.1 Pipeline of methodology

References

Altschul SF, Gish W, Miller W, Myers EW, Lipman DJ (1990) Basic local alignment search tool. J Mol Biol 215:403–410

Ames TD, Breaker RR (2011) Bacterial aptamers that selectively bind glutamine. RNA Biol 8:82–89

Anathraman V, Koonin EV, Aravind L (2002) Comparative genomics and evolution of proteins involved in RNA metabolism. Nucl Acids Res 30:1427–1464

Barrick JE and Breaker RR (2007) The distributions, mechanisms and structures of metabolite-binding riboswitches. Genome Biol R:839

Batey RT, Gilbert SD, Montange RK (2004) Structure of a natural guanine-responsive riboswitch complexed with the metabolite hypoxanthine. Nature 432:411–415

Block KF, Hammond MC, Breaker RR (2010) Evidence for widespread gene control function by the ydaO riboswitch candidate. J Bacteriol 192:3983–3989

Bocobza S, Adato A, Mandel T, Shapiro M, Nudler E, Aharoni A (2007) Riboswitch-dependent gene regulation and its evolution in the plant kingdom. Genes Dev 21:2874–2879

Breaker RR (2011) Prospects for riboswitch: discovery and analysis. Mol Cell Rev 436:867–880

Breaker RR (2012) Riboswitches and the RNA world. Cold Spring Harb Perspect Biol. doi:10.1011/cshperspecta003566

Brochier-Armanet C, Bousseau B, Gribaldo S, Forterre P (2008) Mesophilic Crenarchaeota: proposal for a third archaeal phylum, the Thaumarchaeota. Nat Rev Microbiol 6:245–252

Brown AM, Hoopes SL, White RH, Sarisky CA (2011) Purine biosynthesis in archaea. Biol Direct 6:63

Burns DG, Janssen PM, Itoh T, Kamekura M, Li Z, Bolhuis H, Dyall-Smith ML (2007) Haloquadratum walsbyi gen nov. spec. nov., the square haloarchaeon of Walsby. Intl. J Syst Evol Microbiol 57:387–392

Collins JA, Irnov I, Becker S, Winkler WC (2007) Mechanism of mRNA destabilization by glmS ribozyme. Genes Dev 21:3356–3368

Commichau FM, Dickmanns A, Gundlach J, Filner R, Stulke J (2015) A jack of all trades: the multiple roles of the unique essential second messenger c-di-AMP. Mol Microbiol 97:189–204

Cromie MJ, Shi Y, Latifi T, Groisman EA (2006) An RNA sensor for extracellular Mg^{2+}. Cell 125:71–84

Ferre'-D'Amaré AR, Scott WG (2010) Small self-cleaving ribozymes. CSH Perspect Biol 2: a003574

Dann CE III, Wakeman CA, Sieling CL, Baker SC, Irnov I, Winkler WC (2007) Structure and mechanism of a metal-sensing regulatory mRNA. RNA Cell 130:878–892

Eckburg PB, Bik EM, Bernstein CN, Purdom E, Dethlefsen L, Sargent M, Gill SR, Nelson KE, Relman DA (2005) Diversity of the human intestinal gut flora. Science 308:1635–1685

Eddy SR (2015) Homology searches for structural RNAs: from proof of principle to practical use. RNA 21:605–607

Elkins JG, Podar M, Graham DE, Makarova KS, Wolf Y, Randau L, Hadlund BP, Brochier-Armanet C, Koonin Y, Anderson I (2008) A korarchaeal genome yields insights into the evolution of archaea. Proc Natl Acad Sci USA 105:8102–8107

Fedor MJ (2009) Comparative enzymology and structural biology of RNA self-cleavage. Ann Rev Biophys 38:271

Gaci N, Borrel G, Tottey W, O'Toole PW, Brugere J-F (2014) Archaea and the human gut: new beginnings to an old story. World J Gastroenterol 20:16062–16078

Gao A, Serganov A (2014) Structural insights into recognition of c-di-AMP by the ydaO riboswitch. Nat Chem Biol 10:787–792

Gill SR, Pop M, DeBoy RT, Eckburg PB, Turnbaugh PJ, Samuel BS, Gordon JC, Rehnan DA, Fraser-Ligget CM, Nelson KE (2006) Metagenomic analysis of the human distal gut microbiome. Science 312:1355–1359

Gupta A, Swati D (2016) Exploring riboswitches in archaeal metagenomes. J RNAi Gene Silencing 12:536–543

Guy L, Ettema TJ (2011) The archaeal TACK superphylum and the origin of eukaryotes. Trends Microbiol 19:580–587

Hampel KJ, Tinsley MA (2006) Evidence for pre-organization of the glmS ribozyme ligand binding pocket. Biochemistry 46:7861–7871

Hengge R, Grundling A, Jenal U, Ryan R, Yildiz F (2016) Bacterial signal transduction by cyclic di-AMP and other nucleotide second messengers J Bacteriol 198:15–26

Kim JN, Roth A, Breaker RR (2007) Guanine riboswitch variants from *Mesoplasma florum* selectively bnd 2′ deoxyguanosine. Proc Natl Acad Sci U S A 104:16092–16097

Kim JN, Breaker RR (2008) Purine sensing by riboswitches. Biol Cell 100:1–11

Lazar CS, Podar M, Graham DE, Makarova KS, Wolf Y, Randau L, Hadlund BP, Brochier-Armanet C, Koonin EV, Anderson I (2008) A korarchaeal genome reveals insights into the evolution of the archaea. Proc Natl Sci Acad U S A 107:8102–8107

Lazar CS, Baker BJ, Seitz K, Hyde AS, Dick GJ, Hinrichs K, Teske AP (2016) Genomic evidence of distinct carbon substrate preferences and ecological niches of Bathyarchaeota in estuarine sediments. Environ Microbiol 18:1200–1211

Mandal M, Breaker RR (2004a) Gene regulation by riboswitches. Nat Rev Mol Cell Biol 5:452–463

Mandal M, Breaker RR (2004b) Adenine riboswitch and gene activation by disruption of a transcription terminator. Nat Struct Mol Biol 11:29–35

Mandal M, Boese B, Barrick JE, Winkler WC, Breaker RR (2003) Riboswitches control fundamental biochemical pathways in *Bacillus subtilis* and other bacteria. Cell 113:577–586

Mironov AS, Gusarov I, Rafikov R, Lopez LE, Shatalin K, Kreneva RA, Perumov DA, Nudler E (2002) Sensing small molecules by nascent RNA: a mechanism to control transcription in bacteria. Cell 111:747–756

Montange R, Batey RT (2008) Riboswitches: emerging themes in RNA structure and function. Ann Rev Biophys 37:117–133

Nahvi A, Barrick JE, Breaker RR (2004) Coenzyme B_{12} riboswitches are widespread genetic control elements in prokaryotes. Nucl Acids Res 32:143–150

Nakahira Y, Ogawa A, Asano H, Oyama T, Tozawa Y (2013) Theophylline structure dependent regulation by riboswitches. Plant Cell Physiol 54:1724–1735

Nawrocki EP, Eddy SR (2013) Infernal 1.1: 100-fold faster RNA homology searches. Bioinformatics 29:2933–2935

Nawrocki EP, Burge SW, Bateman A, Daub J, Eberhartt RY, Eddy SR, Hodan EW, Gardner PP, Jones TA, Tate J, Finn RD (2014) Rfam 12.0: updates to the RNA families database. Nucl Acids Res. doi:10.1093/nar/gku1063

Nelson JW, Sudarsan N, Furnkova K, Weinberg Z, Wang JX, Breaker RR (2013) Riboswitches in eubacteria sense the second messenger c-di-AMP. Nat Chem Biol 9:834–839

Nudler E, Mironov AS (2004) The riboswitch control of bacterial metabolism. Trends Biochem Sci 29:11–17

Offre P, Spanga A, Schleper C (2013) Archaea in biogeochemical cycles. Annu Rev Microbiol 67:437–457

Orell A, Frols S, Albers S-V (2013) Archaeal biofilms: the great unexplored. Ann Rev Microbiol 67:337–354

Ren A, Patel DJ (2014) c-di-AMP binds the ydaO riboswitch in two pseudo-related-symmetry-related pockets. Nat Chem Biol 10:780–787

Rodionov DA, Vitreschak AG, Mironov AA, Gelfand MS (2003) Regulation of lysine biosynthesis and transport genes in bacteria: yet another RNA riboswitch? Nucl Acids Res 31:6748–6757

Romling U (2008) Great times for small molecules: c-di-AMP, a second messenger candidate in bacteria and archaea. Sci Signall 1. doi:10.1126/scisignal.133pe39

Roth A, Breaker RR (2009) The structural and functional diversity of metabolite binding riboswitches. Ann Rev Biochem 78:304–334

Seehafer C, Kalweit A, Steger G, Graf S, Hamman C (2011) From alpaca to zebrafish: hammerhead ribozymes wherever you look. RNA 17:21–26

Serganov A, Patel DJ (2007) Ribozymes, riboswitches and beyond: regulation of gene expression without proteins. Nat Rev Genet 8:776–790

Serganov A, Yuan YR, Pikovskaya O, Polonskaia A, Malinina L, Phan AT, Hobartner C, Micur R, Breaker RR (2004) Structural basis for discriminative regulation of gene expression by adenine- and guanine-sensing mRNAs. Chem Biol 11:1729–1741

Sharma M, Bulusu G, Mitra A (2009) MD simulation of a ligand-bound and ligand-free aptamer: molecular level insights into the binding and switching mechanism of *add*A riboswitch. RNA 15:1673–1692

Smith KD, Strobel SA (2011) Interactions of the c-di-AMP with its second messenger ligand. Biochem Soc Trans 39:647–651

Spang A, Hatzenpichler R, Brochier-Armanet C, Raffei T, Tischler P, Spiek E, Streit W, Stahl DA, Wagner M, Schleper C (2010) Distinct gene set in two different lineages of ammonia-oxidizing archaea supports the phylum Thaumarchaeota. Trends Microbiol 18:331–340

Spang A, Saw JH, Jorgensen SL, Zaremba-Niedzwiedzka K, Markjin J, Lind AE, Eijk RV, Schleper C, Guy L, Ettema TJ (2015) Complex archaea that bridge the gap between prokaryotes and eukaryotes. Nature 521:173–179

Sudarsan N, Barrick JE, Breaker RR (2003) Metabolite-binding RNA domains are present in the genes of eukaryotes. RNA 9:6544–6547

Sudarsan N, Lee ER, Weinberg Z, Ming RH, Kim JN, Link K, Breaker RR (2008) Riboswitches in eubacteria sense the second messenger cyclic di-GMP. Science 321:411–413

Vanwonterghem I, Evans PN, Parks DH, Jensen PD, Woodenroft BJ, Hoogenholtz P, Tyson GW (2016) Methylotrophic methanogenesis discovered in the archaeal phyla Verstraetearchaeota. Nat. Microbiology 1:1–9

Vitreschak AG, Rodionov DA, Mironov HH, Gelfand MS (2004) Riboswitches: the oldest mechanism for the regulation of gene expression. Trends Genet 20:44–50

Wachter A, Tunc-Ozdemir M, Grove BC, Green PJ, Shintani DK, Breaker RR (2007) Riboswitch control of gene expression in plants by splicing and 3′ end processing of mRNAs. Plant Cell 19:3437–3450

Winkler WC, Nahvi A, Breaker RR (2002) Thiamine derivatives bind messenger RNAs directly to control gene expression. Nature 419:952–956

Winkler WC, Nahvi A, Roth A, Collins JA, Breaker RR (2004) Control of gene expression by a natural metabolite-responsive ribozyme. Nature 428:281–286

Woese CR, Fox GE (1977) Phylogenetic structure of prokaryotic domain: the primary kingdoms. Proc Natl Acad Sci U S A 74:5088–5090

Woese CR, Kandler O, Wheelis ML (1990) Towards a natural system of organisms: proposal for the domains archaea, bacteria and eukarya. Proc Natl Acad Sci U S A 87:4576–4579

Xin Y, Hamelberg D (2010) Deciphering the role of glucosamine-6-phosphate in the action of glmS ribozyme. RNA 16:2455–2463

Yadav S, Swati D, Chandrashekharan H (2015) Thiamine pyrophosphate riboswitch in some representative plant species: a bioinformatics study. J Comp Biol 22:1–9

Zaremba-Niedzwiedzka K, Caceras EF, Saw JH, Backstrom D, Juzokaite L, Vancaster E, Seitz KW, Ananthraman K, Starnowski P, Kjeldsen KU, Stott MB, Nunoura T, Banfield JF, Schramm A, Baker KJ, Spang A, Ettema TJ (2017) Asgard archaea illuminate the eukaryotic cellular complexity. Nature 1:1–9

Zuker M (2003) Mfold webserver for nucleic acid folding and hybridization prediction. Nucl Acids Res 31:3406–3415

URLs

MFOLD http://www.bioinfo.rpi.edu/applications/mfold/rna/form1.cgi
RNAfold (a part of Vienna Package) http://rna.tbi.univie.ac.at/cgi-bin/RNAfold.cgi
Infernal http://infernal.janelia.org/
RNABOB http://www.eddylab.org/software/rnabob/
NCBI Genome (http://www.ncbi.nlm.nih.gov/genome/)

DNA Damage Repair in Archaea

Qunxin She, Xu Feng and Wenyuan Han

1 Introduction

In their pioneer research of classifying organisms using 16S rRNA gene sequences, Woese and Fox (1977) identified a new form of organisms consisting of extremophilic microbes, including extremely thermophilic organisms like *Sulfolobus* and *Pyrococcus*, halophilic organisms and methane-producer, and this eventually led to the classification of Archaea, the third domain of life (Woese et al. 1990). Comparative genomics of the microbial genome data has revealed that the eukaryal feature for information-processing machineries in Archaea: their replication machinery is of eukaryotic type, whereas the proteins involved in RNA transcription and protein translation are more closely related to the eukaryotic counterparts (Edgell and Doolittle 1997; Olsen and Woese 1997). These analyses have also led to the identification of proteins implicated in several different pathways of DNA repair (Fig. 1), including base excision repair (BER), nucleotide excision repair (NER), homologous recombination repair (HRR), and translesion DNA synthesis (TLS).

As for the information-processing machineries, a number of archaeal DNA repair proteins are also more closely related to those responsible for DNA repair in eukaryotes. For the historical reasons as described above, all current model archaea are extremophiles, and extremophilic archaea represent excellent models for investigating DNA damage repair mechanisms because these organisms are chronically under energy stress (Valentine 2007). As a result, these archaea have to develop more efficient means to deal with DNA damage. In addition, thermophilic organisms are particularly prone to DNA damage since spontaneous hydrolysis of purine and pyrimidine bases, sugar phosphate cleavage and deamination of cytosine in DNA accelerate as temperature increases. Nevertheless, spontaneous mutation

Q. She (✉) · X. Feng · W. Han
Department of Biology, University of Copenhagen, BioCenter,
Ole Maaloes Vej 5, 2200 Copenhagen N, Denmark
e-mail: qunxin@bio.ku.dk

G. Witzany (ed.), *Biocommunication of Archaea*,
DOI 10.1007/978-3-319-65536-9_18

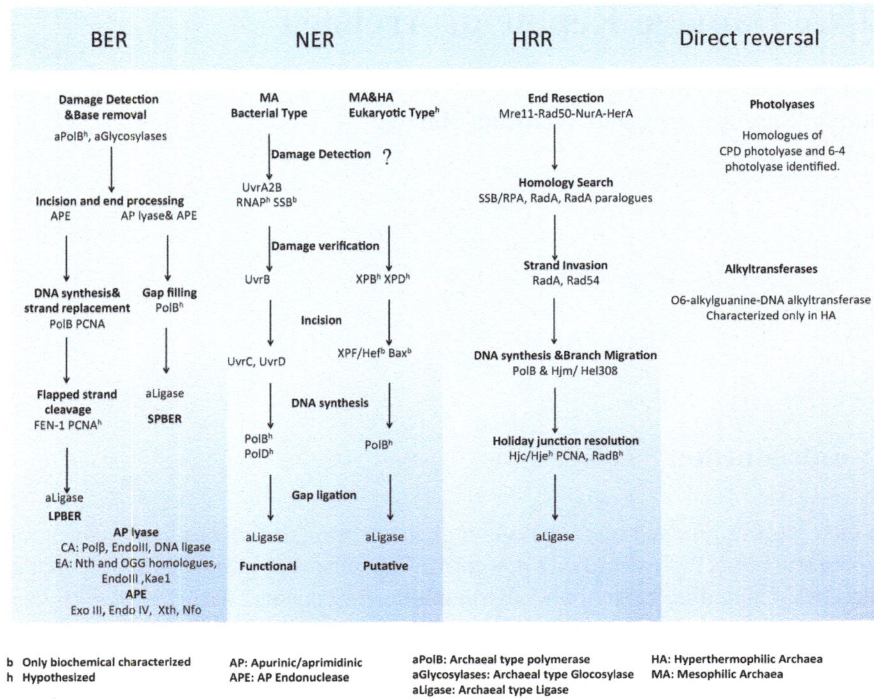

Fig. 1 Known and predicted DNA repair pathways in Archaea. *BER* Base excision repair; *NER* Nucleotide excision repair; *HRR* Homologous recombinational repair

rates in these organisms are maintained to a level comparable to that occurred in the mesophilic bacterium *Escherichia coli* (Grogan et al. 2001). This suggests that archaea possess more efficient DNA repair systems. When archaeal genome sequence became available about 20 years ago, a number of archaeal homologs of DNA repair proteins were identified and immediately characterized. More recently, a few novel enzymes were characterized and implicated in DNA repair in archaea. Furthermore, an emerging picture of archaea-specific DNA damage response (DDR) regulation is taking shape in which novel factors are implicated in DDR. Here we summarize the current knowledge of DNA repair in Archaea obtained from archaeal DNA repair research in the past 20 years.

2 DNA Repair Pathways in Archaea

2.1 Direct Reversal of Base Damage

Two categories of DNA damage are subjected to repair via direct reversal repair, i.e. (a) CPDs (cyclobutane pyrimidine dimers) and 6-4 photoproducts (6-4PP) caused by radiation of short wavelength UV lights, and (b) alkylated bases generated in DNA by alkylating agents. Damage reversal in DNA is the most direct mode of DNA repair since only a single enzyme is required in each pathway. As a result, the processes are also generally error-free because the involved reactions are highly specific.

Two distinct enzymes have been identified to directly repair UV irradiated DNA lesions: i.e. CPD photolyase and 4-6PP photolyase. The former only repairs CPD distortions while the latter specifically removes 4-6PP lesions. Some archaea encode both enzymes, such as *Methanobacterium thermoautotrophicum*, *Halobacterium* NRC1 and *S. acidocaldarius*. As for all known UV photolyases, the archaeal enzymes are also capable of directly removing DNA lesions, and the reaction is driven by photon energy derived from the blue light (Kiener et al. 1989; McCready and Marcello 2003; Baliga et al. 2004; Fujihashi et al. 2007; Kiontke et al. 2011; Sakofsky et al. 2011). Genetic study of gene function in archaea further shows that only CPD photolyase is important for DNA repair in these archaea since deletion of the gene coding for 4-6PP photolyase does not impair DNA repair activity in the mutants (Sakofsky et al. 2011). In *E. coli*, UV lights primarily generate CPDs whereas 4-6PPs only comprise a small portion of UV-irradiated DNA lesions. This is probably why CPD photolyase is more important than 4-6PP photolyase in mediating UV resistance in these archaea.

DNA alkylation is a very common lesion in DNA, and it is highly mutagenic. Alkylating agents, such as the commonly used methyl methanesulfonate (MMS), react with DNA to produce various O-alkylated and N-alkylated products. Four main alkylated bases are known, including O^6-alkylguanine, O^4-alkylthyamine, 1-methyladenine and 3-methylcytosine. Enzymes that are capable of correcting these base alkylations have been identified in different bacteria and eukaryotes. To date, only a few archaeal homologs have been characterized. Three of the archaeal enzymes implicated in the direct removal repair have been characterized, including a DNA alkyltransferases of the archaeon *Pyrococcus* sp. KOD1 (Leclere et al. 1998), a O^6-alkylguanine DNA-protein alkyltransferase of *S. solfataricus* (Perugino et al. 2012) as well as the *Ferroplasma acidarmanus* AGTendoV (Kanugula et al. 2005), the last of which is a fusion of the C-terminal active site domain of AGT with an endonuclease V domain. Interesting, AGTendoV repairs O^6-methylguanine lesions in DNA via alkyl transfer action, although the enzyme carries alterations in key amino acid residues in the active site and it lacks the entire N-terminal domain. In addition, the enzyme also shows the EnodV activity. Alkylation lesions are highly toxic at high temperatures since unrepaired alkylated bases are easily converted into an apurinic or apyrimidinic or abasic (AP) site or DNA breaks that in turn can cause DNA fragmentation. Moreover, investigation of the *S. solfataricus* enzyme has revealed that the alkytransferase is degraded in these cells in an MMS concentration-dependent fashion, and the degradation is a direct effect of the protein alkylation (Perugino et al. 2012). Therefore, the mechanism of directly repairing DNA alkylation is well conserved in all three domain of life.

2.2 Base Excision Repair

Two main sources of DNA damage are to be repaired by BER including deamination and hydrolytic depurination of bases as well as alkylated bases that have not been corrected by direct reversal on DNA molecules. These DNA lesions occur spontaneously due to the intrinsic instability of the genetic material, and furthermore, the formation of these DNA lesions can be facilitated by cellular agents and environmental factors such as reactive oxygen species and ionic radiations. A survey on sources of DNA damage has revealed that base deamination and hydrolytic depurination constitute the top causes of DNA damage occurred for human cells every day (Lindahl 1993). The conservation of the mechanisms in repairing these lesions in all three domains of life suggests the pathways could be present already in the last common ancestor.

Deamination of a base on DNA generates a mutated base that should be corrected to prevent mutation. The first step towards the damage correction in BER is to specifically excise the damaged base from the sugar-phosphate backbone by an enzyme named DNA glycosylase. This leads to the formation of an AP site, which is the same as the DNA lesion resulting from hydrolytic depurination, the most common DNA damage source in all cellular organisms. AP sites are repaired either by a short patch or a long patch pathway of BER: in the former it is AP lyase that makes the first incision at the AP site whereas in the latter AP endonuclease is the incision enzyme. A gap of a single nucleotide gap or 2-11 nt is generated in the process, which is filled up by DNA synthesis (reviewed in Krokan and Bjørås 2013).

Bacteria and eukaryotes encode a large number of DNA glycosylases, and several archaeal homologs have identified and characterized, including DNA alkyltransferase, endonuclease III and uracil DNA glycosylases (UDGs). To date, about 20 archaeal UDGs have been characterized, and they fall into four of the six known UDGs families, including Class II, IV, V and VI. All known DNA glycosylases contain two conserved regions: the active site responsible for generating catalytic water molecule and a DNA-binding motif interacting with the DNA minor groove to stabilize the DNA-UDG complex (Kiyonari et al. 2009). Since these UDGs recognize different base lesions for DNA repair, this suggests that UDGs have co-evolved with their organisms in recognizing a specific type of DNA lesion for correction. Indeed, whereas mesophilic archaea code for UDGs of Class II and VI, the enzymes present in extremely thermophilic archaea are of Class IV and V. Biochemical characterization of Class II and VI UDGs shows that, while the former removes T and U from G/T and G/U mismatches, the latter specifically cleaves hypoxanthine from DNAs. For Class IV and V UDGs, only the former class of enzyme appears to be universal in these organisms. Strikingly, thermophilic bacteria also encode Class IV and V UDGs, further supporting the diversification of UDGs among different physiological groups of organisms.

In a recent review on the archaeal BER pathway, both a short and a long patch BER pathways have been described (Grasso and Tell 2014). In the short patch BER, bifunctional enzymes such DNA glycosylases that also exhibit an AP lyase activity make an incision at the 3′ end of the AP site. The resulting nick is then processed by an AP endonuclease (APE) to remove the protruding deoxyribose phosphate. This yields a single nucleotide gap on the DNA, which is then filled up the gap by DNA replication and ligation. In the long patch BER, the AP site is recognized by an AP endonuclease, generating a 3′-hydroxyl group that can be recognized by DNA polymerase to initiate DNA synthesis to displace the DNA strand containing the AP site. The displaced DNA fragment is then removed by a Flag endonuclease and the nick is finally sealed by a DNA ligase.

A number of archaeal enzymes exhibiting an AP lyase activity have been identified. Most of them are DNA glycosylases although some DNA polymerase and DNA ligase are also active in the β-elimination reaction. On the other hand, several archaeal homologs of endonuclease IV (EndoIV) and exonuclease III (ExoIII) show APE activity among which the *S. islandicus* ExoIII and EndoIV have been characterized both in vitro and in vivo (Yan et al. 2016). The comparative analyses have revealed that the *S. islandicus* EndoIV enzyme is ca. 30-fold more active than the ExoIII enzyme in vitro whereas genetic analysis of the corresponding gene knockout mutants has confirmed that EndoIV plays a more important role in BER than ExoIII in this archaeon because Δ*endoIV* is much more sensitive to the alkylating agent MMS than Δ*exoIII*.

In addition, *P. furiosus* UDG and APE may form a complex with to search for mutated bases since both enzymes interact with proliferation cell nuclear antigen, the replication clamp. Once a base lesion is encountered, the mutated base is removed by the action of UDG, and subsequently, APE makes an incision on the phosphodiester backbone, producing a 3′-OH end. Since the enzyme can enlarge the AP site by its exonuclease activity, it is reasoned that *P. furiosus* most likely to use a long patch BER to repair DNA damage. Nevertheless, whether both mechanisms function in a given archaeon remains to be experimentally tested.

Another conserved endonuclease that may play an important role in mutated bases is a specific endonuclease called endonuclease V (EndoV), which cleaves the second phosphodiester bond on the 3′-side of the deaminated base site. Several archaeal EndoV enzymes have been expressed as recombinant proteins and characterized. Endo V enzymes of *Archaeoglobus fulgidus* (Liu et al. 2000) and *P. furiosus* (Kiyonari et al. 2014) exhibit a specific endonuclease activity for dI-containing substrates in vitro, suggesting they likely function in DNA repair in these archaea. Interestingly, some EndoVs work both on DNA and on RNA, implying that they can have multiple functions in archaea.

2.3 Nucleotide Excision Repair

Nucleotide excision repair (NER) pathways remove a wide spectrum of DNA lesions including bulky adducts, pyrimidine dimers, intrastrand crosslinks and helix-distorting lesions from DNA. The process involves a coordinated action of DNA damage detection, recruitment of NER protein onto the damage site, opening up the dsDNA at the damage site by NER helicases, excision of a patch of nucleotide containing the damaged site by NER nucleases, and finally, the resulting single-stranded DNA gap is fulfilled by DNA synthesis. Two main NER pathways are known: global genomic repair (GGR) and transcription-coupled repair (TCR), which differ in mechanisms of DNA lesion recognition. The latter depends on stalling of RNA polymerase during transcription to initiate the repair pathway, while the former detects DNA lesions by scanning for abnormal regions on dsDNA using specific NER proteins and the process is independent of transcription (Kisker et al. 2013).

While the general mechanism of NER is conserved in bacteria and eukaryotes, the proteins responsible for each NER step in the two domains of life show no sequence homology. Bacterial NER proteins are named as UvrABCD since they were identified as the proteins responsible for repairing DNA damage caused by irradiation of ultraviolet lights (reviewed in Kisker et al. 2013). The functions of each Uvr proteins in *E. coli* are known. In global genomic repair, two UvrA and two UvrB proteins form a complex to scan DNA for distortions. Once encountered, the protein complex remains bound at the damaged site to verify the DNA lesion. Then UvrB helicase open up the double stranded DNA, allowing UvrC nuclease to make nucleolytic incisions on either side of the lesion. Finally, UvrD DNA helicase removes the damaged DNA strand, and the resulting single stranded DNA gap is to be filled up by DNA synthesis with the nick finally sealed by DNA ligase. Although there are only a few mesophilic archaea that encode homologs of the bacterial UvrABCD proteins, it has been demonstrated that these NER homologs function in DNA repair in these archaea (Crowley et al. 2006), and they exhibit transcription-coupled DNA repair activity (Stantial et al. 2016). Nevertheless, the archaeal UvrABC system is likely to be gained from bacteria via horizontal gene transfer since it is absent from most archaeal lineages.

On the other hand, archaea in general code for homologs of eukaryotic NER proteins, including XPF and XPG nucleases and XPB and XPD helicases. The names of these NER proteins are derived from the genetic disease Xeroderma Pigmentosum since deficiencies in any of these proteins lead to the skin disease in humans. A number of archaeal homologs of eukaryotic NER proteins have been characterized. Archaeal XPD and XPB proteins exhibit helicase activity in vitro and unwind DNA from the 3′ side. Further, archaeal XPB homologs form complexes with novel nucleases, such as Bax1 in *Sulfolobus* species, and the complex acts on the 5′ side of the DNA bubble. Therefore, it has been suggested that archaeal XPB and XPD helicases work at both sides of the damaged DNA to open the bubble, followed by cleavage of Bax1 and another nuclease to remove the damaged DNA (Rouillon and White 2011). These putative archaeal NER proteins provide good models for studying functions of their eukaryotic homologs in DNA repair since several eukaryotic NER proteins have multiple functions, e.g. XPB and XPD are components of TFIIH, an essential transcriptional factor in eukaryotes.

Nevertheless, genetic studies reveal that archaeal XP homologs may not have a main role in correcting DNA lesions since deletion of individual *xpb* and *xpd* genes does not impair DNA damage repair in two thermophilic archaea (Fujikane et al. 2010; Zhang et al. 2013). In addition, proteins that are responsible for damage detection in eukaryotes, including CSA/CSB and XPA, among which CSA and CSB represent proteins linked to Cockayne syndrome whereas XPA is associated with Xeroderma Pigmentosum, do not have a recognizable homolog in Archaea. Taken together, whether or not these archaeal XP proteins function in DNA repair in archaea remain to be further investigated.

2.4 Homologous Recombination Repair

Double strand break (DSB) is the most deleterious type of DNA damage in any cells because a single DSB is sufficient to kill the cell deficient in repairing such a DNA lesion. Two distinct strategies are known to repair DSBs: non-homologous end joining (NHEJ) and homologous recombination repair (HRR).

NHEJ is defined as a means of DNA repair in which two DSB ends are joined together by direct ligation. The pathway relies on the specific DNA repair protein Ku70 and Ku80, DNA ligase IV and other associated factors (Chiruvella et al. 2013). The prototypical NHEJ protein is Ku, which is a heterodimer of Ku70 and Ku80. Upon the occurrence of DSBs, the Ku protein complex binds to the ends of DSBs. The DNA ligase is then recruited to join the DNA ends together. Since the Ku factor allows random joining of any two ends of DSBs, mutation occurs upon the end-joining of ends derived from different DSBs. Interestingly, it has been documented that *H. volcanii* efficiently employs NHEJ to repair DSB ends with micro-homology even though no archaea are known to code for any Ku protein homolog (Delmas et al. 2009).

For HRR, different pathways are found to operate in bacteria and eukaryotes (Symington 2014). The two main pathways in bacteria are RecBCD and RecFOR, but no homologs of each complex have been identified in any archaea. By contrast, homologs of eukaryotic HR proteins Rad50 and Mre11 have been identified in all archaeal genomes. In eukaryotes, the Rad50 and Mre11 proteins, in association with a third protein partner (Xrs2 in yeast, Nbs1 in vertebrates), are responsible for the resection of DSB ends. In many archaea, genes coding for Rad50 and Mre11 form an operon with two other genes coding for HerA and NurA that have helicase and nuclease activity, respectively (Constantinesco et al. 2002, 2004). Subsequently, it was found that the archaeal Rad50 and Mre11 form a complex and interact with HerA and NurA to resect 5′ strand at a DNA end, generating 3′ single-stranded DNA that can be used by the archaeal RadA recombinase to catalyze strand invasion and exchange (Hopkins and Paull 2008).

The second step involves homologous DNA pairing and strand exchange. Enzymes promoting the process belongs the ubiquitous RecA family of recombinase, and the archaeal one is named RadA. RadA forms nucleoprotein filaments with the single-stranded DNA generated by archaeal resection enzyme complex in the initiation step, which invade a homologous DNA duplex for yield a three- or four-stranded intermediates termed Holliday junction (HJ). DNA synthesis is invoked to make an identical copy of DNA, producing the two identical copies of DNA in the HJ. In Archaea, Hjm helicase and Hjc nuclease probably perform branch migration and resolve HJ, respectively (White 2011). Mre11, Rad50, HerA, NurA, RadA and Hjm are indispensable for cell survival in Sulfolobus (Huang et al. 2015), indicating that HR activity is essential for Sulfolobus.

Homologous recombination (HR) mediates genetic exchange between identical or similar DNA molecules that provides the basis for meiosis and various mechanisms of DNA repair. When DNA replication forks encounter DNA damage, in the form of template breaks or lesions, HR provides repair by utilizing an undamaged DNA strand to reactivate the stricken replication fork. In this regard, HR mediated repair of stalled DNA replication forks at a high fidelity whereas end-joining processes that result in loss of base pairs, generating mutations. In general, HR repair is kept under strict control in all cellular organisms to balance the need to re-start stalled replication forks versus the potential for generation of genetic diversity.

2.5 *Mismatch Repair in Extremely Thermophilic Archaea Involves Novel Enzymes*

Another surprising finding in archaeal comparative genomics is the lack of the conventional mismatch DNA repair system (MutS and MutL proteins) in many extremely thermophilic archaeal genomes since the corresponding MMR proteins are well conserved in bacteria and eukaryotes and mesophilic archaea. However, extremely thermophilic archaea must encode some unique DNA repair mechanisms to ensure the low spontaneous mutation rates observed for hyperthermophiles as discussed above. More recently, an important research breakthrough has been made in studying *P. furiosus* DNA repair enzymes. A special endonuclease termed EndoMS is capable to cleave mismatched bases on double strand DNAs, including G/T, G/G, T/T, T/C and A/G mismatches. The cleavage produces 5′-protruding ends with the mismatched base pair in the central position, which can represent the initial step of a novel mismatch repair process (Ishino et al. 2016). Further researches are required to prove or disprove the hypothesis.

2.6 *Translesion DNA Synthesis*

During chromosome replication, unrepaired DNA lesions lead to stalled replication forks since the replicase cannot make any DNA synthesis on a damaged template. Nevertheless, DNA lesions can be bypassed by a special class of enzyme called error-prone DNA polymerases, and the process is termed as translesion DNA synthesis (TLS). The archaeal Pol Y family enzymes are of this type, and they are good models for studying the mechanisms of bypass DNA synthesis at damaged sites. Unlike other DNA polymerases, TLS DNA polymerases are able to efficiently bypass various types of damaged template bases, but they also show a low processivity and are error-prone in DNA synthesis since they add a preferred dNTP at any damaged sites (Yang and Woodgate 2007). Interestingly, although error-prone in vitro, Y-family DNA polymerases show a minimal contribution to spontaneous mutation. This suggests that TLS DNA polymerases are effectively excluded from replication forks during normal DNA synthesis. Two Sulfolobus Pol Y enzymes have been investigated extensively and genetic investigation of the gene coding for the Pol Y in *S. acidocaldarius* indicates that the enzyme suppresses certain mutations at particular sites and promotes other mutations elsewhere (Sakofsky et al. 2012), suggesting that TLS can have multiple functions affecting the genome stability in archaea.

3 DNA Damage Response and Regulation

DNA damage repair is an expensive process as it involves numerous genes coding for different DNA repair enzymes except for direct removal of DNA damage where only a single enzyme is required. Therefore, it is only economic for the cell to maintain a low level of expression for DNA repair proteins under physiological condition. Upon DNA damage, bacteria and eukaryotes are known to employ a regulatory mechanism called DNA damage response (DDR) (Giglia-Mari et al. 2011) to activate a large number of genes to up regulate the synthesis of DNA repair proteins and to inhibit the initiation of chromosomal DNA replication, cell cycle progression at the same time. Only after all DNA lesions have been repaired properly, cell cycle will be resumed.

The first step in DDR is the recognition of DNA lesion. In Bacteria, processing of DNA lesion or stalled replication fork will lead to the accumulation of ssDNA, which then will be coated by RecA to form nucleoprotein filaments. The ssDNA bounded RecA could function as a co-protease for the self-cleavage of LexA, the repressor of bacterial SOS promoters. Inactivation of LexA repressor allows the up regulation of over 40 genes involving DNA repair cell cycle regulation (Michel 2005). In eukaryotes, a mechanism named DNA damage checkpoint is used at different stages of cell cycle, involving several DNA damage checkpoint systems (Maréchal and Zou 2013). Two main pathways are know to mediate DDR in eukaryotes: The ataxia telangiectasia mutated (ATM) and the ATM and Rad3-related (ATR) systems. The former is activated by the DSB-MRN (Mre11-Rad50-Nbs1) DNA-protein complex whereas the latter is activated in response to persistence of RPA-bound single-stranded DNAs.

DNA damage response has been investigated only for *Halobacterium* NRC1 (Baliga et al. 2004) and *S. solfataricus* (Gotz et al. 2007; Frols et al. 2007) in archaea. Strikingly, although the haloarchaeon contains genes coding for both the bacterial and the eukaryotic NER pathways, neither set of genes exhibits a significant up regulation of gene expression. In the crenarchaeon, only the eukaryotic NER genes are present but none of them show any DNA damage response, either. This led to the conclusion that these archaea might not show any DNA damage response. Nevertheless, further characterization of organisms of *Sulfolobus* genus has revealed that a number of genes do show elevated levels of expression after UV radiation or other DNA damage treatment. First, UV radiation induces the formation of a special pilus that is essential for cell aggregation of the archaeon (Frols et al. 2008). Since disruption of pilus formation by gene deletion strongly reduces the UV resistance of the mutants, the UV-pilus has been implicated in mediating DNA damage resistance (Ajon et al. 2011). Second, two genes responsible for intercellular DNA transfer have been identified and gene deletion experiments show that loss of the DNA transfer also eliminates the DNA damage resistance (van Wolferen et al. 2015). These findings have two profound implications: (a) It suggests that *Sulfolobus* possesses a novel DNA repair pathway involving UV-inducible pilus formation and intercellular DNA transfer, and (b) this archaeon does show DNA damage response.

Indeed, a number of other genes show differential expression upon DNA damage treatment, and these genes are regarded as DDR genes in analogy to bacterial SOS genes. Two of them are putative transcriptional factors that show strong up regulation, i.e. TFB3 and Cdc6-2. The former is one of the three basal transcription factor TFB while the latter is an ortholog of Cdc6/Orc1 proteins, the archaeal replication initiator (Frols et al. 2007; Gotz et al. 2007). Functional analysis of the three cdc6 genes has confirmed that Cdc6-1 and -3 function in replication initiation whereas Cdc6-2 does not (Samson et al. 2013). In consistence with their function in DNA replication, Cdc6-1 and -3 are expressed to a high level in exponentially growing cells whereas the expression of Cdc6-2 in higher in stationary phase cells of *Sulfolobus* (Robinson et al. 2004). Because the expression of Cdc6-2 is strongly elevated upon UV radiation and treatment with nitroquolonine-1-oxide, a UV-mimic DNA damage agent, this protein has been implicated in DDR (Han et al. 2017). The assumption has been confirmed by genetic analysis of cdc6-2 deletion mutant in which DDR genes no longer show differential expression upon DNA damage treatment (Sun et al. 2017). Therefore, recent researches have revealed that *Sulfolobus* codes for a novel DNA damage response system in which both Cdc6-2 and TFB3 can have very important regulatory functions.

4 Concluding Remarks

The concept of Archaea was first proposed in 1990, and a few years later, genome sequences were determined for a few archaea. Comparative genome analysis has revealed that information-processing machineries encoded in archaeal genomes are of eukaryotic type, including DNA replication, RNA transcription and protein translation. Furthermore, archaeal DNA repair proteins are also either of eukaryotic type or more closely related to their eukaryotic counterparts. This led to a blooming in studying these archaeal homologs of eukaryotic enzymes such as biochemical characterization and crystallographic structural analysis of prominent archaeal DNA repair proteins. The researches have largely confirmed that the predicted DNA repair activity of archaeal enzymes. Furthermore, novel archaeal nucleases implicated in DNA repair have also been identified.

More recently, efficient gene deletion methods have been developed for diverse archaeal models and this allows functional studies of these DNA repair genes to be conducted in these model archaea (Farkas et al. 2013; Peng et al. 2017). An interesting picture is emerging from the genetics of archaeal DNA repair: while BER and HRR are very important for archaeal DNA repair, homologs of eukaryotic NER do not appear to play a role in DNA repair, and the absence of DNA damage-detecting proteins in Archaea supports that archaea XP may have a role other than DNA repair.

Archaea genetics has also led to the discovery of archaea-specific DDR in which Cdc6-2, a DNA replication initiator ortholog functions as a master regulator to regulate DNA damage tolerance. This process also involves TFB3 a basal transcriptional factor. Nevertheless, the DNA damage signal and the DNA repair protein(s) that interact(s) with the signal remain to be identified. Since blocking DNA replication readily induces DNA damage response, a prime candidate for signal transduction could be DSB and proteins interacting with it. In many archaea, there are four HRR proteins, including Mre11, Rad50, HerA and NurA. Most recently, CRISPR-based gene knockout, gene silencing and mutagenesis have been developed for *Sulfolobus* (Peng et al. 2017) and it is also expected that similar genetic tools will soon be developed for other archaeal models. Together, this greatly facilitates the progress of studying DNA damage response regulation and novel DNA repair mechanisms in these fantastic organisms.

References

Ajon M, Frols S, van Wolferen M, Stoecker K, Teichmann D, Driessen AJ, Grogan DW, Albers SV, Schleper C (2011) UV-inducible DNA exchange in hyperthermophilic archaea mediated by type IV pili. Mol Microbiol 82:807–817

Baliga NS, Bjork SJ, Bonneau R, Pan M, Iloanusi C, Kottemann MC, Hood L, DiRuggiero J (2004) Systems level insights into the stress response to UV radiation in the halophilic archaeon Halobacterium NRC-1. Genome Res 14:1025–1035

Chiruvella KK, Liang Z, Wilson TE (2013) Repair of double-strand breaks by end joining. Cold Spring Harb Perspect Biol 5:a012757

Constantinesco F, Forterre P, Elie C (2002) NurA, a novel 5′-3′ nuclease gene linked to rad50 and mre11 homologs of thermophilic Archaea. EMBO Rep 3:537–542

Constantinesco F, Forterre P, Koonin EV, Aravind L, Elie C (2004) A bipolar DNA helicase gene, herA, clusters with rad50, mre11 and nurA genes in thermophilic archaea. Nucleic Acids Res 32:1439–1447

Crowley DJ, Boubriak I, Berquist BR, Clark M, Richard E, Sullivan L, DasSarma S, McCready S (2006) The uvrA, uvrB and uvrC genes are required for repair of ultraviolet light induced DNA photoproducts in *Halobacterium* sp. NRC-1. Saline Syst 2:11

Delmas S, Shunburne L, Ngo HP, Allers T (2009) Mre11-Rad50 promotes rapid repair of DNA damage in the polyploid archaeon *Haloferax volcanii* by restraining homologous recombination. PLoS Genet 5:e1000552

Edgell DR, Doolittle WF (1997) Archaea and the origin(s) of DNA replication proteins. Cell 89:995–998

Farkas JA, Picking JW, Santangelo TJ (2013) Genetic techniques for the archaea. Annu Rev Genet 47:539–561

Frols S, Ajon M, Wagner M, Teichmann D, Zolghadr B, Folea M, Boekema EJ, Driessen AJ, Schleper C, Albers SV (2008) UV-inducible cellular aggregation of the hyperthermophilic archaeon *Sulfolobus solfataricus* is mediated by pili formation. Mol Microbiol 70:938–952

Frols S, Gordon PM, Panlilio MA, Duggin IG, Bell SD, Sensen CW, Schleper C (2007) Response of the hyperthermophilic archaeon *Sulfolobus solfataricus* to UV damage. J Bacteriol 189:8708–8718

Fujihashi M, Numoto N, Kobayashi Y, Mizushima A, Tsujimura M, Nakamura A, Kawarabayasi Y, Miki K (2007) Crystal structure of archaeal photolyase from *Sulfolobus tokodaii* with two FAD molecules: implication of a novel light-harvesting cofactor. J Mol Biol 365:903–910

Fujikane R, Ishino S, Ishino Y, Forterre P (2010) Genetic analysis of DNA repair in the hyperthermophilic archaeon, *Thermococcus kodakaraensis*. Genes Genet Syst 85:243–257

Giglia-Mari G, Zotter A, Vermeulen W (2011) DNA damage response. Cold Spring Harb Perspect Biol 3:a000745

Gotz D, Paytubi S, Munro S, Lundgren M, Bernander R, White MF (2007) Responses of hyperthermophilic crenarchaea to UV irradiation. Genome Biol 8:R220

Grasso S, Tell G (2014) Base excision repair in Archaea: back to the future in DNA repair. DNA Repair (Amst) 21:148–157

Grogan DW, Carver GT, Drake JW (2001) Genetic fidelity under harsh conditions: analysis of spontaneous mutation in the thermoacidophilic archaeon *Sulfolobus acidocaldarius*. Proc Natl Acad Sci U S A 98:7928–7933

Han W, Xu Y, Feng X, Liang Y, Li H, Shen Y, She Q (2017) NQO-induced DNA-less cell formation is associated with chromatin protein degradation and dependent on A0A1-ATPase in *Sulfolobus*. Front Microbiol 8:1480

Hopkins BB, Paull TT (2008) The *P. furiosus* mre11/rad50 complex promotes 5′ strand resection at a DNA double-strand break. Cell 135:250–260

Huang Q, Liu L, Liu J, Ni J, She Q, Shen Y (2015) Efficient 5′-3′ DNA end resection by HerA and NurA is essential for cell viability in the crenarchaeon *Sulfolobus islandicus*. BMC Mol Biol 16:2

Ishino S, Nishi Y, Oda S, Uemori T, Sagara T, Takatsu N, Yamagami T, Shirai T, Ishino Y (2016) Identification of a mismatch-specific endonuclease in hyperthermophilic Archaea. Nucleic Acids Res 44:2977–2986

Kanugula S, Pauly GT, Moschel RC, Pegg AE (2005) A bifunctional DNA repair protein from *Ferroplasma acidarmanus* exhibits O6-alkylguanine-DNA alkyltransferase and endonuclease V activities. Proc Natl Acad Sci U S A 102:3617–3622

Kiener A, Husain I, Sancar A, Walsh C (1989) Purification and properties of *Methanobacterium thermoautotrophicum* DNA photolyase. J Biol Chem 264:13880–13887

Kiontke S, Geisselbrecht Y, Pokorny R, Carell T, Batschauer A, Essen LO (2011) Crystal structures of an archaeal class II DNA photolyase and its complex with UV-damaged duplex DNA. EMBO J 30:4437–4449

Kisker C, Kuper J, Van Houten B (2013) Prokaryotic nucleotide excision repair. Cold Spring Harb Perspect Biol 5:a012591

Kiyonari S, Egashira Y, Ishino S, Ishino Y (2014) Biochemical characterization of endonuclease V from the hyperthermophilic archaeon, *Pyrococcus furiosus*. J Biochem 155:325–333

Kiyonari S, Tahara S, Uchimura M, Shirai T, Ishino S, Ishino Y (2009) Studies on the base excision repair (BER) complex in *Pyrococcus furiosus*. Biochem Soc Trans 37:79–82

Krokan HE, Bjørås M (2013) Base excision repair. Cold Spring Harb Perspect Biol 5:a012583

Leclere M, Nishioka M, Yuasa T, Fujiwara S, Takagi M, Imanaka T (1998) The O 6-methylguanine-DNA methyltransferase from the hyperthermophilic archaeon *Pyrococcus* sp. KOD1: a thermostable repair enzyme. Mol Gen Genet 258:69–77

Lindahl T (1993) Instability and decay of the primary structure of DNA. Nature 362:709–715

Liu J, He B, Qing H, Kow YW (2000) A deoxyinosine specific endonuclease from hyperthermophile, *Archaeoglobus fulgidus*: a homolog of *Escherichia coli* endonuclease V. Mutat Res 461:169–177

Maréchal A, Zou L (2013) DNA damage sensing by the ATM and ATR kinases. Cold Spring Harb Perspect Biol 5:a012716

McCready S, Marcello L (2003) Repair of UV damage in *Halobacterium salinarum*. Biochem Soc Trans 31:694–698

Michel B (2005) After 30 years of study, the bacterial SOS response still surprises us. PLoS Biol 3:e255

Olsen GJ, Woese CR (1997) Archaeal genomics: an overview. Cell 89:991–994

Peng N, Han W, Li Y, Liang Y, She Q (2017) Genetic technologies for extremely thermophilic microorganisms of Sulfolobus, the only genetically tractable genus of crenarchaea. Sci China Life Sci 60:370–385

Perugino G, Vettone A, Illiano G, Valenti A, Ferrara MC, Rossi M, Ciaramella M (2012) Activity and regulation of archaeal DNA alkyltransferase conserved protein involved in repair of DNA alkylation damage. J Biol Chem 287:4222–4231

Robinson NP, Dionne I, Lundgren M, Marsh VL, Bernander R, Bell SD (2004) Identification of two origins of replication in the single chromosome of the archaeon *Sulfolobus solfataricus*. Cell 116:25–38

Rouillon C, White MF (2011) The evolution and mechanisms of nucleotide excision repair proteins. Res Microbiol 162:19–26

Sakofsky CJ, Foster PL, Grogan DW (2012) Roles of the Y-family DNA polymerase Dbh in accurate replication of the Sulfolobus genome at high temperature. DNA Repair 11:391–400

Sakofsky CJ, Runck LA, Grogan DW (2011) Sulfolobus mutants, generated via PCR products, which lack putative enzymes of UV photoproduct repair. Archaea 2011:864015

Samson RY, Xu Y, Gadelha C, Stone TA, Faqiri JN, Li D, Qin N, Pu F, Liang YX, She Q, Bell SD (2013) Specificity and function of archaeal DNA replication initiator proteins. Cell Rep 3:485–496

Stantial N, Dumpe J, Pietrosimone K, Baltazar F, Crowley DJ (2016) Transcription-coupled repair of UV damage in the halophilic archaea. DNA Repair (Amst) 41:63–68

Sun M, Feng X, Liu Z, Han W, Liang Y, She Q (2017) A Orc1 paralogue functions as a global regulator in mediating DNA damage response in Archaea (Submitted for publication)

Symington LS (2014) End resection at double-strand breaks: mechanism and regulation. Cold Spring Harb Perspect Biol 6:a016436

Valentine DL (2007) Adaptations to energy stress dictate the ecology and evolution of the Archaea. Nat Rev Microbiol 5:316–323

van Wolferen M, Ma X, Albers SV (2015) DNA processing proteins involved in the UV-induced stress response of Sulfolobales. J Bacteriol 197:2941–2951

White MF (2011) Homologous recombination in the archaea: the means justify the ends. Biochem Soc Trans 39:15–19

Woese CR, Fox GE (1977) Phylogenetic structure of the prokaryotic domain: the primary kingdoms. Proc Natl Acad Sci U S A 74:5088–5090

Woese CR, Kandler O, Wheelis ML (1990) Towards a natural system of organisms: proposal for the domains Archaea, Bacteria, and Eucarya. Proc Natl Acad Sci U S A 87:4576–4579

Yan Z, Huang Q, Ni J, Shen Y (2016) Distinct catalytic activity and in vivo roles of the ExoIII and EndoIV AP endonucleases from *Sulfolobus islandicus*. Extremophiles 20:785–793

Yang W, Woodgate R (2007) What a difference a decade makes: Insights into translesion DNA synthesis. Proc Natl Acad Sci U S A 104:15591–15598

Zhang C, Tian B, Li S, Ao X, Dalgaard K, Gokce S, Liang Y, She Q (2013) Genetic manipulation in *Sulfolobus islandicus* and functional analysis of DNA repair genes. Biochem Soc Trans 41:405–410

Archaeal Lipids as an Adaptation to Higher Temperatures?

David Penny

1 Introduction

Since the acceptance of the Woese and Fox (1977) hypothesis it has been agreed that prokaryotes (akaryotes in our terminology) are divided into Archaea and Bacteria. Do archaea form different lipids because they were initially adapted to life at higher temperatures? Life as we know it is from −15 to 122 °C (McKay 2014), so it is an interesting question. A similar question is the temperature at which life arose, was it at low temperature, moderate temperatures, or at the highest temperatures. For convenience (Moulton et al. 2000), we suggest these temperatures to be; cold (from under ice to about 20 °C), intermediate (from 20 to 80 °C), and high (above 80 °C). However, the exact points are arbitrary, but the basic principle still holds. The original equation for Gibbs free energy (ΔG) was $\Delta G = \Delta H - T\Delta S$. So the question is whether the origin of life is more of a kinetic problem (ΔH) or more of an entropy problem ($T\Delta S$) (which becomes more difficult as the temperature increases). For example, Attwater et al. (2013) do their experiments in ice, and Martin et al. (2014) favor a higher temperature; above the boiling point of water. Perhaps higher temperatures are the most popular system for the origin of life—most researchers see 'life' as a kinetic problem, not as an 'order' problem—which is enhanced (made more difficult) by higher temperatures.

Several theories are available for the origin of akaryotes (prokaryotes) from a LUCA that had more properties of eukaryotes (Penny 2017, and see Table 1); there are good reasons to consider eukaryotes as 'ancestral', but that does not alter the basic question. Perhaps the first hypothesis is by Patrick Forterre (1995) who proposed that Bacteria (at least) were the result of a high temperature adaptation (although the original paper included both Bacteria and Archaea). This might

D. Penny (✉)
Institute of Fundamental Sciences, Massey University,
Palmerston North, New Zealand
e-mail: Penny@massey.ac.nz

G. Witzany (ed.), *Biocommunication of Archaea*,
DOI 10.1007/978-3-319-65536-9_19

Table 1 Some properties inferred to be in LECA (the Last eukaryotic common ancestor)

Nucleus/cytoplasm subdivision	✓
Nuclear substructures (nucleolus, etc.)	✓
Larger genome size	✓
Exon/Intron structure of genes	✓
Spliceosomal proteins with RNA catalysis	✓
Capping of mRNA	✓
Mitosis and meiosis	✓
Centrioles and centrosomes	✓
Linear chromosomes and histones	✓
Many centers of DNA replication	✓
Apoptosis/argonautes/dicer	✓
Endoplasmic reticulum (and substructures)	✓
Membrane lipids	✓
Epigenetics	✓
Flagella/Cilia	✓
Large genomes	✓
Mitochondria (endosymbionts)	✓

account for the second observation of Fuerst (2013) who points out that the deeply-diverging Planctomycete Bacteria appear to have apparent 'remnants' of a nucleus, this is a second option. A third option is that Bacteria are derived from cytoplasmic viruses that obtained more and more processed genes from a crown group (proto-eukaryotic) cell (Penny et al, 2014). A fourth hypothesis is that the lipids of Archaea (in particular, see Bousseau et al. 2008) are a prior adaptation to a higher temperature). This is the option discussed here. Several lineages subsequently adapted to lower temperatures (López-Garcia et al. 2015). There are probably more, but this is sufficient here to maintain the idea, and the focus here is on the fourth theory (though that does not alter the basic question about the temperature at which life ultimately arose). It is assumed here that the original Last Universal Common Ancestor (LUCA) had a standard membrane—there were many intermediate steps towards LUCA.

2 Archaeal Lipids

So the question here is about archaeal lipids. They are based in the same general principles as other lipids, but have unique characteristics of their own. On the 'same principles' aspect they have 'lipid-like' properties and water-soluble ends of the molecules, thus they have a bipolar aspect that is used to form the membranes. Perhaps the main point is that the lipids of Archaea are different to those of other organisms (Bacteria and Eukaryotes). A surprising aspect about archaeal lipids is that they are unusually temperature-tolerant (Koga (2012)—hence the interest in their properties.

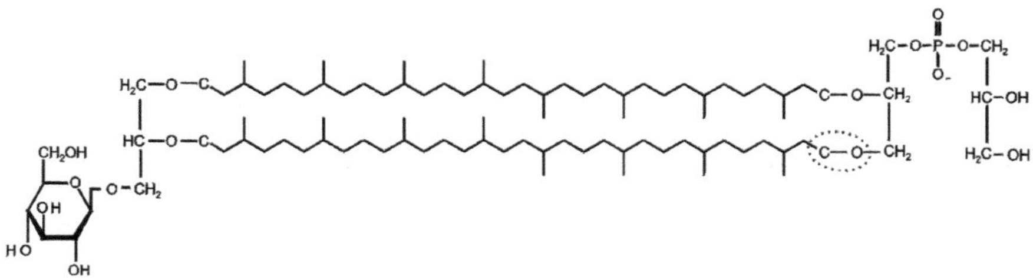

Fig. 1 Archaeal lipids are basically ether-linked isoprenoids. They have the same basic isoprenoid structure, even though the protein layer varies (and it differs from the Bacteria). The ether linkage us shown by a *dotted circle*. Some isoprenoid structures are split in *middle*, there is a wide variety of isoprenoid-based lipids (based on Albers et al. 2000)

Basically, the Archaea have a chain of isoprenoids as their lipids, and are ether linked (not ester linked) to the usual glycerol molecule (Albers and Meyer 2011; Jain et al. 2014), see Fig. 1. Apart from that, the lipids are otherwise variable, sometimes they are continuous isoprenoids, sometimes that apparently split in the middle into two halves (this, in bacteria and eukaryotes, is the more 'normal' situation). Archaeal lipids also have an additional difference, in that the phosphate and hydrocarbons are attached in different places on the glycerol, leading to a glycerol-1-phosphate backbone in Archaea (not the more usual glycerol-3-phosphate, Jain et al. 2014). This doesn't have a huge effect on the function of the lipid, but it is a distinguishing feature of archaeal lipids—this stereochemistry isn't found in any other organisms. Basically, the archaeal lipids are derived as isoprenoids, and which do occur in all three domains of life. So all three domains of life have isoprenoids, but only Archaea has then as lipids. The ether linkage is also found, to a limited extent in Bacteria and Eukaryotes, but not as the main component. Basically, the 'head group' is a glycerol molecule, and the side chains are different.

The question here is whether the early Archaea were initially adapted to higher temperatures. At present the Archaea occupy a range of niches, and at different temperatures, but what was the ancestral condition? This is the question raised here and also by de Wit et al. (2015) and by Lombard and Moreira (2011), even though they do not derive eukaryotes from akaryotes. It is an important question.

The prediction here is that it is desirable to study the temperatures at which membranes can form. The prediction here is that 'membranes' can form at higher temperatures if an isoprenoid is used. Possibly the study of 'black lipid membranes' is a good starting point, and has a long history (Mueller et al. 1962), it is a rather good question that can really only be studied by experiment. Do these membranes form at higher temperatures? The study could be extended to 'Supported lipid bilayers' and 'tethered bilayers' as well. This would be a good starting point. Schouten et al. (2002) do attempt to determine sea temperatures by determining the properties of membranes.

A related question might be the thermo-stability of Coenzyme M (Balch and Wolfe 1979). Is it more stable at higher temperatures, or is it just able to handle a different reaction? This is an additional study that could be investigated. However, the 'origin of life' is a separate question, because we infer that life (as we know it) was a sequential process that was amenable to 'steps along the way'.

3 Discussion

Probably the best answer is 'nobody knows for sure', but that does not stop ideas/hypotheses. We need to do tests of membrane formation at a range of temperatures. Perhaps we favor a lower temperature origin of life. We have shown, for example, that RNA does not fold well at the highest temperatures (Moulton et al. 2000), at least, in the absence of proteins. (The idea of an RNA-world assumes that the first macromolecules were RNA—before proteins and DNA—though it does not exclude other earlier macromolecules.) The experimental results with RNA were carried out at 1 atmosphere pressure, but it does appear that this is not a problem—experiments at high pressures do not appear to alter the main result (Lepper et al. 2017).

Some researchers like to think that the 'tree of life' is well-known, including that the Archaea have at least two clades—one of which leads to the Eukaryotes. However, Mossel and Steel (2004) have shown (mathematically) that the Markov models that we use (and love) are prone to error at the deeper divergences. They lose information (exponentially with time), but they do gain information linearly with the number of sequences. So we have an exponential decay 'fighting' a linear process, and there is no doubt which must win for the deeper divergences. For example, we do not know the orders of the branching of animals, but we appear to go much deeper than this in time, and infer the branching of Eukaryotes and of Archaea. So many biologists appear to reconstruct the deeper convergences, but even though the main animal groups are uncertain. The net result of this observation is that we must keep an open mind about the Archaea.

There are additional differences in the external proteins that make up the outer membrane of the archaeal cell, but the emphasis here is on the membranes. The question of archaeal membranes being adapter to higher temperatures has been

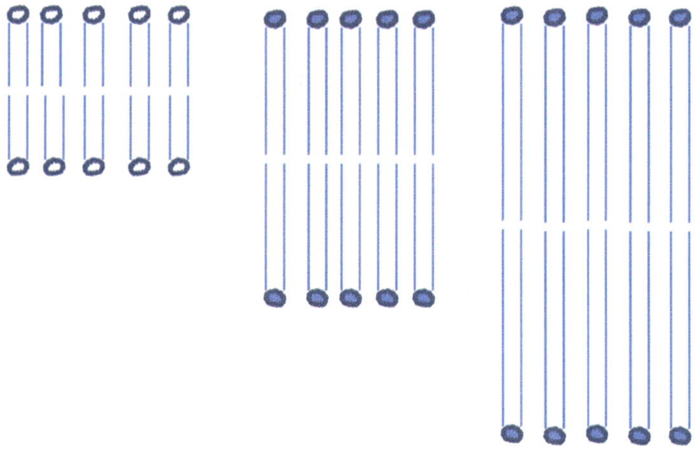

Fig. 2 The origin of life may have been easier with shorter lipids—even though the thickness of the membrane now appears more constant in width. Basically, we expect that the first membranes (during the origin of life) to have been not as wide as at present. The image assumes that non-Archaeal lipids came first!

raised before (e.g. Albers et al 2000). The aim here is to raise questions for future study, and not to come to any conclusions (yet). The ideas presented here are all subject to further testing, and the future is very positive. The main thing is that we should not come to premature decisions (Fig. 2).

References

Albers S-J, van de Vossenberg JLCM, Driessen AJM, Konings WN (2000) Adaptations of Archaeal membranes to heat stress. Front Biosci 5:d706–d803

Albers S-J, Meyer BH (2011) The archaeal cell envelope. Nature Rev Microbiol 9:414–426

Attwater J, Wochner A, Holliger P (2013) In-ice evolution of RNA polymerase activity. Nature Chem 5:1011–1018

Balch WE, Wolfe RS (1979) Specificity and biological distribution of coenzyme M (2-mercaptoethanesulfonic acid). J Bacteriol 137:256–263

Boussau B et al (2008) Parallel adaptations to high temperatures in the Archaean eon. Nature 456:942–945

de Wit G, Danial JSH, Kukura P, Wallace MI (2015) Dynamic label-free imaging of lipid nanodomains. Proc Natl Acad Sci USA 112:12299–12303

Forterre P (1995) Thermoreduction, a hypothesis for the origin of prokaryotes C. R Acad Sci 318:415–422

Fuerst J (2013) The PVC superphylum: exceptions to the bacterial definition? Antonie Van Leeuwenhoek 104:451–466

Jain S, Caforio A, Driessen AJM (2014) Biosynthesis of archaeal membrane ether lipids. Front Microbiol 5:1–16 (article 641)

Koga Y (2012) Thermal adaptation of the archaeal and bacterial lipid membranes. *Archaea* 789652

Lepper CP et al (2017) The effects of pressure (0.1 – 200 MPa)and pH on the hydrolysis of cytosine at 373 K: implications for nucleotide stability around deep-sea blacksmokers. Proc Natl Acad Sci USA (submitted)

Lombard J, Moreira D (2011) Origins and early evolution of the mevalonate pathway of isoprenoid biosynthesis in the three domains of life. Mol Biol Evol 28:87–99

López-Garcia P, Zivanovic Y, Deschamps P, Moreira D (2015) Bacterial gene import and mesophilic adaptation in archaea. Nat Rev Microbiol 13:447–456

Martin WF, Sousa FL, Lane N (2014) Energy at life's origin. Science 344:1092–1093

McKay CP (2014) Requirements and limits for life in the context of exoplanets. Proc Natl Acad Sci USA 111:12628–12633

Mossel E, Steel MA (2004) A phase transition for a random cluster model on phylogenetic trees. Math Biosci 187:189–203

Moulton V et al (2000) RNA folding argues against a hot start origin of life. J Mol Evol 51: 416–421

Mueller P, Rudin DO, Tien HI, Wescott WC (1962) Reconstitution of cell membrane structure in vitro and its transformation into an excitable system. Nature 194:979–980

Penny D (2017) Evolution now. Xlibris, Australia

Penny D, Collins LJ, Daly T, Cox SJ (2014) The relative ages of Eukaryotes and Akaryotes. J Mol Evol 79:228–239

Schouten S, Hopmans EC, Schefuss E, Sinninghe Damsté JS (2002) Distributional variations in marine crenarchaeotal membrane lipids: a new tool for reconstructing ancient sea water temperatures? Earth Planet Sci Lett 204:265–274

Woese CR, Fox GE (1977) Phylogenetic structure of the prokaryotic domains: the primary kingdoms. Proc Natl Acad Sci USA 74:5088–5090

Printed by Printforce, the Netherlands